UPGRADING PETROLEUM RESIDUES AND HEAVY OILS

CHEMICAL INDUSTRIES

A Series of Reference Books and Textbooks

Consulting Editor

HEINZ HEINEMANN
Berkeley, California

1. *Fluid Catalytic Cracking with Zeolite Catalysts,* Paul B. Venuto and E. Thomas Habib, Jr.
2. *Ethylene: Keystone to the Petrochemical Industry,* Ludwig Kniel, Olaf Winter, and Karl Stork
3. *The Chemistry and Technology of Petroleum,* James G. Speight
4. *The Desulfurization of Heavy Oils and Residua,* James G. Speight
5. *Catalysis of Organic Reactions,* edited by William R. Moser
6. *Acetylene-Based Chemicals from Coal and Other Natural Resources,* Robert J. Tedeschi
7. *Chemically Resistant Masonry,* Walter Lee Sheppard, Jr.
8. *Compressors and Expanders: Selection and Application for the Process Industry,* Heinz P. Bloch, Joseph A. Cameron, Frank M. Danowski, Jr., Ralph James, Jr., Judson S. Swearingen, and Marilyn E. Weightman
9. *Metering Pumps: Selection and Application,* James P. Poynton
10. *Hydrocarbons from Methanol,* Clarence D. Chang
11. *Form Flotation: Theory and Applications,* Ann N. Clarke and David J. Wilson
12. *The Chemistry and Technology of Coal,* James G. Speight
13. *Pneumatic and Hydraulic Conveying of Solids,* O. A. Williams
14. *Catalyst Manufacture: Laboratory and Commercial Preparations,* Alvin B. Stiles
15. *Characterization of Heterogeneous Catalysts,* edited by Francis Delannay
16. *BASIC Programs for Chemical Engineering Design,* James H. Weber
17. *Catalyst Poisoning,* L. Louis Hegedus and Robert W. McCabe
18. *Catalysis of Organic Reactions,* edited by John R. Kosak
19. *Adsorption Technology: A Step-by-Step Approach to Process Evaluation and Application,* edited by Frank L. Slejko
20. *Deactivation and Poisoning of Catalysts,* edited by Jacques Oudar and Henry Wise

21. *Catalysis and Surface Science: Developments in Chemicals from Methanol, Hydrotreating of Hydrocarbons, Catalyst Preparation, Monomers and Polymers, Photocatalysis and Photovoltaics,* edited by Heinz Heinemann and Gabor A. Somorjai
22. *Catalysis of Organic Reactions,* edited by Robert L. Augustine
23. *Modern Control Techniques for the Processing Industries,* T. H. Tsai, J. W. Lane, and C. S. Lin
24. *Temperature-Programmed Reduction for Solid Materials Characterization,* Alan Jones and Brian McNichol
25. *Catalytic Cracking: Catalysts, Chemistry, and Kinetics,* Bohdan W. Wojciechowski and Avelino Corma
26. *Chemical Reaction and Reactor Engineering,* edited by J. J. Carberry and A. Varma
27. *Filtration: Principles and Practices: Second Edition,* edited by Michael J. Matteson and Clyde Orr
28. *Corrosion Mechanisms,* edited by Florian Mansfeld
29. *Catalysis and Surface Properties of Liquid Metals and Alloys,* Yoshisada Ogino
30. *Catalyst Deactivation,* edited by Eugene E. Petersen and Alexis T. Bell
31. *Hydrogen Effects in Catalysis: Fundamentals and Practical Applications,* edited by Zoltán Paál and P. G. Menon
32. *Flow Management for Engineers and Scientists,* Nicholas P. Cheremisinoff and Paul N. Cheremisinoff
33. *Catalysis of Organic Reactions,* edited by Paul N. Rylander, Harold Greenfield, and Robert L. Augustine
34. *Powder and Bulk Solids Handling Processes: Instrumentation and Control,* Koichi Iinoya, Hiroaki Masuda, and Kinnosuke Watanabe
35. *Reverse Osmosis Technology: Applications for High-Purity-Water Production,* edited by Bipin S. Parekh
36. *Shape Selective Catalysis in Industrial Applications,* N. Y. Chen, William E. Garwood, and Frank G. Dwyer
37. *Alpha Olefins Applications Handbook,* edited by George R. Lappin and Joseph L. Sauer
38. *Process Modeling and Control in Chemical Industries,* edited by Kaddour Najim
39. *Clathrate Hydrates of Natural Gases,* E. Dendy Sloan, Jr.
40. *Catalysis of Organic Reactions,* edited by Dale W. Blackburn
41. *Fuel Science and Technology Handbook,* edited by James G. Speight
42. *Octane-Enhancing Zeolitic FCC Catalysts,* Julius Scherzer
43. *Oxygen in Catalysis,* Adam Bielański and Jerzy Haber
44. *The Chemistry and Technology of Petroleum: Second Edition, Revised and Expanded,* James G. Speight

45. *Industrial Drying Equipment: Selection and Application,* C. M. van't Land
46. *Novel Production Methods for Ethylene, Light Hydrocarbons, and Aromatics*, edited by Lyle F. Albright, Billy L. Crynes, and Siegfried Nowak
47. *Catalysis of Organic Reactions*, edited by William E. Pascoe
48. *Synthetic Lubricants and High-Performance Functional Fluids*, edited by Ronald L. Shubkin
49. *Acetic Acid and Its Derivatives*, edited by Victor H. Agreda and Joseph R. Zoeller
50. *Properties and Applications of Perovskite-Type Oxides*, edited by L. G. Tejuca and J. L. G. Fierro
51. *Computer-Aided Design of Catalysts*, edited by E. Robert Becker and Carmo J. Pereira
52. *Models for Thermodynamic and Phase Equilibria Calculations*, edited by Stanley I. Sandler
53. *Catalysis of Organic Reactions*, edited by John R. Kosak and Thomas A. Johnson
54. *Composition and Analysis of Heavy Petroleum Fractions*, Klaus H. Altgelt and Mieczyslaw M. Boduszynski
55. *NMR Techniques in Catalysis,* edited by Alexis T. Bell and Alexander Pines
56. *Upgrading Petroleum Residues and Heavy Oils*, Murray R. Gray

ADDITIONAL VOLUMES IN PREPARATION

Methanol Production and Use, edited by Wu-Hsun Cheng and Harold H. Kung

Catalytic Hydroprocessing of Petroleum and Distillates, edited by Michael C. Oballah and Stuart S. Shih

The Chemistry and Technology of Coal: Second Edition, Revised and Expanded, James G. Speight

Lubricant Base Oil and Wax Processing, Avilino Sequeira, Jr.

Catalytic Naphtha Reforming: Science and Technology, edited by George J. Antos, A. M. Aitani, and J. M. Parera

UPGRADING PETROLEUM RESIDUES AND HEAVY OILS

Murray R. Gray

University of Alberta
Alberta, Canada

CRC Press is an imprint of the
Taylor & Francis Group, an **informa** business

CRC Press
Taylor & Francis Group
6000 Broken Sound Parkway NW, Suite 300
Boca Raton, FL 33487-2742

First issued in paperback 2019

ISBN-13: 978-0-8247-9211-4 (hbk)
ISBN-13: 978-0-367-40207-5 (pbk)

Library of Congress Cataloging-in-Publication Data

Gray, Murray R.
Upgrading petroleum residues and heavy oils / Murray R. Gray.
p. cm. -- (Chemical industries; 56)
Includes bibliographical references and index.
ISBN 0-8247-9211-4
1. Petroleum--Refining. I. Title. II. Series: Chemical industries; v. 56.
TP690.G69 1994
665.5'3--dc20 94-803
CIP

Visit the Taylor & Francis Web site at
http://www.taylorandfrancis.com

and the CRC Press Web site at
http://www.crcpress.com

To Dinah

Preface

As the world's supply of light, sweet crude oil is depleted, the stocks of heavy oils and bitumens become more and more important as a component in supplying the demand for fuels and petrochemical feeds. Similarly, the need to process the residue fractions of petroleum has also increased in importance. Thirty years ago, the process options were limited; residues were used for coking, manufacture of asphalts, or refinery fuel. The combination of higher fuel prices and environmental constraints has given rise to a new generation of process alternatives and focused attention on the efficient upgrading of residues.

In order to meet the challenge of converting the high molecular weight components of residues and heavy oils into distillable fractions, one requires a knowledge of the chemistry of these feeds, their physical and thermodynamic properties, and the available conversion processes. This book provides current information on both the fundamentals of residue composition and behavior, and a critical review of the available process technologies. The book is intended for chemists in refinery laboratories, for process and design engineers, and for researchers who are interested in upgrading heavy hydrocarbons. Much of the content is aimed at a general technical audience, so that the book can serve as an introduction to the field of upgrading, as well as a reference source. The

Preface

As the world's supply of light, sweet crude oil is depleted, the stocks of heavy oils and bitumens become more and more important as a component in supplying the demand for fuels and petrochemical feeds. Similarly, the need to process the residue fractions of petroleum has also increased in importance. Thirty years ago, the process options were limited; residues were used for coking, manufacture of asphalts, or refinery fuel. The combination of higher fuel prices and environmental constraints has given rise to a new generation of process alternatives and focused attention on the efficient upgrading of residues.

In order to meet the challenge of converting the high molecular weight components of residues and heavy oils into distillable fractions, one requires a knowledge of the chemistry of these feeds, their physical and thermodynamic properties, and the available conversion processes. This book provides current information on both the fundamentals of residue composition and behavior, and a critical review of the available process technologies. The book is intended for chemists in refinery laboratories, for process and design engineers, and for researchers who are interested in upgrading heavy hydrocarbons. Much of the content is aimed at a general technical audience, so that the book can serve as an introduction to the field of upgrading, as well as a reference source. The problems at the end of each chapter are intended to illustrate important concepts and are suitable for self-study.

Chapters 1 through 4 review the chemistry, properties, and reactions of residues and distillates fractions. Current methods of estimating phase behavior, thermodynamic properties, and transport properties are presented. The kinetics of the reactions of petroleum fractions are discussed in detail, with an emphasis on how the observed kinetics are determined by the underlying behavior of complex mixtures of components.

Chapters 5 through 8 cover the processing of residues. These chapters go beyond the description of the process flowsheets to discuss kinetics, reactor hydrodynamics, operability, and environmental impact. The emphasis is on covering the major classes of processes, rather than describing every patented technology in detail. All of the processes presented are in use at the commercial scale, or have been demonstrated at the semi-commercial scale within a refinery operation. The most emphasis is placed on hydrogen addition processes (Chapter 5) because this approach to upgrading is economically attractive and has been developing most rapidly. The coverage of fluid catalytic cracking, or FCC (Chapter 7), is brief, with an emphasis on modifications to accommodate residues. Comprehensive coverage of FCC technology for gas oil feeds is available elsewhere. The final chapter discusses the secondary and tertiary hydrotreating of cracked distillates.

Ackowledgements

My work on heavy oil upgrading over the past ten years has been aided by enthusiastic graduate students (Paula Jokuty, Farhad Khorasheh, and Lyle Trytten), post-doctoral fellows (Jacek Thiel, John Choi, and Nosa Egiebor) and coinvestigators from industry (Emerson Sanford, Ed Chan, Marten Ternan, Roger Kirchen, Andrzej Krzywicki, Oscar Sy, Bruce Sankey, Umesh Achia, and Alberto Ravella). My continuing interactions with staff at Syncrude Research in Edmonton have been particularly stimulating and productive. Generous funding from Alberta Oil Sands Technology and Research Authority, Syncrude Canada Ltd., Imperial Oil Ltd. and the Natural Sciences and Engineering Research Council of Canada has supported my research work on upgrading over the past nine years.

While the manuscript was in preparation, several people agreed to review chapters for technical accuracy and content. I am indebted to Walter Bishop, Roger Kirchen, Alan Mather, Emerson Sanford, Milan Selucky, Marten Ternan and Sok Yui for their helpful comments and suggestions. Michelle Portwood was delightful to work with, as she spent a summer working on the figures. Much of the appearance of the book was molded by her careful attention to format and detail. The support and assistance of Cindy Heisler enabled me to complete the book while at the same time attending to my other duties. Finally, I would like to thank my wife, Dinah, for her support and encouragement throughout this project.

Murray R. Gray

Contents

Contents

CHAPTER 1

Heavy Oil and Residue Properties and Composition

1.1 Definition of Petroleum Residues, Heavy Oils and Bitumens

Residue, or residuum, is operationally defined as the fraction of petroleum, heavy oil, or bitumen that does not distill under vacuum. Residues have atmospheric equivalent boiling points over 525° C, and in a refinery would be produced as the bottom product from a vacuum distillation column. Many conventional crude oils from around the world contain 10—30% residue (see Nelson, 1958, or Aalund, 1983, for extensive tables of crude oil assay data, including residue contents). These residues were often used to manufacture asphalt, depending on the properties of the residue fraction and the local demand for asphalts.

As conventional crude oils have becomes more expensive, interest in processing heavier feeds has increased. These heavier feeds, however, have residue contents of 40% or more, which may require processing in order to find a market. The upgrading of these residues, and residues from conventional crudes, into distillable fractions is the subject of this book. Processing of heavy oils and bitumens tends to go hand in hand with a need for upgrading of residue, so that most process examples will deal with fractions from these materials.

1.2 Physical Properties

Bitumens and heavy oils are defined in terms of their physical properties. Bitumen is defined as "Any natural mixture of of solid and semi–solid hydrocarbons", but this description is too broad to be of use in classifying oils. UNITAR defined a bitumen based on viscosity and density, as follows:

Table 1–1
UNITAR Definition of Heavy Oils and Bitumens

	Viscosity mPa·s	Density g/cm³	API Gravity
Heavy Oil	10^2–10^5	0.934–1.0	20–10
Bitumen	$> 10^5$	>1.00	< 10

Density and API gravity are reported at a standard temperature of 15.6°C. The API gravity is an expanded density scale which is defined as follows:

$$^{o}API = \frac{141.5}{\text{specific gravity at } 15.6^{0}C} - 131.5 \qquad (1.1)$$

The API scale gives water a value of 10, while methane is 340 (as liquid at its saturation pressure). Petroleum and bitumen residues and refinery process streams can have gravities that are negative. The UNITAR definition deals with naturally occurring materials. From a processing point of view, we are most interested in the residue fraction rather than the naturally occurring material, so that the classification as bitumen or heavy oil is not significant.

Another characterization factor which appears in the refining literature is the Watson characterization factor:

$$K_w = \frac{T_b^{1/3}}{SG} \qquad (1.2)$$

where T_b is the mean average boiling point in °R (English units only; a definition of this "mean average" is given in Chapter 2), and SG is the specific gravity at 15°C. The value of K_w ranges from < 10 for very aromatic oils, to 15 to paraffinic materials.

Table 1–2 compares light oil to typical bitumens and heavy oils, while Table 1–3 gives properties of residue fractions.

Table 1–2
Comparison of Light and Heavy Crudes

	Light Crude	Cold Lake	Atha–basca	Morichal
API Gravity	38	10	9	4.9
Sulfur, wt%	0.5	4.4	4.9	4.1
Nitrogen, wt%	0.1	0.4	0.5	0.8
Metals, wppm	22	220	280	863
Viscosity, $m^2/s \times 10^6$ at 40° C	5	5000	7000	
Vacuum Resid 525° C+, Liquid Vol%	11	52	52	80[1]

	Heavy Arabian	Hondo	Maya	Gach Saran
API Gravity	12.6	13.4	9.4	15.6
Sulfur, wt%	4.23	5.10	4.42	2.60
Nitrogen, wt%	0.26	0.70	0.52	0.41
Metals, wppm	115	372	496	144
Vacuum Resid 525° C+, Weight%	51	46	59	50

(wppm = parts per million by weight)
(Data from Gray *et al.*, 1991 and Dolbear *et al.*,1987)
1. Properties are for a vacuum residue, not an as–produced oil

Although Cold Lake oil is often referred to as heavy oil, Table 1–2 shows that Athabasca bitumen has very similar properties, and that both can be classified as bitumens by the UNITAR definition. The properties of the residues in Table 1–3 are clearly much more important, from the point of view of upgrading, than those of whole crude oils prior to distillation. These feeds are extremely complex mixtures which contain hundreds of thousands of components. The individual components cannot all be measured, so a variety of characteristics of an oil are measured to give an indication of its value upon processing.

Table 1–3
Properties of Vacuum Residue Feedstocks
(Data from Wenzel, 1992; Han *et al.*, 1992)

	Content +500°C	°API	S wt%	CCR[1] wt%	Metals ppm
Middle East					
Arabian Heavy	97	3.9	4.5	16	252
Venezuela					
Bachaquero	93	5.3	3.36	21.4	720
Boscan	81	5.1	5.88	19.1	1655
Morichal	85	5.4	3.91	19.9	620
Canada					
Athabasca	97	2.1	6.18	21.4	490
Cold Lake	98	2.1	6.15	22.8	470
USA					
Hondo	92	6.8	6.53	14.2	555
Russia					
Export Blend	98	7.1	3.21	18.7	240
China					
Daqing	98	21	0.14	7.3	16
Shengli	98	14.4	1.35	13.2	48
Refinery Streams					
FCC[2] recycle	5.5	–3.5	2.61	5.2	<5
Solvent Deasph—alted Bottoms	99	0	6.3	29.3	323

1. CCR = Conradson carbon residue (see Section 1.2.5)
2. FCC = Fluid catalytic cracking (see Chapter 7)

1.2.1 Hydrogen to carbon ratio

In the range of fuel materials from natural gas to coal, the molar ratio of hydrogen to carbon gives an indication of heating value and combustion properties.

Table 1—4
H/C Ratios for Fuels

Fuel	H/C Ratio
Methane	4.0
Gasoline, Diesel	1.9
Light Crude	1.8
Bitumen	1.4—1.6
Coal	0.5—0.8

Residue	H/C Ratio	Density kg/m^3
Athabasca (424° C+)	1.46	1019
Arabian Heavy (343° C+)	1.53	993
Cold Lake (424° C+)	1.40	994
Lloydminster (424° C+)	1.47	1018
Peace River (424° C+)	1.42	1010

Unlike coals, the H/C ratio varies little between residues. The H/C ratio of the distillates must be increased for use as a diesel of jet engine transportation fuel, so that these feeds must either be disproportionated to remove a carbon—rich stream, or hydrogenated. The objective of H/C ratio is a simple concept which is useful for thinking about process objectives, but even when hydrogen is added catalytically the change in the total hydrogen content is quite small.

1.2.2 Heteroatom content

The heteroatoms are elements in the residue other than carbon and hydrogen. The typical range of values for residues is:

Sulfur	2 — 7 wt%
Nitrogen	0.2 — 0.7 wt%
Oxygen	≃1 wt%
Vanadium	100 — 1000 ppm
Nickel	20 — 200 ppm

All of these values are higher than the range acceptable for distillate fractions in a conventional refinery, particularly sulfur, and therefore these feeds require processing to upgrade their characteristics. Table 1—5 gives some examples of the elemental composition of residues.

Table 1–5
Elemental Composition of Residue Feedstocks

	Heteroatom Content (weight percent)			Metals in +525° C cut (ppm)	
	S	N	O	Ni	V
Athabasca[1]	5.14	0.56	1.17	150	290
Bachaquero[2]	3.39			100	880
Cerro Negro[2]	4.50			200	1040
Cold Lake[1]	5.10	0.45	0.97	200	490
Gach Saran	2.60	0.41	nd	40	110
Heavy Arabian	4.23	0.26	nd	30	90
Hondo	4.42	0.70	nd	90	280
Lloydminster[1]	4.69	0.53	0.99	140	190
Maya	4.42	0.52	nd	80	410
Peace River[1]	7.02	0.63	1.09	130	410
Tia Juana[2]	3.17			100	760
Zuata[2]	4.17			160	820

1. 424° C+ fraction
2. 540° C+ fraction

(Data from Gray *et al.*,1991, Dolbear *et al.*,1987, Carbognani *et al.*, 1987)

1.2.3 Class fractionation

Class fractions are groups of compounds separated from an oil on the basis of combined solubility and adsorption characteristics. The terminology is derived from the study of coal and asphalt materials, and is based on solubility in solvents of decreasing polarity or solvent power.

Solubility fractions:

Coke and solids	Insoluble in tetrahydrofuran (THF)
Preasphaltenes	Soluble in THF and insoluble in benzene
Asphaltenes	Soluble in benzene and insoluble in n–pentane
Maltenes	Soluble in n–pentane

The maltenes can be further separated into column chromatography fractions:

Resins	Maltenes adsorbed by silica gel or by clay from a solution in n–pentane
Oils	Maltenes not adsorbed from n–pentane
Aromatics	Oils adsorbed from a solution of n–pentane by a silica/alumina column
Saturates	Oils not adsorbed

The presence of solids or coke depends on the production and processing history of the sample. The remaining fractions are routinely determined for residues, by a sequence of separations using dilute solutions in the appropriate solvents and adsorption columns. A common acronym for this separation protocol is SARA (Saturates, Aromatics, Resins, and Asphaltenes). Asphaltenes are determined as either n–pentane insolubles or n–heptane insolubles. As for all of these fractions, the definition of the separated material is a blend of chemistry and technique. A class fraction is mainly defined by the procedures followed in its isolation. For example, C_5 asphaltenes are not the same as C_7 asphaltenes; the yield from a given oil is higher and the properties differ. A detailed description of the various separation schemes is given by Speight [1991].

Comparison of the data for 424°C+ and 525°C+ compositions in Table 1–6 clearly shows the trend in class composition with boiling point; more resins and less oils. Asphaltenes are found only in the residue fraction. The class designations are indirect labels in terms of chemical structure and composition. The saturates fraction is most descriptive, because it contains only aliphatic compounds. The aromatics contain a variety of aromatic compounds with saturated groups attached. The resins are higher in heteroatoms, and have a higher concentration of aromatic carbon, while the asphaltenes are the highest molecular weight fraction, and contain most of the polar compounds. The distinctions between aromatics and resins, and between resins and asphaltenes, therefore, are not clear. Despite this, the asphaltenes have attracted a tremendous amount of interest and attention.

The original methods for preparing these fractions involved simple precipitation and chromatography on large open columns at low pressure. More recent adaptations of this analysis include automated chromatography and detection using silica rods and flame ionization (Iatroscan), and High–Performance Liquid Chromatography (HPLC) using high–pressure columns. The problem with the latter method is quantitation of the complex fractions as they elute from the column.

Table 1–6
Class Fractions in Residues

	Weight %			
	Satur–ates	Aroma–tics	Resin	Asph–altene
Athabasca 424° C+	10.3	5.3	62.3	22.2
Athabasca 525° C+		6.6[1]	54.6	38.8
Bachaquero 540° C+	8.9	54.5	25.3	11.3
Cerro Negro 540° +	1.1	50.6	29.5	18.8
Cold Lake 424° C+	12.3	6.7	58.1	22.9
Cold Lake 525° C+		4.9[1]	55.5	36.1
Gach Saran		64.7[1]	28.5	6.8
Heavy Arabian		59.9[1]	27.5	12.6
Hondo		43.9[1]	40.2	13.9
Lloydminster 424+	15.4	6.4	58.2	19.9
Lloydminster 525° C+		10.8[1]	52.4	36.9
Maya		48.9[1]	25.9	25.2
Peace River 424° +	7.0	4.2	63.3	25.5
Peace River 525° C+		6.4[1]	57.8	35.8
Tia Juana 540° C+	5.4	57.7	25.6	10.8
Zuata 540° C+	2.8	50.0	31.8	15.4

1. Oils fraction

(Data from Gray *et al.*,1991, Dolbear *et al.*,1987, Carbognani *et al.*,1987)

1.2.4 Boiling cuts

The distillation curve of any hydrocarbon mixture is important because it indicates the quantities of useful fractions to a refiner. The commonly used fractions for refinery processing are listed in the following table:

Table 1—7
Boiling Ranges from Petroleum and Bitumen

Name	Range, °C	Use
Light straight run gasoline	32—104	Gasoline
Naphtha	82—204	Reformed for gasoline
Kerosine	165—282	"
Light Gas Oil (LGO)	215—337	Diesel fuel
Heavy Gas Oil (HGO)	320—426	Catalytic cracker feed or
Vacuum Gas Oil (VGO)	398—565	hydrocracker feed
Residue, Resid, Residuum	565+	Asphalt, coker, or hydrocracker feed

The overlaps between these ranges indicate variations in refinery specifications and practice. For convenience, we will use a common four—fraction designation for distillate cuts:

Naphtha	Initial Boiling Point to 177°C
Middle Distillate	177—343°C
Gas Oil	343—525°C
Residue	525°C +

The words for these refinery fractions are descriptive in some cases, as in middle distillate or residue. Naphtha comes from the Greek word for petroleum, and kerosine from the word for wax. Gas oils were first produced in the nineteenth and early twentieth century as by—products from coal gasification plants, which supplied city gas mains.

Distillation curves for hydrocarbon mixtures are given as the fraction by weight or volume which distills in standard equipment as a function of temperature (see Chapter 2). Examples are given in Table 1—8. The fraction boiling between the ranges listed above is the yield of that range.

Fractionation of oils to equivalent boiling temperatures beyond 525—540°C has been demonstrated using short—path distillation, high—temperature gas chromatography, and supercritical fluid chromatography. These methods rely on volatilization, therefore they cannot guarantee reaching the endpoint of a residuum. They have not been scaled up as yet, therefore they do not correspond to separations that can be achieved on a large scale.

Table 1—8
Distillation curves for bitumens

Athabasca Vol %	T_b, °C	Peace River Vol %	T_b, °C
Initial	257	Initial	65
10	326	6.5	234
20	383	12.6	284
30	434	23.8	352
40	483	33.7	419
50	530	42.5	486
End Pt.	536	48.3	538

The heteroatoms in the residues are not distributed evenly with boiling point; in general the concentration of S, N, O, V, and Ni increases with boiling point (or molecular weight). High boiling fractions, therefore, have more heteroatoms on a weight basis and on a basis of heteroatoms per molecule. Data from Boduszynski [1987] illustrate this trend very well; fractions of atmospheric residues (343° C+ including both gas oil and residue) were separated up to an equivalent atmospheric boiling point of 760° C. The overall heteroatom contents are given in Table 1—9, while the distribution of each element with boiling point is given in Figures 1—1 through 1—5.

Table 1—9
Elemental Composition of Atmospheric Residues Before Fractionation
(Data from Boduszynski, 1987)

Feed	Element S wt%	N wt%	O wt%	V ppm	Ni ppm
Arabian Heavy	4.36	0.25	nr	80	25
Boscan	5.94	0.79	nr	1400	125
Kern River	1.2	0.96	0.73	38	81
Maya	4.17	0.47	nr	340	72
Offshore Calif.	4.41	0.83	0.60	364	97

nr = not reported

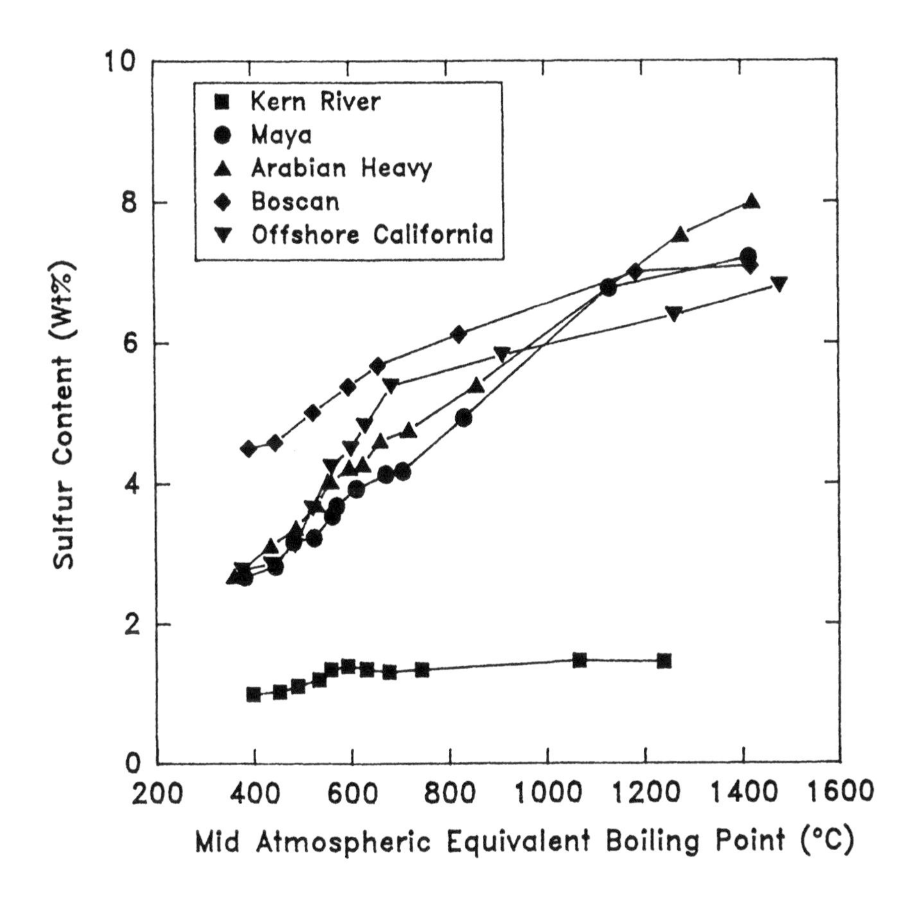

Figure 1—1
Concentration of sulfur in heavy oils as a function of equivalent boiling point at 101.3 kPa (Data from Boduszynski, 1987)

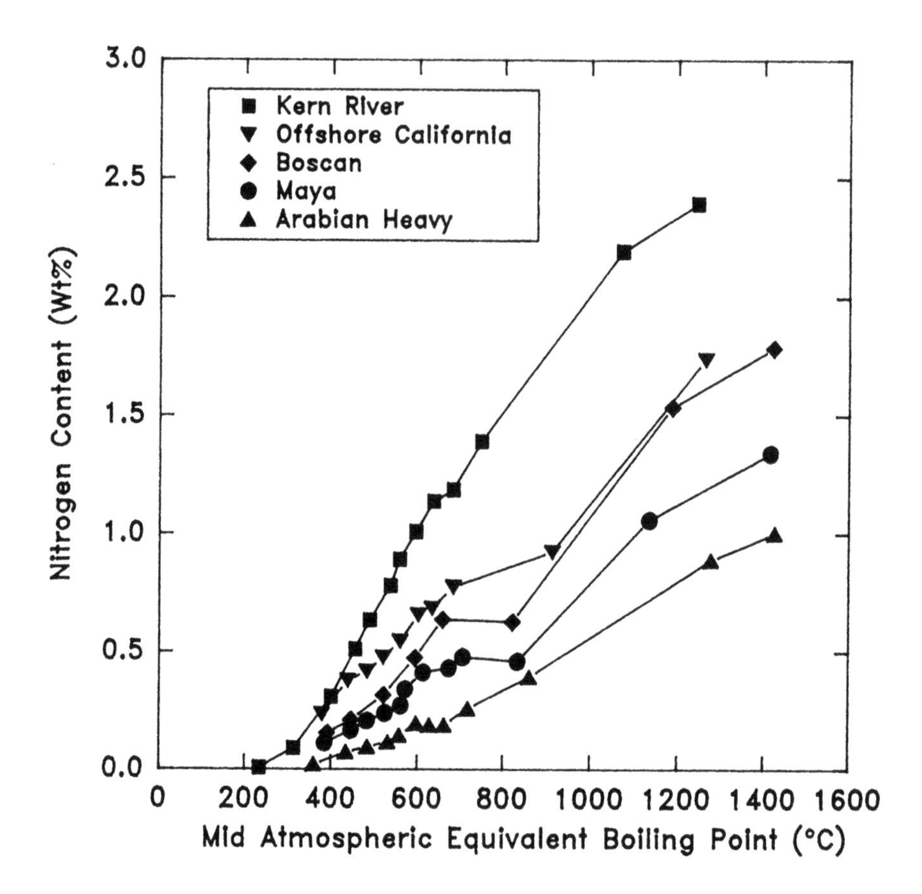

Figure 1–2
Concentration of nitrogen in heavy oils as a function of equivalent boiling point at 101.3 kPa (Data from Boduszynski, 1987)

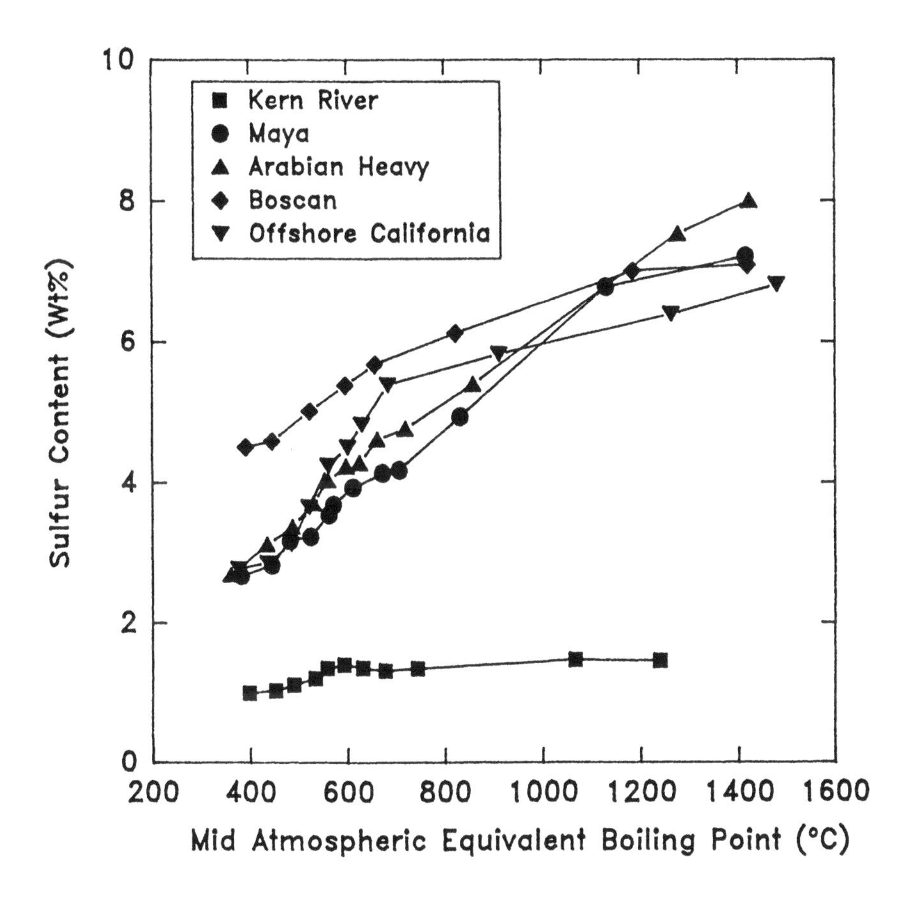

Figure 1—1
Concentration of sulfur in heavy oils as a function of equivalent boiling point at 101.3 kPa (Data from Boduszynski, 1987)

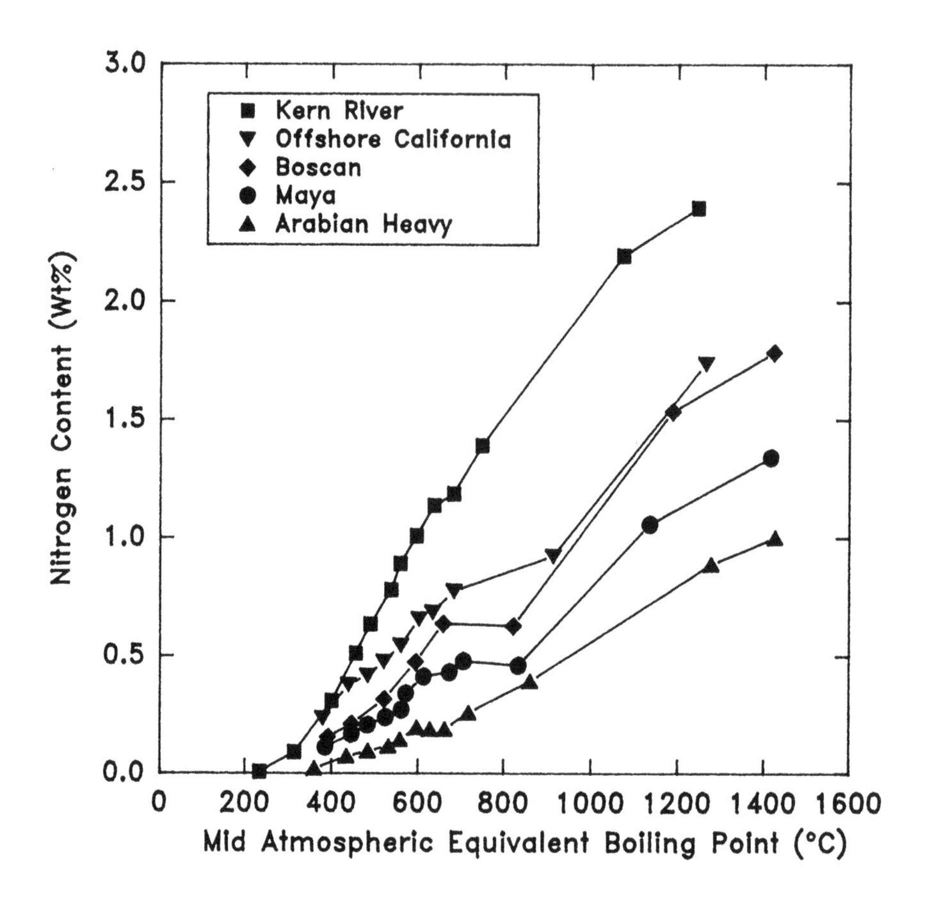

Figure 1—2
Concentration of nitrogen in heavy oils as a function of equivalent boiling point at 101.3 kPa (Data from Boduszynski, 1987)

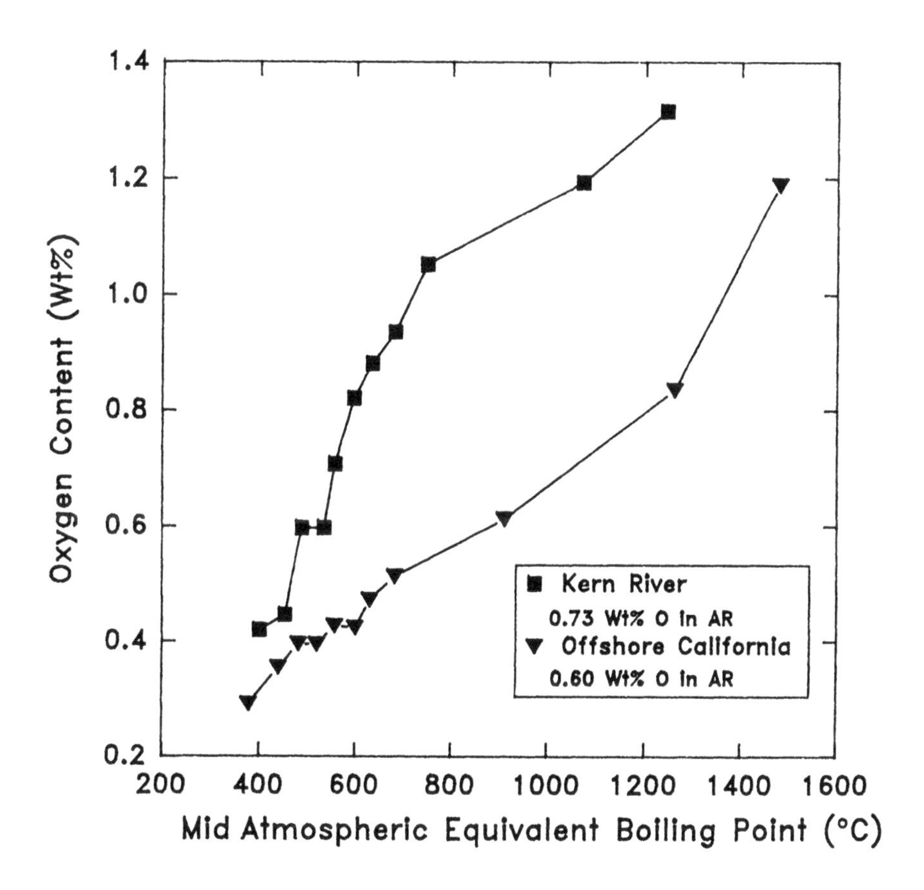

Figure 1–3
Concentration of oxygen in heavy oils as a function of equivalent boiling point at 101.3 kPa (Data from Boduszynski, 1987)

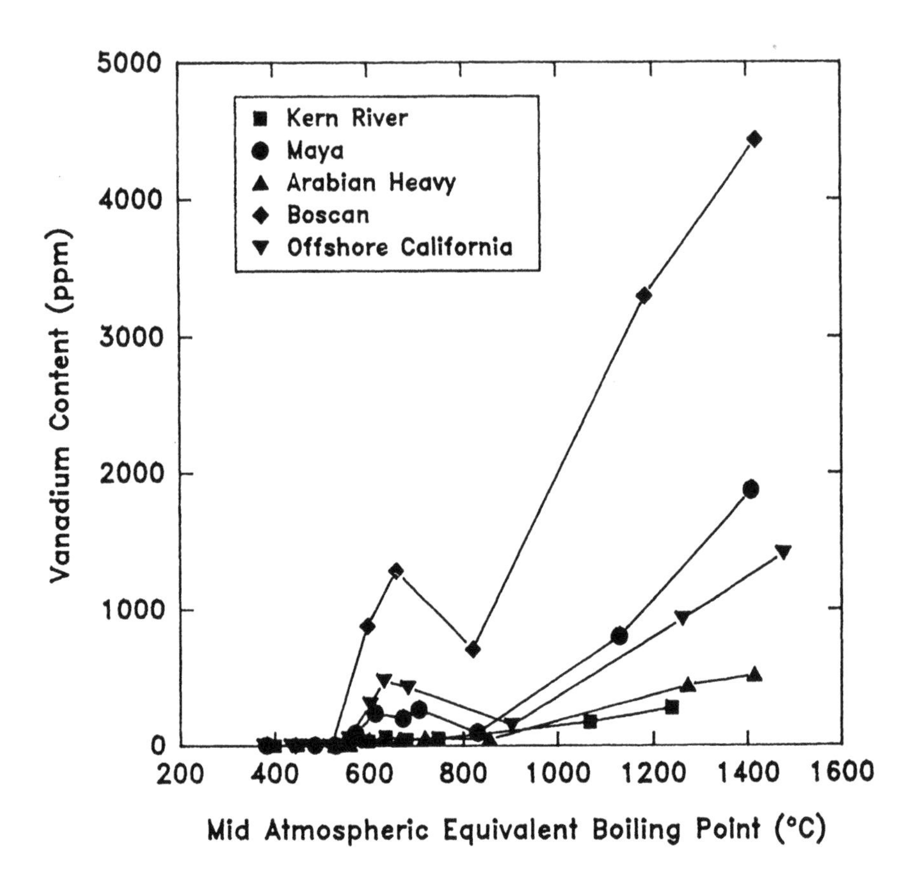

Figure 1–4
Concentration of vanadium in heavy oils as a function of equivalent boiling point at 101.3 kPa (Data from Boduszynski, 1987)

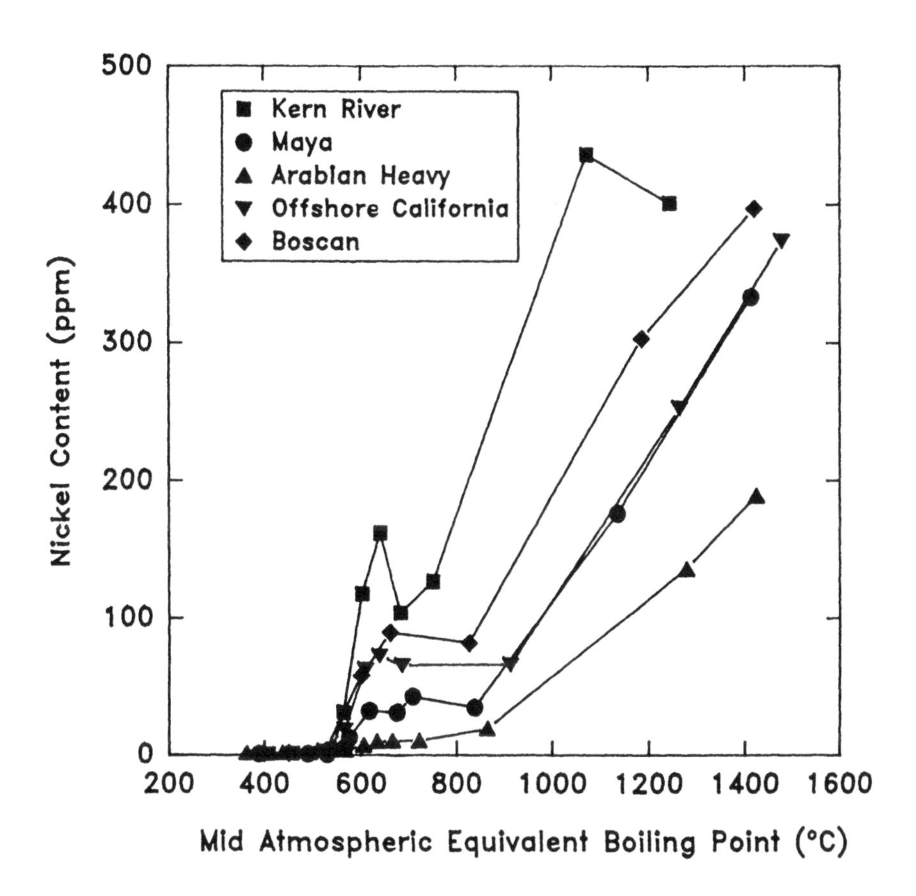

Figure 1–5

Concentration of nickel in heavy oils as a function of equivalent boiling point at 101.3 kPa (Data from Boduszynski, 1987)

1.2.5 Other Properties

A variety of other physical or chemical properties may be quoted or specified for a hydrocarbon mixture. These analyses refer to a standard method such as an ASTM (American Society for the Testing of Materials) code, and are related to the end use of the material. For example, pour point is the lowest temperature where the liquid will flow under standard conditions, providing a measure of cold—weather service for fuels.

An important parameter for upgrading is carbon residue, usually given as Conradson carbon residue (CCR), Ramsbottom carbon residue (RCR), or Micro carbon residue (MCR). The carbon residue is the amount of solid left behind when a sample is pyrolyzed in an inert gas. The names refer to the specific equipment and conditions employed. The most common method in use today is MCR, because it can be done in a computerized micro—balance (ASTM Method D4530).

1.3 Chemical Composition of Residues

The chemical species in bitumen and heavy oil are thought to be heavily degraded by bacterial action. The alkanes have, therefore, been removed along with the lighter boiling fractions. The remaining oil has a low concentration of paraffinic groups, and many of these are attached to other entities in the oil. The high heteroatom content is also a consequence of this ancient bacterial attack.

The composition of residues from conventional crudes depends on both the original oil and the subsequent processing. For example, many conventional crudes give some nondistillable material, which may or may not be similar to a natural bitumen. Some crudes, such as Daqing, can give a residue with a high wax content which corresponds to long—chain paraffins. Refinery processing can give rise to residues with very different properties.

The following sections outline the main building blocks which are combined to give the high boiling compounds found in residue fractions. The complexity of residue fractions derives from the combination of simple groups to give complex molecules and countless isomers.

1.3.1 Hydrocarbon groups

1.3.1.1 Aromatic structures

These aromatic rings are normally substituted with alkyl groups, or bridges to other aromatics.

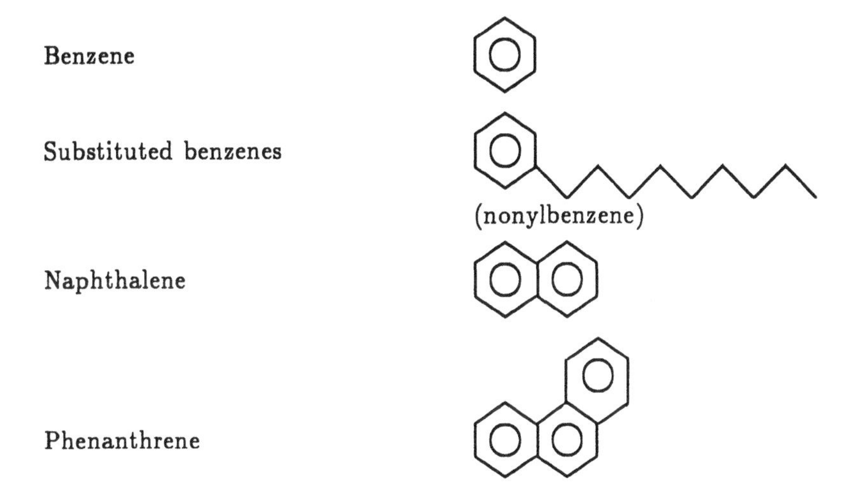

More highly condensed aromatic ring groups, corresponding to pyrene or chrysene (4 rings), are less abundant in unprocessed residues than groups with one, two, and three rings [Strausz, 1989]. Refinery processing, particularly over acidic catalysts, can promote reactions that build higher polynuclear aromatics containing up to 8—10 rings [Boduszynski and Sullivan, 1989], so that product streams may contain higher aromatics than raw feeds.

1.3.1.2 Hydroaromatics (Hydrogenated aromatics)

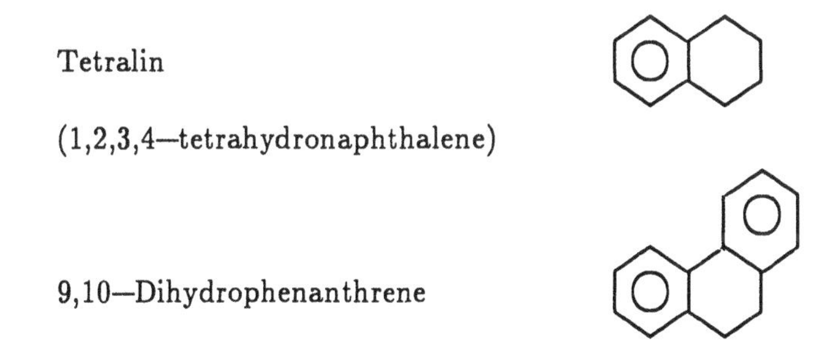

The hydroaromatics occur in the natural oil, as well as in streams that have been hydrogenated. The ability of compounds such as naphthalene to accept hydrogen to form tetralin is useful in supplying hydrogen from the liquid phase. Many hydroaromatics in feed residues are likely derived from naphthene rings. Interconversion of aromatics, hydroaromatics, and naphthenes is active in some upgrading processes [Peters *et al.*, 1992].

1.3.1.3 Naphthenes (Cyclic Saturates)

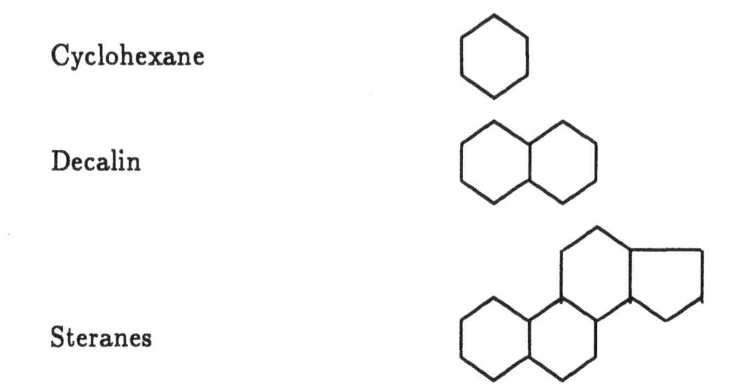

The steranes are derived from the bacteria which attacked the original oil. These materials are abundant in crude oils, and serve as markers for past biological activity [Strausz, 1989]. Naphthenic compounds containing one to six fused rings have been identified in a variety of residues and gas oils [Strausz, 1989; Boduszynski, 1988].

1.3.1.4 Paraffins

Straight—chain paraffins may be present in residue depending on its origin, or alkane groups may be present as side chains, as in nonylbenzene. In oils that have been subjected to bacterial degradation, the surviving paraffins are branched in various configurations:

The former structure is an isoprenoid, while the latter is a T—branched alkane [Gough and Rowland, 1990].

1.3.2 Heteroatomic groups

1.3.2.1 Sulfur

Sulfur is found as thiophene homologs, which can be resistant to further processing, and as sulfides which are more easily removed.

Dibenzothiophene

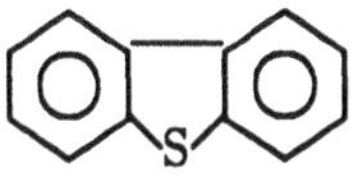

Sulfides or Thioethers

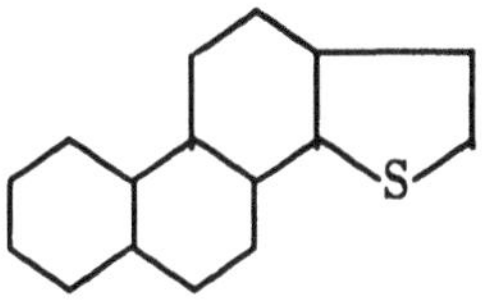

$R_1-CH_2-S-CH_2-R_2$

Disulfide $R_1-CH_2-S-S-CH_2-R_2$

Sulfoxide $R_1-CH_2-S(=O)-CH_2-R_2$

Thiols or Mercaptans R–SH

The sulfides are susceptable to oxidation to sulfoxides, which accounts for the presence of the latter group.

A series of alkyl–substituted thiophenes, benzothiophenes and cyclic thioethers were isolated from Athabasca bitumen [Payzant *et al.*, 1988] by showing a consistent pattern of substitution adjacent to the sulfur atom. This pattern of substitution has implications both for catalytic conversion of sulfur compounds, and the origin of these compounds from linear unsaturated alkanes.

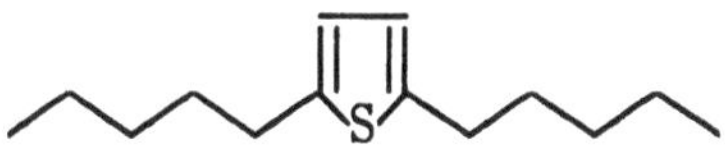

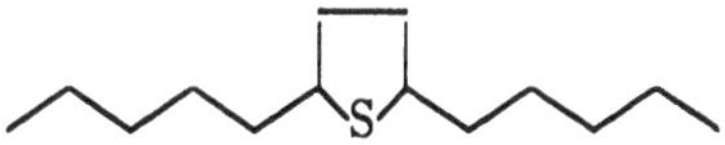

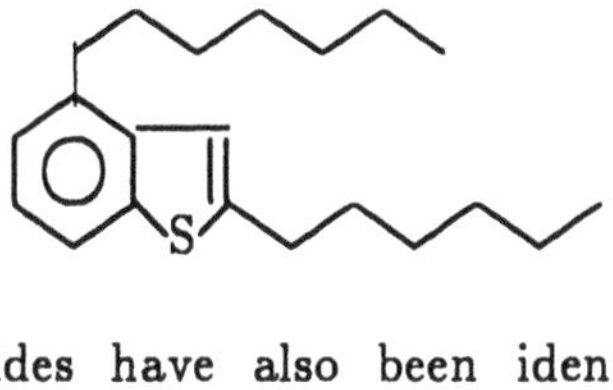

Sulfides have also been identified as bridging between sterane side groups and an alkane backbone [Adam *et al.*, 1992], suggesting that these groups may be responsible for holding together some high–molecular components in residues.

1.3.2.2 Nitrogen

Nitrogen is present in two main forms; the nonbasic derivatives of pyrrole, and the basic derivative of pyridine. Both types of nitrogen are highly resistant to removal.

Nonbasic nitrogen types:

Pyrrole

N
H

Indole

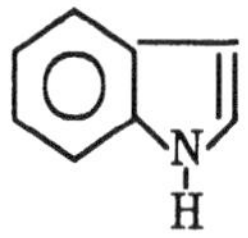

N
H

N—substituted Indole

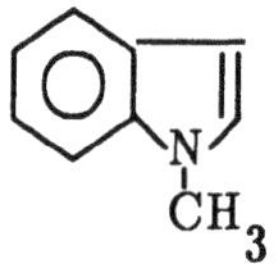

N
CH_3

Carbazole

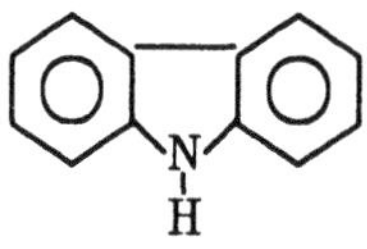

N
H

Basic nitrogen types:

Pyridine

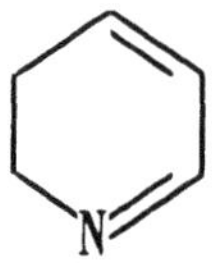

N

Quinoline

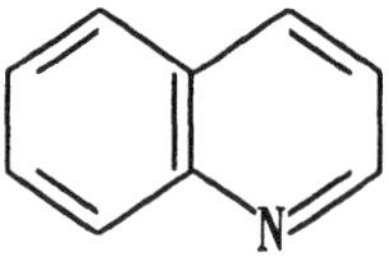

N

Acridine

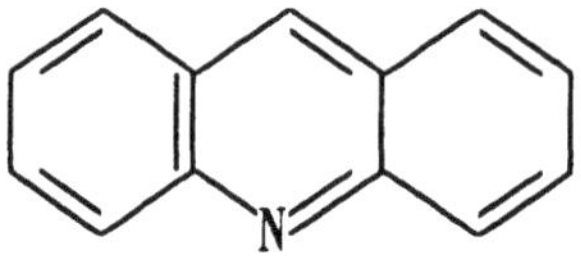

N

Very weak bases:

Amide

O
||
R—C—NH_2

Quinolone

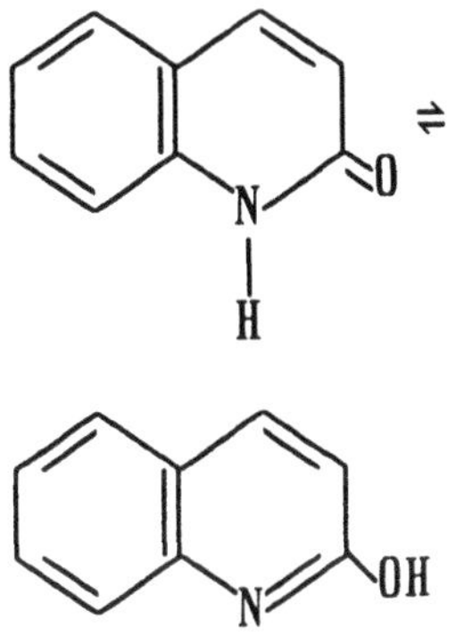

$$\text{(2-quinolone, N–H)} \rightleftharpoons \text{(2-hydroxyquinoline, OH)}$$

Higher–ring benzologs of carbazole and pyridine have also been identified in many gas oils, residues, and asphaltenes, extending up to pentacyclics (dibenzoquinoline) and hexacyclics (dibenzocarbazole) [Mojelsky *et al.*, 1986a; Schmitter *et al.*, 1984; Boduszynski, 1988].

1.3.2.3 Oxygen

The oxygen compounds are the least studied of the constituents of petroleum, although the oxygen compounds in Athabasca bitumen have been studied by Strausz and coworkers [Strausz, 1989]. Like sulfur, the oxygen is present in furan homologs, ethers, and as other polar functional groups.

Benzofuran	(benzofuran ring, O)
Carboxylic acid	$R{-}\overset{O}{\overset{\|\|}{C}}{-}OH$
Ketone	$R_1{-}\overset{O}{\overset{\|\|}{C}}{-}CH_2{-}R_2$
Aromatic hydroxyl (Phenol)	(benzene ring–OH)

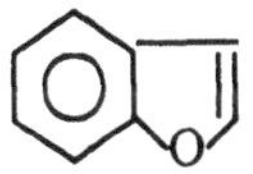

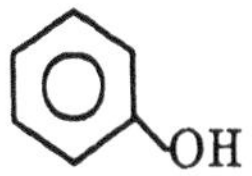

1.3.2.4 Metals (Iron, Nickel, Vanadium)

Metals occur in two organic forms in crude oils and bitumen: porphyrin metals, which are chelated in porphyrin structures analogous to chlorophyll, and nonporphyrin metals which are thought to be associated with the polar groups in the asphaltenes. The porphyrin ring is based on pyrrole groups which complex the metal atom (M):

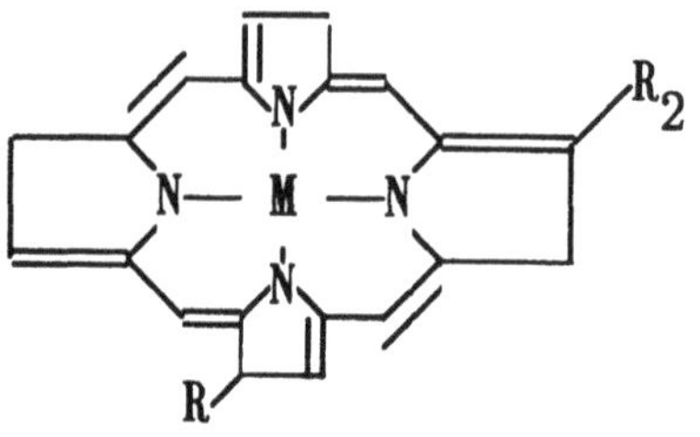

Data for metal compounds:

Metal Compounds in Alberta Bitumens
(Data from Strong and Filby, 1987)

Property	Atha—basca	Peace River	Cold Lake
V, ppm	196	180	191
Ni, ppm	75	62	63
V—porpyrins, ppm	93	92	77
V—porphyrins, %V	47%	51%	40%

1.3.3 Asphaltene composition and structure

The term "asphaltene" was introduced in section 1.2.3 to describe all material that was soluble in benzene (or toluene) and insoluble when diluted with a large excess of n—alkane (pentane or heptane). This experimental definition encompasses a heterogeneous and complex fraction of materials. Complete coverage of the extensive literature on asphaltenes is beyond the scope of this work, and the excellent summary by Speight [1991] is recommended for more details. This section will review the major points on the composition of asphaltenes.

One view of asphaltenes is that they are ordered structures within the liquid phase. The nonporphyrin component of the metals is thought to be associated with the asphaltenes, which form micelles. These micelles constitute a second phase within the oil, which can precipitate from the oil upon dilution with aliphatic solvents such as pentane or gas condensate. A conjectural structure is illustrated in Figure 1—6. This type of associated structure has been invoked to explain the wide range of molecular weights that have been reported for asphaltenic materials. What is most important from the point of view of upgrading processes is that micellar structures (as in Figure 1–6) unfold and break up as the temperature is raised. Using neutron scattering, Overfield *et al.* [1988] found that molecular weight dropped from an apparent value of > 500,000 at 25° C in toluene solution to a limiting value of 6000 at 250° C due to disruption of the aggregates. Such aggregates may be very important for the properties of asphaltic materials at low temperature, but not at temperatures above 250° C.

The composition of residues can also be viewed as a copolymer of the various structural subunits discussed above. The various aromatic, aliphatic, and heteroatomic groups are combined fairly randomly to make up the observed molecular weight. A hypothetical asphaltene molecule is shown in Figure 1–7 to illustrate the linked molecular building blocks. This view is only partly compatible with the "micelle" model, though such copolymers may indeed form aggregated structures at low temperatures.

The heterogeneity of the asphaltenes is nicely illustrated by the results of Cyr *et al.* [1987], who separated asphaltene fractions by gel–permeation chromatography. For an average molecular weight of 3600, five fractions were isolated with an abundance of 11 to 30% of the sample by weight, and molecular weights ranging from 1200 to 16,900. The heaviest fraction (MW=16,900) had the lowest aromatic carbon fraction (35%) and the most aliphatic side chains. The lightest fraction (MW=1200) was most aromatic (48% of the carbon in aromatic rings) and it also had the lowest concentration of long side chains.

1.4 Chemical Analysis of Residues

The mixtures of components in heavy distillates and residues are so daunting that components are only very rarely separated for identification and analysis. The constituent building blocks described in Section 1.3 were identified by a variety of chemical methods that liberated fragments for identification. Methods such as chromatographic separation, pyrolysis and selective oxidation, followed by specific methods for identifying chemical structures (e.g. mass spectroscopy) showed that the carbon structures were found in both the distillates and the residue.

Process studies require analysis of hundreds of samples, so that the routine analytical methods must be much faster and easier to interpret. In addition, the methods must be amenable to material balance calculations, so that methods that isolate only some chemical fragments of residue molecules are unsatisfactory. Standard refinery analysis, therefore, emphasizes distillation data and measurements on distillates relevant to product specifications, e.g. S and N content, viscosity, pour

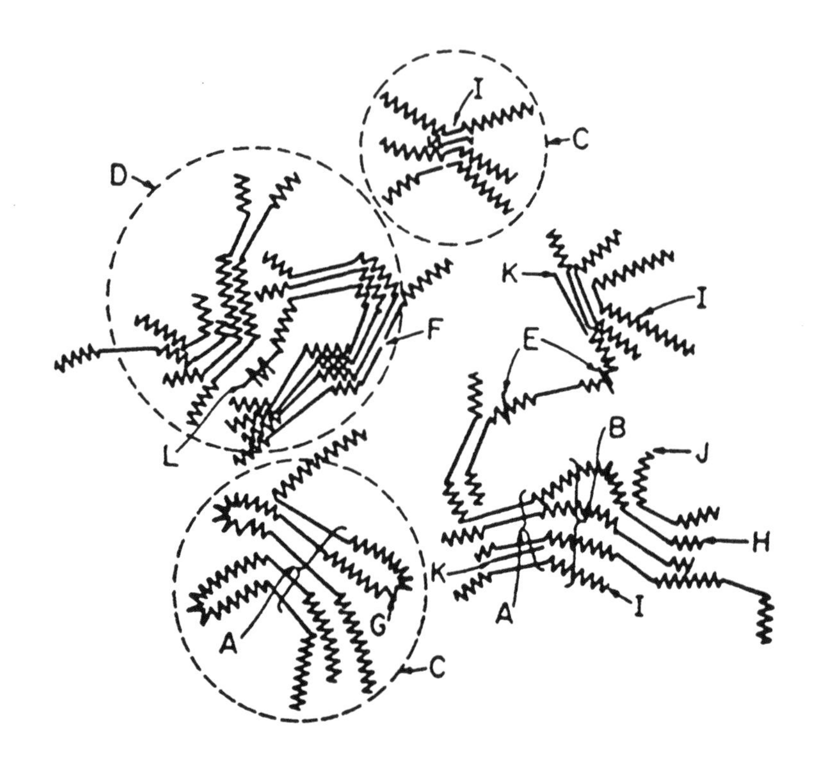

Figure 1—6
Macrostructure of asphaltenes

A. Crystallite; B. Chain Bundle; C. Particle
D. Micelle; E. Weak link; F. Gap & hole
G. Intracluster; H. Intercluster; I. Resin
J. Single layer; K. Porphyrin;L. Metal (M)
(Reprinted from J.P. Dickie and T.F. Yen, "Macrostructures of the asphaltic fractions by various instrumental methods", Anal. Chem. vol 39, p. 1847, 1967. Copyright by the American Chemical Society)

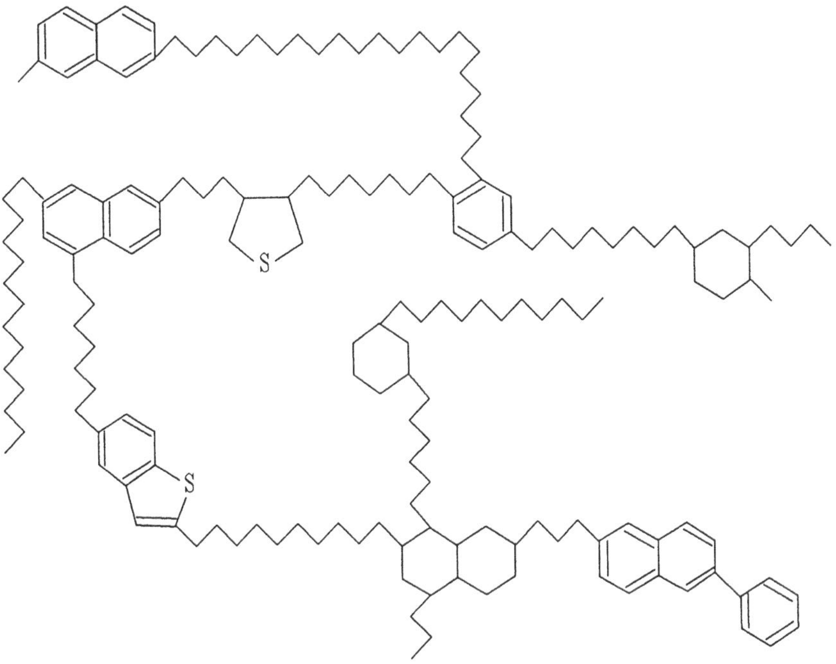

S

S

Figure 1—7
Chemical model for Athabasca asphaltene [Strausz, 1989b]

point, octane number, or cetane number. The only methods suitable for residue, i.e. elemental analysis and ASTM methods such as Micro carbon residue, give little information on the organic structural components.

1.4.1 Chemical group analysis by NMR

One approach to characterizing complex residue mixtures is to use averaged data, based on simple measurements of the whole sample. The most common analysis of this type is proton and ^{13}C—NMR, which gives the distribution of hydrogen and carbon types in a sample. Figure 1—8 illustrates the carbon types that appear in an NMR spectrum. The common notation is to designate carbons and hydrogens according to their proximity to an aromatic ring: α is immediately adjacent, β is one removed, and γ is two positions (or more) away from the ring. Unlike pure materials, the spectra consist of broad bands (Figure 1—9). From a knowledge of where signals appear from pure compounds, these spectra can be divided into bands and integrated.

Table 1—10
Proton NMR Band Assignments for Petroleum Fractions
(Data from Khorasheh *et al.*, 1987 and Petrakis and Allen, 1987)

Band	Range ppm	Hydrogen Types
1	0.5—1.0	γ—methyl
2	1.0—2.0	β, naphthenic H
3	2.0—4.5	α—H, amine, CH_2 α to polars
4	6.3—8.3	aromatic H, amide, phenol
5	8.3—9.0	phenanthrene hindered H

The NMR data can be used at several levels. The simplest is to calculate aromatic H and aromatic C fractions in a sample. The aromatic carbon content of residues and heavy oils covers a range from 20% of the carbon up to 40%, as listed in Table 1—12 for selected residues. Extreme process conditions can increase this level up to 75% of the total carbon.

The second level is to integrate portions of the ^{13}C—NMR spectrum to determine paraffinic, naphthenic, and aromatic carbon concentrations in a sample [Young and Galya, 1984]. The most complex approach is to

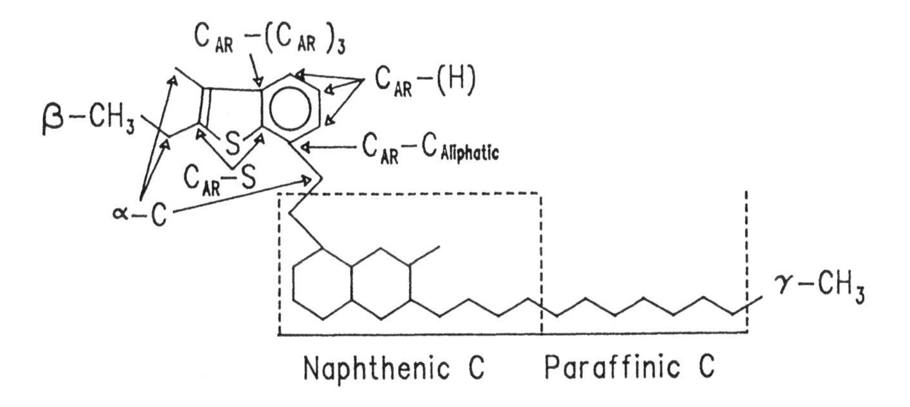

Figure 1–8
Major carbon structural types in molecules of petroleum residue, heavy oil and bitumen

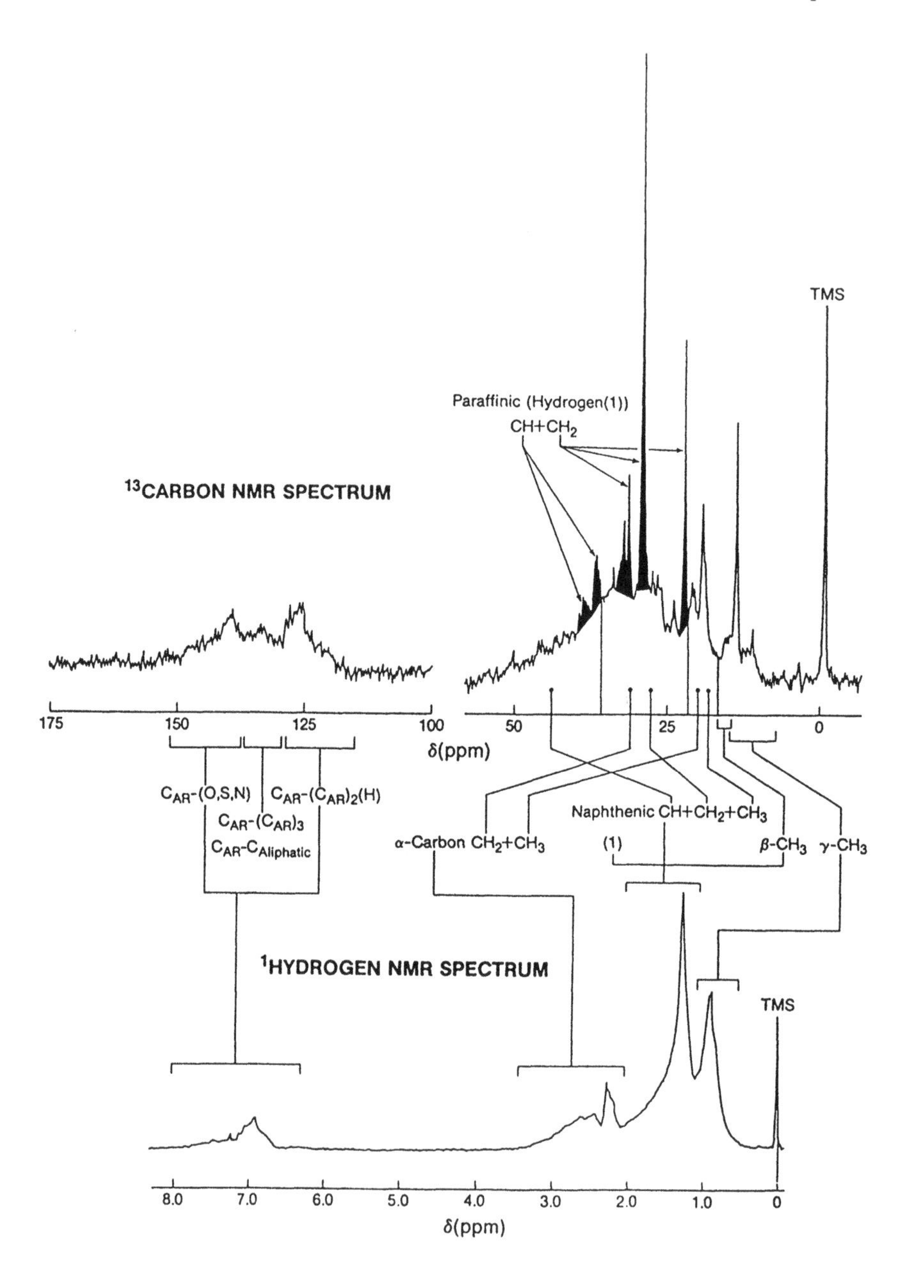

Figure 1—9

Interpretation of ^{1}H— and ^{13}C—NMR spectra of heavy petroleum fractions *(Reprinted from M.R. Gray "Lumped kinetics of structural groups: Hydrotreating of heavy distillate", Ind. Eng. Chem. Res. vol 29, p. 506, 1990. Copyright by the American Chemical Society)*

Table 1—11

^{13}C NMR Band Assignments for Petroleum Fractions

(Data from Khorasheh *et al.*, 1987)

Band	Range, ppm	Carbon Structures
1	11—15	γ—methyl
2	15—18	β—methyl
3	18—22.5	α—methyl, naphthenic methyl
4	22.5—37	methylene (CH_2)
5	37—60	methyne (CH)
6	100—129.5	$C_{aromatic}$—H
7	129.5—140	$C_{aromatic}$—C
8	140—160	$C_{aromatic}$—O,N,S

Table 1—12

Aromatic Carbon Content of Selected Residues by ^{13}C—NMR

(Data from Beret and Reynolds, 1990)

Residue	Fraction of Carbon in Aromatic Rings
Athabasca	0.316
Athabasca 525° C+	0.355
Boscan 343° C+	0.279
Cold Lake 525° C+	0.357
Duri 538° C+	0.224
Kern River 510° C+	0.330
Khafji 525° C+	0.358
Kuwait 538° C+	0.374
Maya 510° C+	0.401
Orinoco 525° C+	0.378

combine NMR data with elemental analysis to calculate group concentrations [Khorasheh *et al.*, 1987]. An example is given in Table 1—13. These concentrations can then be used to monitor chemical changes during processing.

Table 1—13
Structural Characteristics of Topped Bitumens (424° C+)
(Data from Gray *et al.*, 1991)

	Concentration, mol group/100g			
	ATH	**CL**	**LL**	**PR**
424—525° C Fraction				
Total α—carbon	0.58	0.65	0.58	0.58
Total paraffinic C	0.38	0.78	0.74	0.59
Total naphthenic C	3.41	3.26	3.63	2.87
Total aromatic C	1.99	1.74	1.52	2.15
525° C + Fraction				
Total α—carbon	0.37	0.51	0.59	0.62
Total paraffinic C	0.70	0.86	0.77	0.73
Total naphthenic C	2.67	2.43	2.59	2.48
Total aromatic C	2.32	2.49	2.24	2.38

ATH = Athabasca; CL = Cold Lake; LL = Lloydminster; PR = Peace River

Many early studies used NMR and elemental analysis data to construct average molecular structures as a representation of feeds and products [Petrakis and Allen, 1987]. The approach of trying to represent a complex mixture containing many chemical structures by a single average is subject to a number of errors and biases, and is only recommended as a means of visualizing chemical pathways [Boduszynski, 1984]. One method under development that avoids the limitation of the average molecule approach is to represent the oil fractions as ensembles of representative molecules. This Monte Carlo approach is computationally intensive, but has significant promise for studies of reaction pathways [Neurock *et al.*, 1990; Trauth *et al.*, 1993].

More sophisticated NMR methods can be used to identify the carbon or hydrogen types giving rise to a given signal. By selecting sequences of radio—frequency pulses, signals from selected carbons can be inverted. For example, CH_2 and CH signals can be inverted so that they appear below the baseline. Two dimensional NMR methods can be used to determine how carbon signals in the ^{13}C–NMR spectrum are correlated with proton signals in the ^{1}H–NMR. For example, these 2–dimensional methods can link the carbon and hydrogen signals from methyl groups. Petrakis and Allen [1987] give a review of the theory of NMR spectroscopy as applied to fossil fuels, including the more advanced pulse–sequence and 2–dimensional methods.

1.4.2 Average molecular weight and molecular weight distribution

Determination of molecular weight (MW) is important for residues as an alternative to distillation. The interconversion of boiling point and molecular weight is presented in quantitative terms in Chapter 2, but in general an increase in molecular weight gives an increase in boiling point. The methods for determining average molecular weight and molecular weight distributions are closely related to methods for analysing polymers.

Dissolution of a polymer in a solvent changes the vapor pressure and freezing point of the solvent. The most commonly used method is the thermoelectric measurement of vapor pressure (ASTM D2503), also called vapor–pressure osmometry (VPO). The sample is dissolved in a suitable solvent such as benzene. A drop of solution is suspended on a thermistor side by side with a drop of pure solvent in a chamber saturated with solvent vapor. The reduction in vapor pressure caused by the solute induces condensation on the drop of solution, which is then measured as a difference in temperature. High molecular–weight materials tend to give data that depend both on the solvent and the sample concentration.

The molecular weight measured by VPO is a molar average or number average, from the calculation of the molarity of the solute in the solvent. The ASTM D2503 method can be applied to samples with average molecular weights in the range 200 to 3000, although equivalent methods have been used to measure asphaltenes with MW over 10,000 and polymer MW up to 40,000 [Kroschwitz, 1985]. The lower limit is set by the requirement that the sample be nonvolatile at the temperature of the chamber (37° C).

The distribution of molecular weight in a nonvolatile residue sample can, in principle, be determined by gel permeation chromatography (GPC). Briefly, a sample is pumped through a chromatographic column as a dilute solution in a strong solvent, such as THF or toluene. The molecules diffuse into the column packing depending on their molecular weight; high–molecular–weight material diffuses the least, while low–molecular–weight material diffuses the most. The eluant from the column, therefore, appears in order of reverse molecular weight from high to low (Figure 1—10). An appropriate detector measures the quantity of sample in the eluant as a function of time; the time axis is converted to molecular weight by repeating the procedure with analytical standards of known molecular weight. The distribution of molecular weights can then be used to calculate averages as follows:

$$M_n = \Sigma\, w_i / \Sigma\, (w_i/M_i) \tag{1.3}$$

$$M_w = \Sigma\, (w_i M_i) / \Sigma\, w_i \tag{1.4}$$

$$M_z = \Sigma\, (w_i M_i^2) / \Sigma\, (w_i M_i) \tag{1.5}$$

where M_n is the number—average or molar—average molecular weight, M_w is the weight—average and M_z is the z—average molecular weight. M_i is the molecular weight of the fraction with weight fraction w_i. As illustrated in Figure 1—10 the M_n value is most influenced by low—molecular weight components, while M_z is most affected by heavy species.

GPC, in principle, can be used to track the changes in molecular weight distribution of residues as they undergo processing. Serious practical problems arise, however, in analyzing a heterogeneous oil fraction as compared to a synthetic organic polymer. A distribution curve is only useful if the data are quantitative. Detectors for liquid chromatrography respond to the properties of the eluant solution from the column, such as UV absorbance or refractive index. In the case of residues, the chemical composition changes with molecular weight, so that the response factor of a detector will also change. At present, no detectors are capable of giving a response proportional to the true concentration of complex hydrocarbons.

The calibration of molecular weight from a GPC column is also fraught with difficulty. The retention of a molecule in the gel is influenced by its molecular volume and to a lesser extent its polarity. Analytical standards are normally polystyrene of well—defined molecular weight, but polystyrene has a different relationship between molecular volume and mass than does an asphaltene material, and it also differs in chemical composition and polarity. Consequently, the calibration of GPC columns for analysis of residues must be considered approximate.

This problem is illlustrated in the data of Table 1—13, which compares data from vapor pressure osmometry (number—average molecular weight) to GPC results using toluene as a solvent. The calibration of the masses does not correspond at all; the number—average

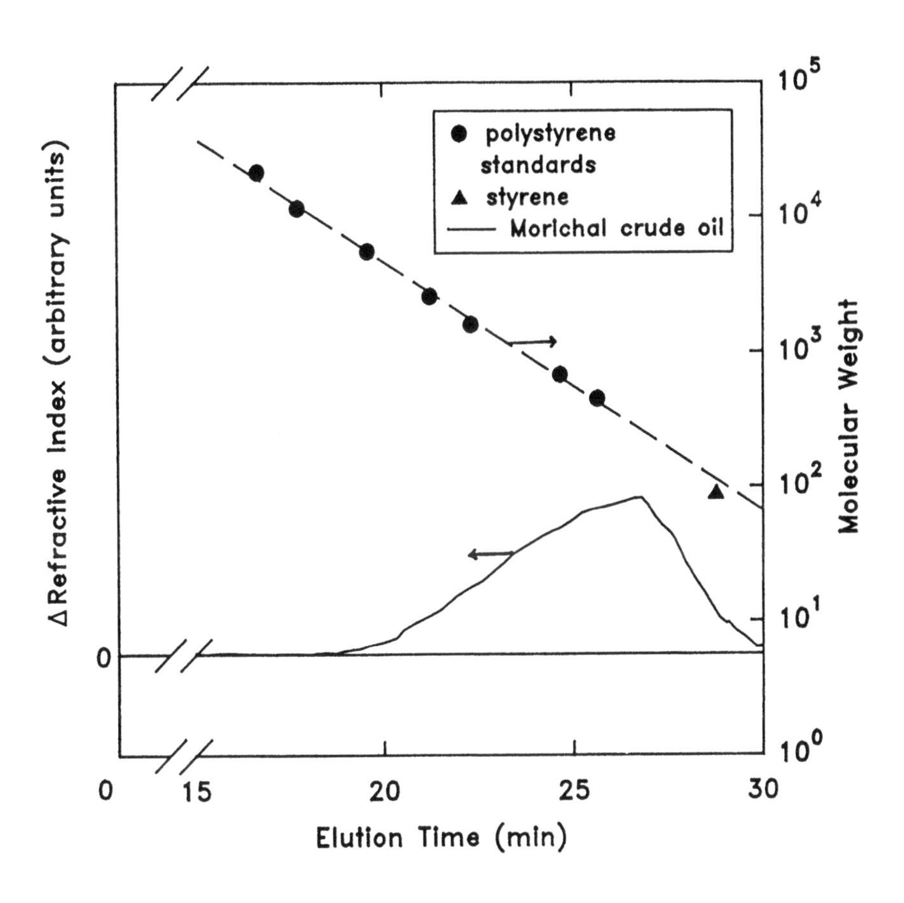

Figure 1—10

Gel Permeation Chromatogram of Morichal Crude (Data from Sanchez *et al.* [1984], using chloroform solvent and μ Styragel columns with polystyrene calibration standards. From these data for Morichal M_n = 393, M_W = 1196, and M_Z = 3500) molecular weights by GPC are consistently much lower than the result from VPO. In the gas oil cut the M_n from VPO was comparable to the M_z from GPC. The molecular weight distributions from GPC, therefore, are semi–quantitative and should only be used carefully in comparing closely related process samples.

Table 1–13
Molecular Weight of Residue Fractions
(Data from Gray *et al.*, 1992b)

		Molecular Mass (Da)			
		ATH	**CL**	**LL**	**PR**
424–525° C Fraction					
Avg. MW by VPO		460	490	420	470
MW by GPC	M_n	290	330	340	290
	M_w	370	420	430	380
	M_z	450	500	510	450
525° C + Fraction					
Avg. MW by VPO		1690	1325	1295	1250
MW by GPC	M_n	960	950	960	820
	M_w	3220	3080	3140	2990
	M_z*E–3	1.49	1.31	1.37	1.53

ATH = Athabasca; CL = Cold Lake; LL = Lloydminster; PR = Peace River

1.4.3 Mass spectrometry of residues

Application of mass spectrometry to 525° C+ residues is hampered by low volatility and resolution. The molecules in a sample must be volatilized in order to obtain molecular ions at the detector; but residue samples cannot be expected to volatilize even under vacuum. Complex molecules give correspondingly complex mass spectra under fragmentation conditions, so that the spectra give little useful information, and the resolution of the detectors does not allow determination of elemental composition at high masses.

The only technique of mass spectrometry that gives useful information under these limitations is field–ionization mass spectrometry (FIMS), which is designed to minimize the decomposition of the parent molecule due to electron bombardment. This technique is designed to register parent molecules without fragmentation. With a mass limit ca. 1800, this technique can be used to analyze material recovered by high—vacuum distillation techniques, and give data both on molecular weight and molecular classes (nitrogen compounds, saturates etc.) [Boduszynski, 1987]. Since residues may include molecular weights over 20,000, the FIMS technique cannot be used for routine analysis of these materials.

1.5 Environmental Properties of Heavy Oils and Residues

Each step in the production, refining, and distribution of petroleum products has an impact on the environment. Steam injection production processes for heavy oil can affect local supplies of ground and surface water, both through consumption of boiler feedwater and through production of oily waters and emulsions. Mining of tar sands disturbs surface land use, and the subsequent extraction generates large volumes of tailings with a high water content [MacKinnon, 1989]. The aqueous streams from such production operations are usually toxic to aquatic organisms, and potentially mutagenic [Birkholtz *et al.*, 1987; von Borstel and Mehta, 1987].

The environmental impacts of upgrading of residue are due to the transport of raw materials and products, the operation of upgrading processes, and the end use of the products and by—products of the oil. Although occupational health within a refinery operation is of concern, the discussion of the environmental aspects of upgrading will focus on materials that pass the plant boundary as products, waste streams and emissions to air and water. In particular, the prevention of sulfur dioxide releases to the atmosphere, during processing and end use of petroluem products, drives the design of upgrading plants and the specifications for upgraded products. Sulfur removal, therefore, will be a recurring theme throughout this book. In this chapter, data are presented on the biological effects of heavy oils and residues themselves. Environmental aspects of upgrading and the final products will be presented in Chapters 5—8.

1.5.1 Acute toxicity

Heavy oils, bitumens, and residues have a potential effect on the environment when they are released during transportation or at a plant site. Consequently, the effects of these materials on aquatic organisms, soil, and animals is of concern. The acute toxicity (i.e. the immediate toxic effect) of high boiling petroleum fractions is very low. Immediate toxic effects in marine and soil environments are due to compounds dissolved in water, such as phenols, naphthenic acids, and C_6—C_8 aromatics [Birkholtz *et al.*, 1987]. The latter compounds, benzene, toluene, ethylbenzene and xylenes are often referred to as "BTEX". The solubility of heavy oil, bitumen, and residues in water is very low, so that acute toxicity is minimal. The ubiquitous use of petroleum asphalt as a binder in road pavement and roofing products emphasizes the high stability of high–boiling hydrocarbons toward biological degradation, and their low toxicity. Addition of asphalt at 1000 kg/ha to agricultural soil in France, for example, did not affect plant growth [Gudin and Syratt, 1975]. Emulsions of asphaltic hydrocarbons have been used as a soil conditioner to improve water retention [Voets *et al.*, 1973].

The most serious effects of accidental spills are due to indirect effects, such as fouling of sea bottoms, beaches, and soil, with a concomitant reduction in biological activity. Dense oils sink to the bottom after a spill, and affect bottom dwelling organisms. The effect of a spill on soil depends on the penetration into the root zone and the resulting loss of soil structure and moisture supply [DeJong, 1980].

Oral toxicity of oil is a major concern in marine spills, due to ingestion by seabirds and aquatic mammals. In the case of residues and heavy oils, the very low water solubility minimizes the effect of ingesting these materials. Acute toxicity of residue fractions in mammals is very low, both via ingestion and skin contact [Stubblefield *et al.*, 1989

1.5.2 Mutagenicity and carcinogenicity

A greater concern with high–boiling fractions is the potential for mutagenic and carcingenic effects on animal cells. The classic carcinogen, benzo[a]pyrene, is found in heavy petroleum fractions along with chemically related polynuclear aromatics (PNA) with known mutagenic activity [Guerin *et al.*, 1978]. Tests of mutagenicity on bacteria (the Ames test), however, show very low mutagenic activity in natural petroleum fractions [Greenly *et al.*, 1982]. Any observable mutagenic activity was associated with the neutral fraction from column chromatography, containing PAH and neutral heterocycles of nitrogen, sulfur, and oxygen.

Tests of carcinogenicity have focused on the ability to induce tumors in the skin of mice. Regular application of petroleum blends and Wilmington crude did not induce tumors at all, in contrast to coal liquids which induced both tumors and mortality [Greenly *et al.*, 1982].

The lack of toxicity of the natural petroleum fractions, despite the presence of benzo[a]pyrene and related compounds, must be attributed to the extensive alkyl sustitutuion of aromatic rings in petroleum materials. In contrast to coal–derived liquids, aromatic groups such as PAH are present as a variety of alkyl–substituted isomers, while the parent hydrocarbons are seldom present at significant concentration. This substitution must account for the low mutagenicity and carcinogenicity of the natural crude oils and petroleum residues. Consequently, the only human health concerns associated with handling of unprocessed residues are burns from handling hot material, and chronic inhalation of vapors [Bright *et al.*, 1982].

1.6 Product Specifications for Bitumen, Heavy Oil, and Residue

Residues are often produced and consumed within a refinery or in proximate plants, in which case transportation and product specifications are not an issue. When a residue or heavy oil is to be sold on the open market, however, it must be transportable. This transport often requires that the product be pipelined to the purchaser, or shipped by tanker. In the former case, the viscosity of the oil must be low enough to pump it through a pipeline at ambient temperatures, about 250 cs at winter temperatures of 2—3° C in northern latitudes, or about 100 cs at 15° C.

One method, used at Cold Lake and Loydminster in Western Canada, is to dilute the oil with gas condensate until the viscosity is acceptable. The blend is then sold based on the content of condensate and oil. The main limitations to this approach are the supply of gas condensate, and precipitiation of asphaltenes in the pipeline due to the dilution with an aliphatic solvent.

An alternative is mild thermal treatment, intended only to reduce the viscosity. These viscosity reduction or visbreaking processes will be discussed in more detail in a later chapter. A different approach to reduction of viscosity of heavy oil, which is under development, is to emulsify it with water.

The most expensive alternative is to produce a synthetic crude oil, which meets some of the requirements of a conventional refinery. This alternative was adopted by Syncrude and Suncor in Alberta. Typical specifications for synthetic crude oils are listed in Table 1—14, and these specifications provide a target for upgrading processes in general when the intent is to sell the products on the open market.

Upgraded products must give a naphtha cut which has a nitrogen content low enough for catalytic reforming, a middle distillate cut which is suitable for use as a diesel blendstock, and a gas oil cut which can be used in a fluid—catalytic cracker. These downstream uses for the synthetic crude limit the acceptable nitrogen and sulfur contents, and constrain the cetane number of the mid—distillate. The upgrading to produce a synthetic crude gives a distribution of chemical types which is quite distinct from a virgin oil, even when the boiling distribution and heteroatom levels are comparable. This molecular difference between synthetic oils and conventional crudes limits the capacity of conventional refineries to accept synthetic crude oil. Conventional refineries must

TABLE 1—14
Typical Synthetic Crude Oil Compositions

Property	Syn—crude (1)	Husky Oil (2)	AOSTRA (3)	Alberta Light Crude (4)
Gravity, °API	32.4	32	32	38.5
Sulfur, Wt%	0.11	0.15		0.4
Nitrogen, ppm	570	500		1400
Naphtha (82—177° C)				
Vol%	16	8		25
Nitrogen ppm	1.1	1.0	10	
Sulfur ppm	2.2		50	
Mid—Distillate (177—343° C)				
Vol%	49	40(min)		40
Sulfur wt%	0.04	0.2	0.2	
Cetane #		40	40	
Gas Oil (343° C+)				
Vol%	33	40		24
Sulfur wt%	0.24	0.5	0.5	
Nitrogen wt%	0.14	0.1	0.1	
Basic N ppm	150	275		

1. Assay of synthetic crude, from Hyndman and Liu [1987].
2. Minimum specifications for Husky Upgrader, from Jeffries and Gupta [1986].
3. Minimum AOSTRA Synthetic Crude specifications, from Colyar *et al.*[1989].
4. Typical light crude composition, from Van Driesen *et al.*[1979].

blend the synthetic crude with conventional crude oil. The synthetic crude oil gives a poor quality diesel fuel unless extra processing is performed, and the high aromatic carbon content makes the synthetic crude a poor candidate for catalytic cracking. The alternative is to specially design the refinery to accept synthetic crude, as in the case of Shell Scotford at Fort Saskatchewan, AB.

Notation

K_W	Watson characterization factor
M_i	molecular weight of fraction i, Da
M_n	number—average molecular weight, Da
M_w	weight—average molecular weight, Da
M_z	z—average molecular weight, Da
SG	specific gravity at 15°C
T_b	boiling point at 101.3 kPa
w_i	weight fraction of *i*th fraction

Further Reading

AOSTRA Technical Handbook on Oil Sands, Bitumens, and Heavy Oils, L.C. Hepler and C. Hsi, eds., AOSTRA, 1989.

Speight, J.G. The Chemistry and Technology of Petroleum, 2nd Ed., Marcel Dekker, New York, 1991.

Problems

1.1 Two pathways are normally available for upgrading of bitumen; removal of carbon or addition of hydrogen. For a production rate of 5 x 10^3 m^3/d of Athabasca bitumen, calculate the maximum possible yield of synthetic crude by each pathway:

a) Carbon rejection: Coke is formed as a by—product from the bitumen with the following composition:

Carbon	79.99 wt%
Hydrogen	1.66
Sulfur	6.63
Nitrogen	1.9
Ash	6.92
Volatiles	2.9

b) Hydrogen addition: Hydrogen is added to the bitumen to improve the H/C ratio. Additional data:

Bitumen		
	Specific Gravity	1.019 (15° C)
	Carbon	83.3 wt%
	Hydrogen	10.5
	Nitrogen	0.4
	Oxygen	1.2
	Sulfur	4.6
Synthetic Crude		
	Specific Gravity	0.835 (15° C)
	Carbon	86.4 wt%
	Hydrogen	13.0
	Nitrogen	0.1
	Sulfur	0.5

The nitrogen, oxygen, and sulfur are produced as NH_3, H_2O, and H_2S, or are retained in the coke. Ignore the formation of light ends.

1.2 How much hydrogen would be required to totally hydrogenate the 424° C+ fraction of Lloydminster heavy oil (in kg hydrogen/kg oil)? Total hydrogenation would convert all aromatic carbon to aliphatic carbon, convert sulfur to hydrogen sulfide, etc. The carbon content of the bitumen is 84% by weight.

1.3 An upgrader has been proposed to process 30,000 bbl/d of Maya crude. How much sulfur and nitrogen would be produced in upgrading this feed to meet synthetic crude specifications? Assume that the plant produces no solid by—products and emits no sulfur dioxide.

1.4 Estimate the average elemental formula for Morichal crude, i.e. $C_nH_mS_iN_j$ where n, m, i, and j are mols of element per mol of oil. State your assumptions.

CHAPTER

2

Thermodynamic and Transport Properties of Heavy Oil and Bitumen

2.1 Equations of State for Hydrocarbon Mixtures

Process engineering calculations require a variety of properties for a process stream, including pressure—volume—temperature (PVT) relationships, densities, enthalpies, entropies, fugacities, thermal conductivities, etc. The most efficient method is to use an accurate equation of state to calculate the thermodynamic properties, and to use auxiliary correlations for transport properties such as viscosity and thermal conductivity. A number of studies have verified that the equation of state approach can give predictions for petroleum distillates and residues that are more accurate than methods based on K—values or activity coefficients (see, for example, Sim and Daubert [1980]). Thermal properties of real fluids can also be calculated from the equation of state, for example, the departure of the enthalpy from that of an ideal gas can be obtained directly from PVT properties:

$$\frac{H - H^{ID}}{RT} = z - 1 + \frac{1}{RT}\int_{\infty}^{V} [T(\partial P/\partial T)_V - P]\, dV \tag{2.1}$$

where z is the compressibility of the fluid and the superscript "ID" indicates ideal gas properties.

2.1.1 Cubic Equations of State

The equation of state selected for process calculations must be reasonably accurate for liquid and vapor phase mixtures of many components over a wide range of temperature and pressure, and it must be computationally efficient. These requirements have led to the almost universal adoption of cubic equations of state for general process engineering calculations for hydrocarbon gases and liquids. Corrections have been developed to allow use of the equations for hydrocarbon mixtures with carbon dioxide, hydrogen sulfide, and hydrogen. The general form of a cubic equation of state is

$$P = \frac{RT}{v - b} - \frac{a(T)}{F(v)} \tag{2.2}$$

where R is the gas constant and b and a are parameters for a given fluid mixture. The qualitative significant of b is a molecular volume of exclusion, while a represents intermolecular forces. The most widely used equations are the Soave–Redlich–Kwong (SRK) [1972]:

$$P = \frac{RT}{v - b} - \frac{a(T)}{v(v+b)} \tag{2.3}$$

where a(T) is a polynomial function of reduced temperature and acentric factor (ω), and the Peng–Robinson (PR) [1976]:

$$P = \frac{RT}{v - b} - \frac{a(T)}{v(v+b)+b(v-b)} \tag{2.4}$$

The adjustable parameters are evaluated using the characteristics of the critical isotherm, which has an inflection at the critical point, so that

$$[\partial P/\partial v]_{T_c} = [\partial^2 P/\partial v^2]_{T_c} = 0 \text{ when } v = v_c \tag{2.5}$$

where the subscript c indicates values at the critical point. This condition gives two equations in two unknowns from a two–constant cubic equation (i.e. (2.3) or (2.4)) which is sufficient to solve for a and b in terms of critical properties. In addition the critical compressibility is given by:

$$z_c = \frac{P_c\, v_c}{R\, T_c} \tag{2.6}$$

The conditions at the critical point, therefore, are overdetermined and a two–constant equation cannot satisify all three constraints in equations (2.5) and (2.6). In order to ensure that the PVT behavior follows real fluids at least qualitatively, equation (2.5) is used and the equations are only approximate for the value of z_C. A typical critical compressibility for a hydrocarbon is $z_C = 0.27$, while the SRK gives $z_C = 0.333$ and PR gives $z_C = 0.308$. The improved fit in the critical region is an advantage for the PR equation in some applications, and this equation will be used in the following discussion and derivations.

The two–constant equations of state follow the corresponding states principle in the sense that all components have equivalent reduced critical behavior, i.e. the same z_C. This restriction is more accurate than corresponding states itself, which assumes that all compounds have identical PVT behavior at the same values of $T_r = T/T_C$ and $P_r = P/P_C$. Real fluids have differing values of z_C, for example 0.29 for methane, 0.26 for octane, 0.225 for methanol and 0.303 for hydrogen. In practice, however, the PR equation is sufficiently accurate for many process calculations on hydrocarbons and their mixtures.

Using the existence of an inflection point at the critical condition (equation (2.5)), the PR equation (2.4) gives the following constants:

$$a = 0.45724\, \alpha \frac{R^2\, T_c^2}{P_c} \tag{2.7}$$

$$b = 0.07780 \frac{R\, T_c}{P_c} \tag{2.8}$$

The parameter α was inserted to improve the fit of the equation of state to vapor–pressure data away from the critical point:

$$\alpha^{1/2} = 1 + \kappa(1 - T_r^{\,1/2}) \tag{2.9}$$

$$\kappa = 0.37464 + 1.54226\omega - 0.26992\, \omega^2 \tag{2.10}$$

$$\omega \equiv -\log_{10}[P_{sat}/P_c]_{T_r=0.7} - 1.0 \qquad (2.11)$$

The definition of α in equation (2.9) is such that $\alpha = 1.0$ at the critical point for all compounds, so that it has no effect on the critical behavior or the value of z_C. Away from the critical point, the Pitzer acentric factor ω is used to correct for real–fluid PVT behavior. The acentric factor, calculated from the saturation pressure (P_{sat}) at a reduced temperature of 0.7, is available for most compounds of interest.

The strength of the equation–of–state approach is that process calculations for a large number of compounds requires data only for the critical properties T_C and P_C, and the acentric factor ω. By solving for the cubic roots of equation (2.4), with constants from equations (2.7) and (2.8), the molar volumes of the liquid and vapor phases are obtained as the largest and smallest of the three roots. If the fluid is in the single–phase region, then equation (2.4) will give only one real root.

The two–constant equations of state (both SRK and PR) have two key limitations: they give poor predictions of liquid–phase density, and they are inaccurate for many nonhydrocarbons. The SRK and PR equations are so efficient for process calculations, however, that a major focus of the research effort on equations of state has been on correcting the equations. Examples are volume translation methods for improving critical point predictions and liquid densities [Chou and Prausnitz, 1989] and modification of equation (2.10) for the coefficient κ to improve accuracy for non–hydrocarbons such as alcohols, ketones, ethers, and halogenated compounds [Stryjek and Vera, 1986; Proust and Vera, 1989]. Such extended equations of state are used by some process engineering software, such as HYSIM.

2.1.2 Mixing rules for cubic equations

The attractiveness of the cubic equations is partly due to the ease with which they can be extended to calculations for mixtures. In order to use equation (2.4) to calculate v_L and v_G for a mixture, the parameters a and b must be evaluated for the mixture. These values are calculated from the component parameters for the PR equation as follows for an n–component mixture:

$$a = \sum_{i=1}^{n} \sum_{j=1}^{n} x_i x_j a_{ij} \qquad (2.12)$$

$$a_{ij} = (1 - \delta_{ij})\, a_i^{1/2}\, a_j^{1/2} \qquad (2.13)$$

$$b = \sum_{i=1}^{n} x_i b_i \tag{2.14}$$

where x_i is mol fraction of component i in the liquid phase. The parameter δ_{ij} is a binary interaction coefficent to correct for non–ideal binary interactions. Values of δ_{ij} range from 0.0 for pairs of n–alkanes to 0.7 for hydrogen–hydrocarbon binaries. Carbon dioxide, hydrogen sulfide and water have non–zero interaction parameters, in the range 0.13 to 0.5. The extended PR equation for non–hydrocarbons of Stryjek and Vera [1986] requires two interaction parameters per binary pair. For hydrocarbon systems in upgrading applications, however, values of δ_{ij} for hydrogen sulfide–hydrocarbon and hydrogen–hydrocarbon binaries are sufficient.

Once the components and their interactions are characterized, then calculations of PVT properties, phase behavior, and excess thermodynamic properties are straightforward. Details of such calculations are given by Smith and Van Ness [1987]. Such computational details are handled by process engineering computer software once the components in a mixture have been properly characterized.

2.2 Phase Behavior of Undefined Mixtures

Fluids such as crude oil, butumen, and gasoline contain so many components that calculations of properties cannot be performed on a component–by–component basis. Recall that calculations for pure components using equations of state require several parameters:

MW (molecular weight) — converts mass into moles of compound

T_c — critical temperature

P_c — critical pressure

ω — acentric factor, based on boiling point at standard conditions

These parameters have a straightforward physical meaning for pure components; even when compounds do not reach the critical condition due to instability we understand the definition of the critical point for a pure component and can estimate its temperature and pressure.

In order to work with complex mixtures, we must define pseudocomponents which contain a number of related compounds boiling over a narrow range of temperature. These pseudocomponents can then be assigned parameter values for calculating phase behavior. For example, consider a mixture of hydrocarbons boiling between 300 and 325° C. This mixture has an average boiling point ca. 312° C, and so it has an average acentric factor. Similarly, the mixture has an average molecular weight which can be measured by vapor—pressure osmometry. The most common approach, however, is to characterize such mixtures of hydrocarbons in terms of experimental boiling curves, and use these to estimate parameters for calculating phase behavior.

2.3 Measurement of Oil Distillation Characteristics

2.3.1 Batch distillation

Nonequilibrium batch distillation in standardized equipment is most commonly used to measure boiling curves. A typical system is illustrated in Figure 2—1. The vapor temperature and volume distilled are measured as the oil is slowly boiled off. The ASTM (American Society for Testing and Materials) D86 determination for light fractions is at atmospheric pressure, while the ASTM D1160 determination uses vacuum distillation down to 130 Pa to determine vacuum residue content. In either case the raw data are a record of temperature versus volume distilled. The residue is the material remaining in the flask after a run, atmospheric residue from ASTM D86 and vacuum residue from ASTM D1160. Losses (Initial — Collected — Residue) are due to volatile components which are not condensed. In the case of vacuum distillation, the actual temperature is kept below 250° C to prevent decomposition of the oil, but the normal boiling points go up to 525° C.

The measured temperatures are at the operating pressure, so they are then converted to a standard basis of 101.3 kPa, using standard tables or charts. An equation for converting the measured boiling temperture, T, to the normal boiling point, T_b, was given by Maxwell and Bonnett [1955]:

$$T_b = \frac{748.1(A + 0.0002867)}{(1/T) - 0.0002867 + 0.2145(A + 0.0002867)} \tag{2.15}$$

$$A = \frac{5.9082 - \log_{10} P}{2926.8526 - 43 \log_{10} P}; \quad P < 0.0387 \text{ psia} \tag{2.15a}$$

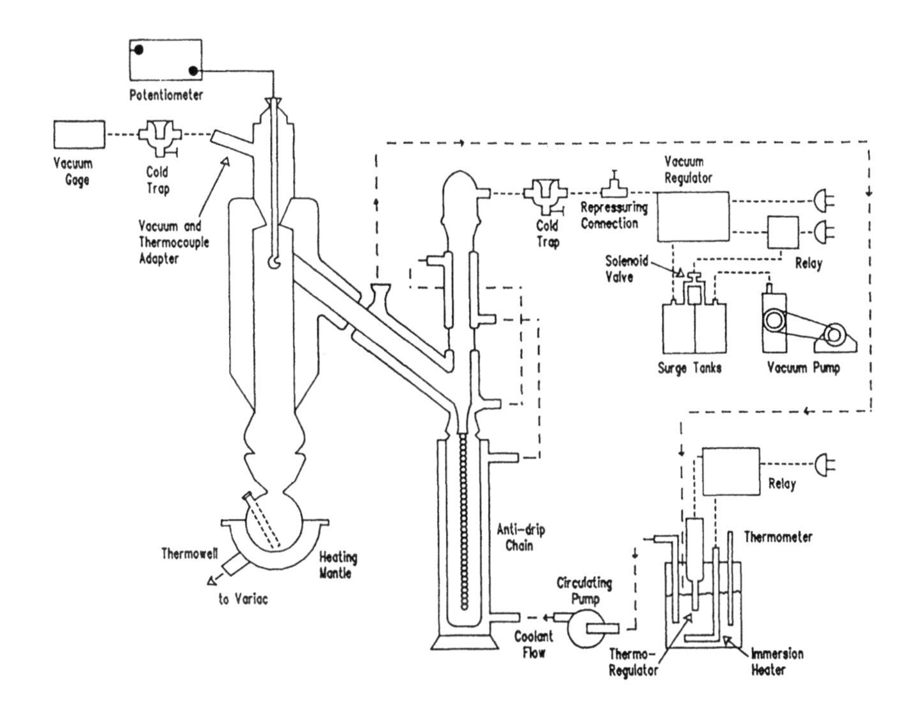

Figure 2–1
Batch distillation equipment for ASTM D1160

$$A = \frac{5.0442 - \log_{10}P}{2499.0330 - 95.76\,\log_{10}P};\ 0.0387 \leq P \leq 14.7 \text{ psia} \tag{2.15b}$$

Here T_b and T are in °F, while P is the measured pressure in the distillation apparatus (psia) as the liquid boils under vacuum. Equation (2.15) is based on conventional petroleum fractions with a Watson characterization factor $K_W = 12$ (Equation 1.2). A correction for samples with $K_W < 12$ is

$$T_{b,actual} = T_{b,\ Eq.(2.15)} - 2.5 \cdot F \cdot (K_w - 12)\log_{10}(P/14.696) \tag{2.16}$$

where $F = 0$ for $T_b \leq 200°F$; $F = -1 + 0.005 \cdot T_b$ for $200 \leq Tb \leq 400°F$; and $F = 1$ for $T_b > 400°F$. Tsonopoulos *et al.* [1986] found that equations (2.15) and (2.16) worked well at pressures below atmospheric even for aromatic coal liquids; therefore, this method can be applied to conventional petroleum fractions as well as upgraded materials. Tsonopoulos *et al.* [1986] give corrections for pressures above atmospheric pressure.

2.3.2 Simulated distillation

The distillation curve can also be determined by simulated distillation using gas chromatography for oils with components boiling between 55 and 538°C (ASTM Method D2887). A sample of the oil is injected into a stream of carrier gas (usually helium) which sweeps it into an analytical column in an oven. The column packing preferentially adsorbs the heavier components of the oil. As the temperature of the column is raised, heavier and heavier components are released from the column and swept into a detector, which senses changes in the thermal conductivity of the gas. For a given program of heating, the retention time in the column is proportional to the normal boiling point of the compounds. Figure 2—2 illustrates such a calibration curve, using a known series of n—alkanes as standards to calibrate the boiling point.

This method of calibration can be used in two ways: one is to relate boiling point to retention time as in Figure 2—2, to report a boiling curve for the mixture. The second is to determine how much sample elutes between each pair of known n—alkanes, to give a determination in terms of carbon number. The chromatographic column is not able to separate the individual components, so that reporting an analysis by carbon number is simply an alternate method of defining pseudocomponents. The temperature—carbon number relationship for the

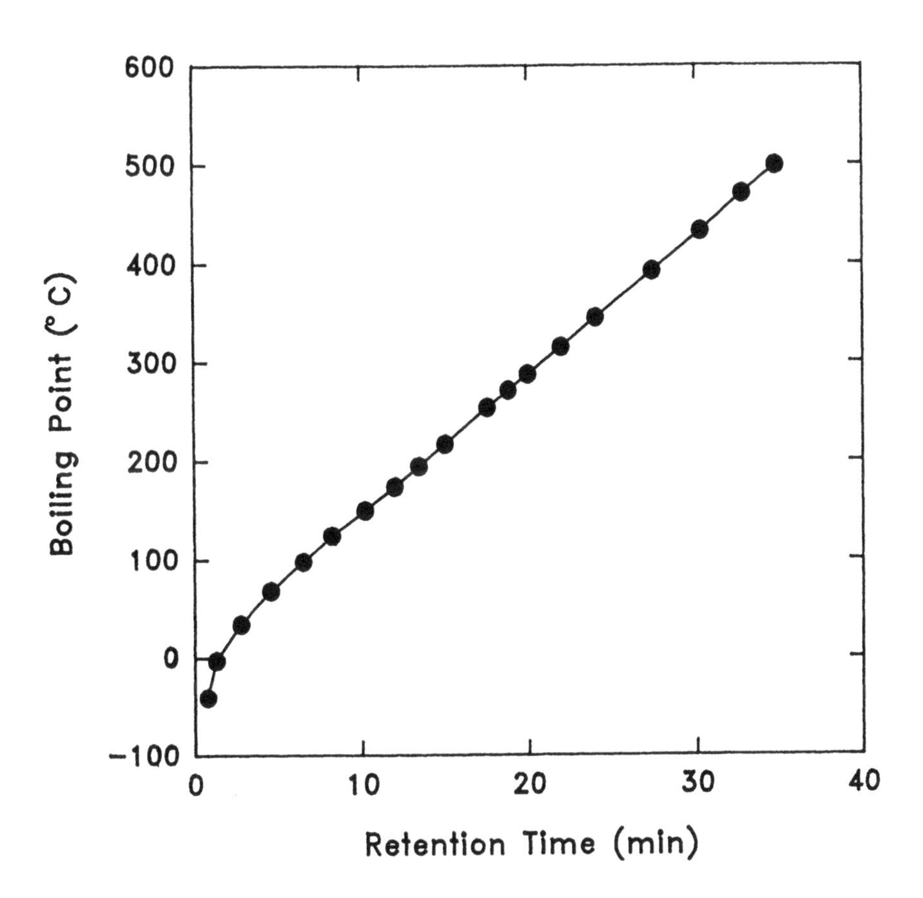

Figure 2–2
Calibration curve for simulated distillation (Data from ASTM, 1990)

n—alkanes is given in Table 2—1.

Table 2—1
Boiling Points of Normal Paraffins

Carbon #	BP, °C	Carbon#	BP, °C
2	—89	24	391
4	0	26	412
6	69	28	431
8	126	30	449
10	174	32	466
12	216	34	481
14	254	36	496
16	287	38	509
18	316	40	522
20	344	42	534
22	369	44	545

Note that the boiling point increases by approximately 10—14° C for each additional carbon.

The carbon number method is most common in the analysis of gas condensate because the psuedocomponents are easily represented by the equivalent n—alkane for calculations by equation of state. For example, a gas condensate boiling between 36 and 174° C is easily represented by the 6 n—alkanes that boil in this range.

The simulated distillation method is convenient because it can be automated, but it suffers from several drawbacks:

1. The sample must be distillable, i.e. no residue fraction can be present. Methods using internal standards have been proposed for residual oils, but have not been generally accepted.
2. Fractions of boiling cuts cannot be collected for subsequent analysis.
3. Aromatic compounds do not follow the calibration of boiling point with retention time based on n—alkanes. Multiring aromatics tend to elute earlier than expected, as illustrated in Figure 2—3. The compounds range from benzene (2) and thiophene (3) which are in good agreement, to four ring aromatics such as pyrene (45) and chrysene (50) which give deviations in NBP of 40—50° C. The calibration method prevents use of ASTM 2887 for highly aromatic oils such as coal liquids. An example high—sulfur coker gas oil gave

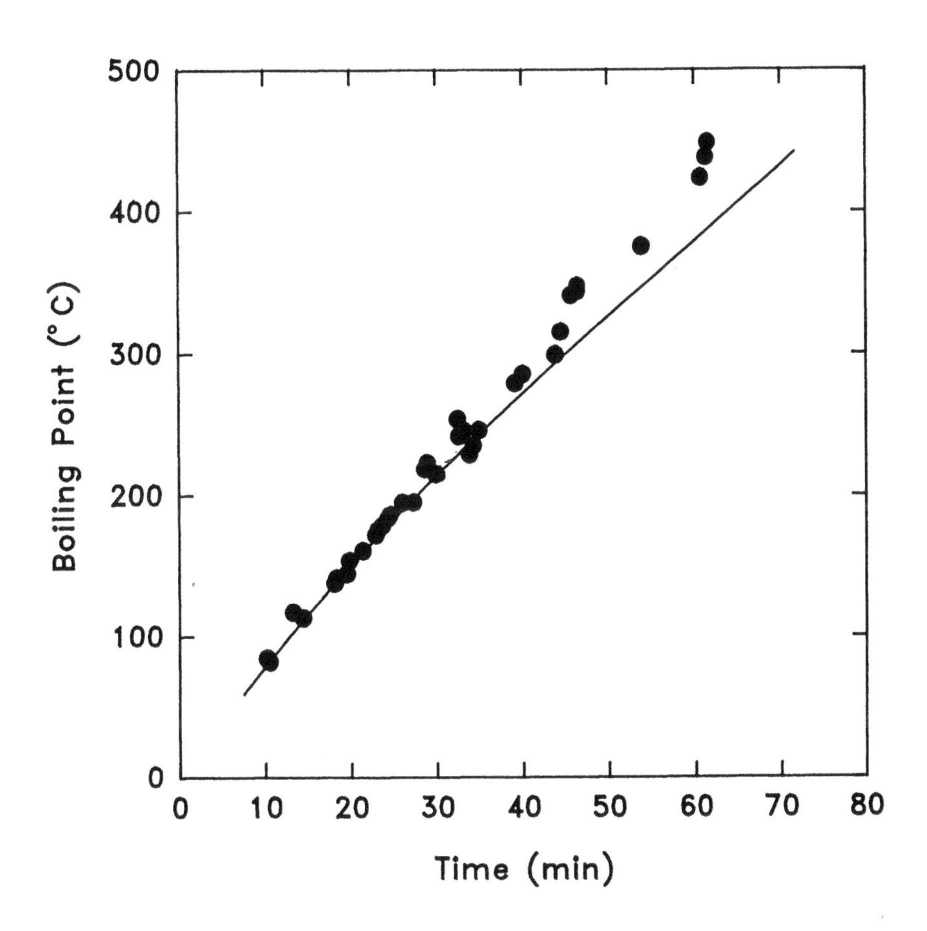

Figure 2–3
Boiling point–retention time relationship for aromatics (Data from ASTM, 1990)

deviations of 7—10° C between the true boiling point and the simulated distillation curves.

2.3.3 True boiling point (TBP) distillation

A more time—consuming method of determining distillation behavior is to use nearly equilibrium conditions. This separation requires a column with about 15 theoretical plates and a reflux ratio of at least 5:1. The ASTM method is D2892. As indicated by its name, the reflux gives true boiling points which, are essential for correlating other properties such as critical properties and molecular weight. The most common approach is to convert ASTM data to TBP curves rather than doing direct measurements (Section 2.3.5).

2.3.4. Equilibrium flash vaporization (EFV) curves

The distillation methods are all incremental; they remove the vapors from the column, leaving a heavier and heavier oil to distill. These measurements do not correspond to liquid—vapor equilibrium of an oil where all of the vapor components are retained in contact with the liquid, as illustrated in Figure 2—4. The measurement of equilibrium flash curves is very time—consuming, so the normal procedure is to calculate equilibrium conditions using pseudocomponents from boiling curves via an equation of state.

2.3.5 Conversion of ASTM boiling point data to true boiling points

Graphical and numerical techniques have been developed to convert ASTM D86 and D1160 data into TBP and EFV data. The TBP curve is needed for defining pseudocomponents to characterize mixtures. Edmister [1988] recommends converting ASTM data on an as—measured basis, i.e. convert vacuum ASTM D1160 data to TBP data at a pressure of 1.33 kPa (10 mm Hg). The TBP data are then converted to 101.3 kPa basis using equation 2.15.

The conversion of the ASTM D1160 boiling point curve to a true boiling point curve assumes that the temperatures at and above the 50% points are identical, so that corrections are needed only for the portion of the curve up to 50% distilled. The temperature differences for each volume (or weight) increment away from the 50% point (i.e. 30%—50% and 50%—70%) are then corrected using Figure 2—5. Apart from the initial range (0—10%) the corrections are quite small, $\lesssim$ 4 C°.

Empirical graphical correlations for converting TBP data to equilibrium flash volume curves are given in the API Technical Data Book [1983], and by Edmister [1988]. The recommended method, however, is to use the equation of state method to calculate any distillation or equilibrium volumes.

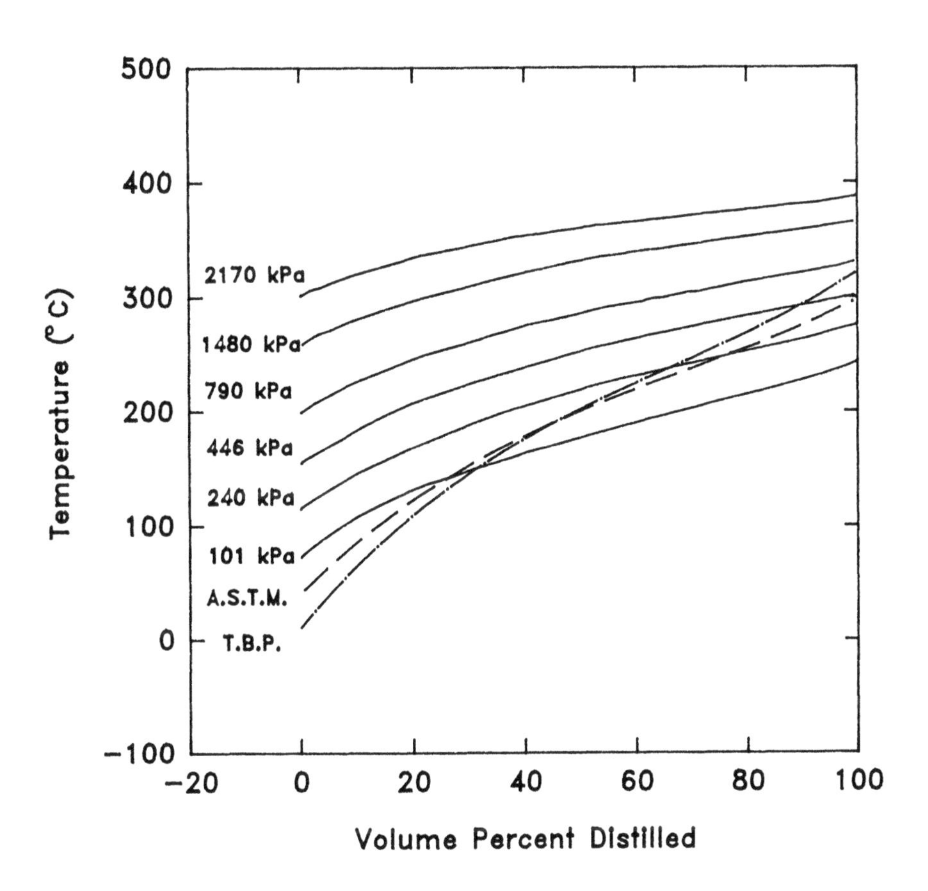

Figure 2–4
Example ASTM, TBP, and equilibrium flash vaporization curves (Data from Edmister, 1988)

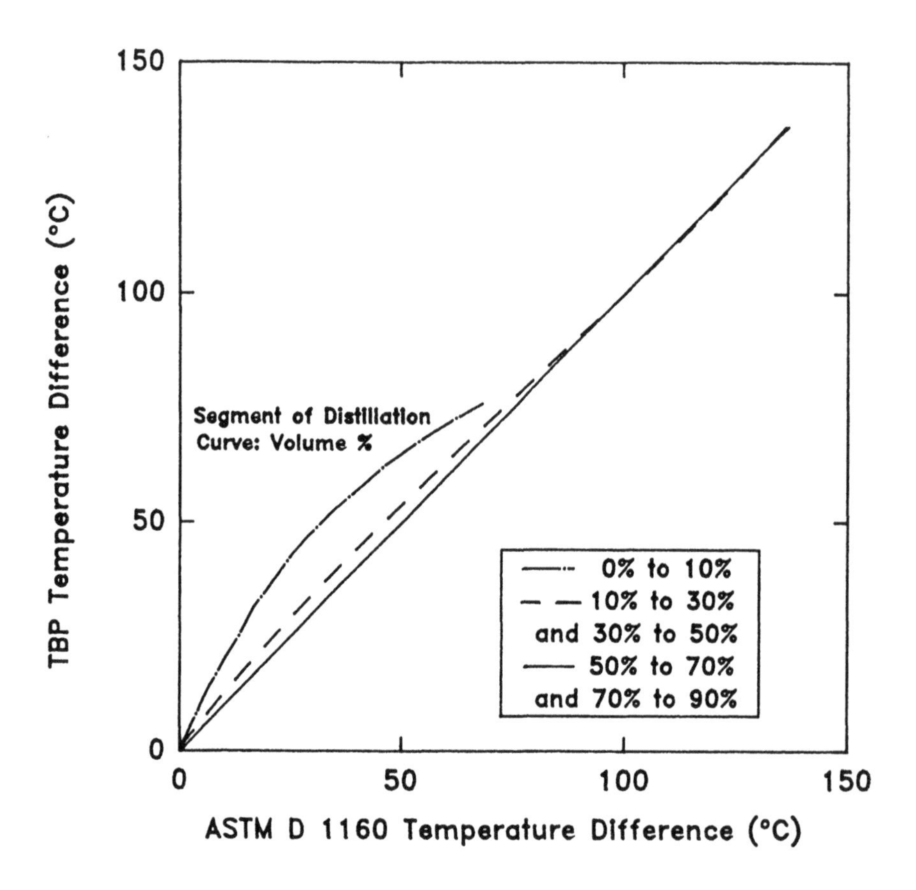

Figure 2–5
Correction of ASTM D1160 distillation data to true boiling point (Data from Edminster, 1988)

2.3.6 Specific Gravity

Specific gravity data are useful both for material balance calculations on petroleum streams, and for characterizing the oil. The combination of specific gravity and boiling point is sufficient to obtain a rough estimate of the molecular weight and thermodynamic properties. The ease of determination of specific gravity makes it a preferred characterization measurement, as compared to molecular weight measurements which are subject to error and bias.

2.4 Calculation of Mean Boiling Points

The boiling curve represents a distribution of properties, but in order to use pseudocomponents we must be able to calculate appropriate average properties. An average of a continuous property, T, over the range from a to b is defined as follows:

$$\overline{T} = \int_a^b T(q)dq/(b-a) \tag{2.17}$$

To obtain the average temperature, we must integrate the area under the curve of temperature (T) versus quantity (q). The quantity, however, could be volume (from a distillation curve), weight, or moles. A different average is obtained from each of these quantities. Consider, for example, a mixture of equal volumes of hexane (MW=86, NBP=69° C, s.g.=0.654) and decane (MW=142, NBP=174° C, s.g.=0.73). The volume average boiling point (VABP) is calculated from the summation form of equation (2.17).

$$\text{VABP} = 0.5 \cdot 69 + 0.5 \cdot 174 = 121.5^\circ \text{C} \tag{2.18}$$

The weight average boiling point (WABP) is calculated from the weight fractions of hexane (0.473) and decane (0.527), to obtain a value of 124.3° C. Similarly the molal average boiling point (MABP) is calculated using the molar quantities:

$$\text{MABP} = x_1T_1 + x_2T_2 = 0.6 \cdot 69 + 0.4 \cdot 174 = 111.3^\circ \text{C} \tag{2.19}$$

The choice of how to define quantity gives a range of 13° C in the average boiling point! This example illustrates how a weight average will be biased toward the densest components, while the molar average is biased toward the lightest components.

In integrating a boiling curve the same considerations will apply. The magnitude of the difference between estimation methods depends on the spread between the components. In a boiling curve this can be defined in terms of the slope of the curve. Consider the following example oil:

Vol. %	Temp., °C
IBP	38
5	54
10	67
20	88
30	103
40	118
50	138
60	159
70	196
80	240
90	311
EP	338

The volume–average boiling point (VABP) can be calculated using 5 evenly spaced points from the distribution, each representing 20% of the total oil.

$$\mathrm{VABP} = \sum_{i=1}^{5} v_i T_i \tag{2.20}$$

$$= (0.2T_{10\%} + 0.2T_{30\%} + 0.2T_{50\%} + 0.2T_{70\%} + 0.2T_{90\%})$$

$$= 162^\circ\mathrm{C}$$

where v_i is the volume fraction. Equation (2.20) is the summation approximation to the integral in equation (2.17). The slope is the change in temperature going from 10 to 90% liquid volume.

$$\mathrm{Slope} = (311-67)/80 = 3.05\ {}^\circ\mathrm{C}/\%\ \mathrm{recovered} \tag{2.21}$$

These data for VABP and slope can then be used to obtain a correction

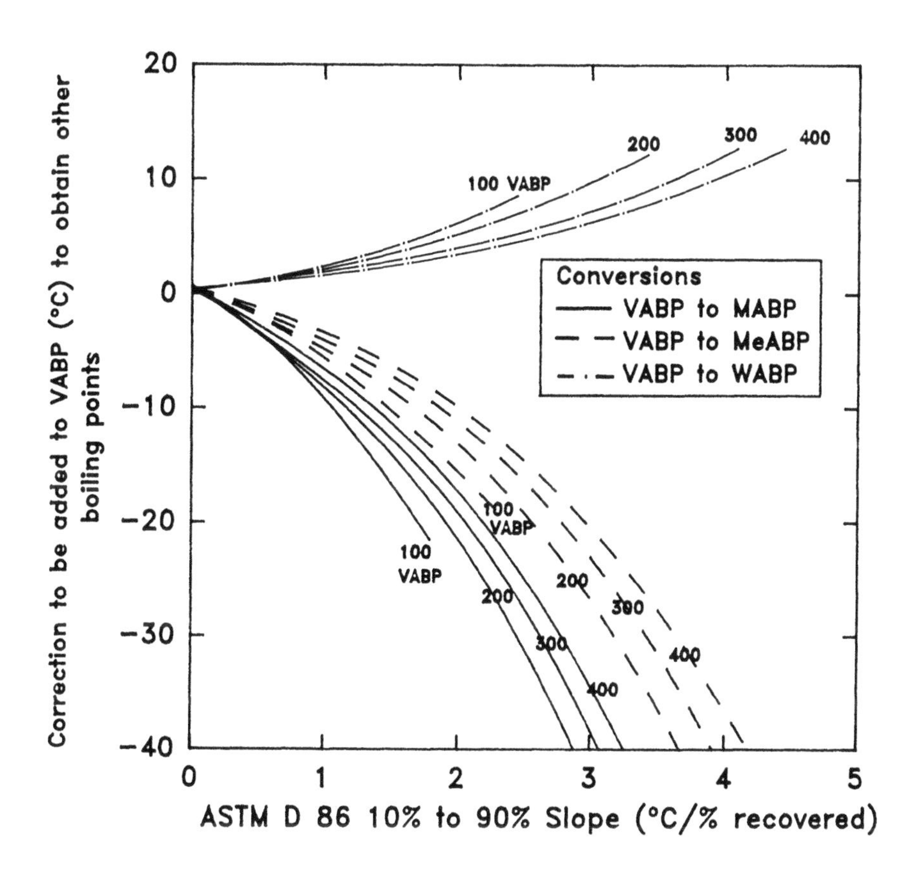

Figure 2–6
Conversion of volume average boiling points for distillates to molal, mean, and weight average boiling points using the slope of the ASTM D86 boiling curve (Adapted from GPSA Data Book, 1987)

value from charts, illustrated in Figure 2—6. The corrected values are given in Table 2—2.

Table 2—2
Example Corrections to Average Boiling Point

Average Basis	Correction, °C	Boiling Point, °C
Volume		162
Molal	—44	118
Mean	—30	132
Weight	+10	172

The narrower the cut, the smaller the value of the slope, and from Figure 2—6 the smaller the correction term. For narrow boiling fractions, on the order of 20—50 C°, the data of Figure 2—6 show that corrections between the different average boiling points are negligible. The mean—average boiling point (MeABP) is used for correlating other properties such as molecular weight:

$$\text{MeABP} = 1/2\ [\ \text{MABP} + (\Sigma\, v_i T_i^{1/3})^3] \tag{2.22}$$

2.5 Estimation of Critical Properties

At the true critical point, the liquid and the vapor become indistinguishable. Both pure components and mixtures exhibit a true critical point. For mixtures, however, the true critical point is not an additive property based on the individual components. Pseudocritical properties are therefore defined for mixtures as follows:

$$P_{pc} = \sum_{i=1}^{n} x_i P_{ci} \qquad T_{pc} = \sum_{i=1}^{n} x_i T_{ci} \tag{2.23}$$

This calculation of pseudocritical values as a weighted sum of the pure components is called Kay's Rule. The pseudocritical temperatures and pressures of mixtures, along with the molecular weight and the acentric factor, can also be estimated for petroleum fractions using correlations with average boiling point and specific gravity (or API gravity). This equation—based method is suitable for computer calculations. Process engineering software programs use data for boiling point and specific gravity (or other combinations of properties) to estimate the missing data for equation of state calculations. The correlations of Twu [1984] use the n—alkanes as reference compounds, and follow the principle of corresponding states. The benefit of this approach is consistency in estimating properties as a function of the dependent variables T_b and specific gravity. The properties of the reference n—alkanes were correlated as follows:

Critical temperature:

$$T_c^\circ = T_b(0.533272 + 0.191017 \times 10^{-3}T_b + 0.779681 \times 10^{-7}T_b^2 - 0.284376 \times 10^{-10}T_b^3 + 0.959486 \times 10^{28}/T_b^{13})^{-1} \quad (2.24)$$

Critical volume:

$$V_c^\circ = [1 - (0.419869 - 0.505839\alpha - 1.56436\alpha^3 - 9481.7\alpha^{14})]^{-8} \quad (2.25)$$

Specific gravity:

$$SG^\circ = 0.843593 - 0.128624\alpha - 3.36159\alpha^3 - 13749.5\alpha^{12} \quad (2.26)$$

Molecular weight:

$$T_b = \exp(5.71419 + 2.71579\theta - 0.286590\theta^2 - 39.8544/\theta - 0.122488/\theta^2) - 24.7522\theta + 35.155\theta^2 \quad (2.27)$$

Critical pressure:

$$P_c^\circ = (3.83354 + 1.19629\alpha^{1/2} + 34.8888\alpha + 36.1952\alpha^2 + 104.193\alpha^4)^2 \quad (2.28)$$

where

$$\alpha = 1 - T_b/T_c^\circ$$

$$\theta = \ln MW^\circ \quad (2.29)$$

Here temperature is R, volume is ft^3/lb—mol, and pressure is in psia.

2.5.1 Critical temperature

The critical properties of each pseudocomponent can then be calculated as a perturbation of the corresponding property of the n–alkane homologous series. The pseudocomponent critical temperature is referenced to the critical temperature (T_c^o) and specific gravity (SG^o) of the n–alkane with the same boiling point:

$$T_c \text{ (psia)} = T_c^o \, [(1+2f_T)/(1-2f_T)]^2 \tag{2.30}$$

$$f_T = \Delta SG_T[-\,0.362456/T_b^{1/2} + (0.0398285 - 0.948125/T_b^{1/2})\Delta SG_T] \tag{2.30a}$$

$$\Delta SG_T = \exp[5(SG^o - SG)] - 1 \tag{2.30b}$$

In these equations the values of T_b and SG are the boiling point (R) and specific gravity of the pseudocomponent respectively.

2.5.2 Critical volume

$$V_c = V_c^\circ \, [(1+2f_v)/(1-2f_v)]^2 \tag{2.31}$$

$$f_v = \Delta SG_v[0.466590/T_b^{1/2} + (-0.182421 + 3.01721/T_b^{1/2})\Delta SG_v] \tag{2.31a}$$

$$\Delta SG_v = \exp[4(SG^{\circ\,2} - SG^2)] - 1 \tag{2.31b}$$

2.5.3 Critical pressure

$$P_c = P_c^\circ \, (T_c/T_c^\circ)(V_c^\circ/V_c)[(1+2f_p)/(1-2f_p)]^2 \tag{2.32}$$

$$f_p = \Delta SG_p[(2.53262 - 46.1955/T_b^{1/2} - 0.00127885\,T_b) + (-11.4277 + 252.14/T_b^{1/2} + 0.00230535\,T_b)\Delta SG_p] \tag{2.32a}$$

$$\Delta SG_p = \exp[0.5(SG^\circ - SG)] - 1 \tag{2.32b}$$

2.5.4 Molecular weight

$$\ln MW = \ln MW^\circ \; [(1+2f_m)/(1-2f_m)]^2 \tag{2.33}$$

$$f_m = \Delta SG_m[x + (-0.0175691 + 0.193168/T_b^{1/2})\Delta SG_m] \tag{2.33a}$$

$$x = |\; 0.0123420 - 0.328086/T_b^{1/2} \;| \tag{2.33b}$$

$$\Delta SG_m = \exp[5(SG^\circ - SG)] - 1 \tag{2.33c}$$

The adjustable parameters were fitted using data for 191 petroleum and coal tar materials, with boiling points ranging up to 715° C. The errors using equations (2.30) to (2.33) were in the range of 0.7 to 3.75% on average, with the best accuracy for critical temperature, and the worst for critical pressure. The second most accurate method was from Kesler and Lee [1976], which is also widely used. Many other correlations using boiling point and specific gravity have been proposed (see, for example, Sim and Daubert [1980], Walas [1985] and Altgelt and Boduszynski [1992]), but they were based on a more limited data set, and were not subjected to the same tests for continuity.

In applying these equation to residues, the boiling point (T_b) is an extended or extrapolated atmospheric equivalent boiling point. For residue materials, specific gravity and measured molecular weight can be used to solve for T_b in equation (2.33).

2.5.5 Acentric factor

Two equations are commonly used to estimate the acentric factor for petroleum fractions, one from Edmister [1958]:

$$\omega = 0.1861 \frac{\ln P_c}{(T_c/T_b - 1)} - 1 \tag{2.34}$$

where P_c is in atmospheres, and the other from Kesler and Lee [1976]:

$$\omega = [\ln P_b/P_c - 5.92714 + 6.09648T_b/T_c + 1.28862 \ln T_b/T_c + 0.43577 (T_b/T_c)^6]/[15.2518 - 15.6875T_c/T_b - 13.4721 \ln T_b/T_c + 0.43577 (T_b/T_c)^6]$$

$$(\text{for } T_b/T_c < 0.8) \tag{2.35}$$

$$\omega = -7.904 + 0.1352K_W - 0.007465K_W^2 + 8.359T_b/T_c + (1.408 - 0.01063K)T_c/T_b$$

$$(\text{for } T_b/T_c > 0.8) \tag{2.36}$$

where K_W is the Watson characterization factor.

2.6 Selection of Pseudocomponents

The selection of psudocomponents is an artistic balancing act between the desire to represent the behavior of the mixture on the one hand, and on the cost of computation and the availability of data on the other. Clearly the selection of hundreds of pseudocomponents is not justified if you have 10 points on a boiling curve, with a specific gravity for each. A reasonable procedure is to define how many cuts, or products, a given oil is to be separated into. If the oil all remains in the liquid phase, then one pseudocomponent may be sufficient. If the oil is to be split into more than one cut, then one or more pseudocomponents should be defined within each cut.

Consider, for example, the boiling curve data for a hydrocracker distillate, given in Table 2—3. Boiling points of pseudocomponents are based on the boiling curve, and selected so that the lighter fractions are properly represented, especially for flash calculation. The midpoint of the boiling range of each pseudocomponent is listed in Table 2—4. Note that the pseudocomponent temperatures are not evenly spaced because the boiling points have been corrected for the slope of the boiling curve. Table 2—5 gives estimated properties for some of these pseudocomponents, estimated from the data in Table 2—3.

Table 2–3
Boiling Curve and Specific Gravity for Example Distillate

Weight Percent	True Boiling Temp. (°C)	Molecular Weight	Liquid Density (kg/m^3)
0.0	21	56.5	590
2.0	71	87.6	697
5.0	117	111	744
10.0	153	131	771
15.0	180	148	788
20.0	211	172	805
30.0	281	224	837
40.0	327	263	856
50.0	361	295	871
60.0	396	332	886
70.0	430	371	899
80.0	459	409	911
90.0	490	454	921
95.0	512	493	928
98.0	531	540	936
100.0	548	581	944

Table 2–4
Product Cuts and Pseudocomponents for Example Distillate

Cut, °C	# of comp	NBP of Pseudocomponents (°C)
<177	13	–2, 11, 23, 40, 53, 66, 81, 95, 110, 123, 137, 151, 165
177–343	12	179, 193, 207, 22, 236, 250, 264, 278, 292, 306, 321, 335
343–525	11	348, 363, 377, 391, 405, 419, 441, 468, 495, 520, 552

Table 2–5
Selected Pseudocomponents and Estimated Properties for Example Distillate

T_b °C	T_c °C	P_c kPa	ω	Molecular Weight
53	221.7	1318	0.2459	79
123	303.5	2840	0.3637	114
207	389.9	2197	0.5222	170
306	480.4	1583	0.7473	245
348	517	1386	0.8498	284
419	578	1114	1.0222	359
520	660	807	1.2549	511

2.7 Critical Properties of Residue Fractions

Methods based on boiling curves are clearly not sufficient for characterizing residue fractions, because they cannot be distilled. In the case of bitumen, the residue accounts for 50% of the raw material, and will therefore have a large effect on phase behavior. Two methods have been suggested, depending on the type of calculation.

2.7.1 Single pseudocomponent

If all of the oil remains in the liquid phase, then it can be treated as a single pseudocomponent. Examples of this type of condition are solubility of light gases in bitumen at temperatures up to 100–150° C. Measurable quantities are specific gravity and molecular weight, so that the boiling point can be back–calculated from equation (2.33), and the pseudocritical properties calculated from equations (2.30)–(2.32). Another approach is to adjust the critical parameters of the pseudocomponent (T_c and P_c) to obtain a good fit to a specific set of experimental data. For example, the Peng–Robinson equation of state gave good predictions for the solubility of methane, carbon dioxide, nitrogen, and ethane in Athabasca bitumen, with T_c = 824K, P_c = 1.267 MPa, $\omega = 1.231$, and average molecular weight = 544 [Lu and Fu, 1989].

2.7.2 Multiple Pseudocomponents by Extrapolation

If the heavier fractions of the oil are vaporizing, then the residue must be more fully characterized. Pedersen *et al.* [1984] outlined several methods for dealing with residue fractions in crude oils with about 10% residue. The best prediction of phase behavior resulted when the residue was split into fractions.

One method is to use the fact that natural hydrocarbons often follow a log—normal distribution of mole fraction, z_n, with carbon number, C_n. Hence, the high boiling residue will follow a logarithmic function as follows:

$$C_n = A + B \ln(z_n) \tag{2.37}$$

where A and B are adjustable constants. A and B are determined from the distillable oil, so that equation (2.29) provides an extrapolation equation. Specific gravities were extrapolated by a similar equation:

$$S(C_n) - S(C_{no}) = D[\ln(C_n) - \ln(C_{no})] \tag{2.38}$$

where $S(C_n)$ is the specific gravity of carbon number fraction C_n, C_{no} is the heaviest distillate fraction for which the specific gravity was measured, and D was a constant. The constraint on the value of D is that the estimated specific gravity for the summed pseudo components (on a weight average basis) must equal the measured specific gravity of the residue fraction.

$$\text{Predicted s.g. of resid} = \frac{\displaystyle\sum_{C_n=C_{no}+1}^{C_{nmax}} z_n \cdot MW(C_n)}{\displaystyle\sum_{C_n=C_{no}+1}^{C_{nmax}} \frac{z_n \cdot MW(C_n)}{S(C_n)}} \tag{2.39}$$

The value of D in equation (2.38) is adjusted until the predicted s.g. from (2.39) agrees with the measured value. $MW(C_n)$ is the molecular weight of carbon number n, from the simple equation:

$$MW(C_n) = 14 \cdot C_n - 4 \tag{2.40}$$

The calculation also requires a maximum value for the carbon number, C_{nmax}. Pedersen *et al.* [1984] used C_{100} as the limiting value. Agreement between the actual and calculated dew points was very good by this method (Figure 2–7).

These results show that the equation of state approach can give good results for petroleum and residue materials. Sim and Daubert [1980] suggested some modifications to the SRK equation to improve these flash calculations, but they also pointed out the crucial need for accurate boiling point data. Another tuning approach for residues is adjustment of the binary interaction parameters (δ_{ij}) for residue fractions; this approach is used in the HYSIM package [Morris *et al.*, 1988].

Process engineering software packages, such as HYSIM, include many of these correlation methods for petroleum fractions, as well as methods for extrapolating residue properties. The responsibility of the process engineer is to ensure that the appropriate estimates and calculation methods are used for the component mixture.

2.7.3 Chemical characterization of residue

The residue fraction is not amenable to characterization by the boiling–point/specific gravity methods simply because the molecules are too heavy to distill without significant decomposition. Several methods, such as short–path distillation, high–temperature gas chromatography, and supercritical fluid chromatography promise to extend the range of boiling points that can be separated; however, unless all of the residue is recovered from such methods a portion of the feed will always be "residual" and will require either extrapolation estimates (section 2.7.2) or some other approach. To the extent that these separation methods can provide boiling/specific gravity data, they will reduce the importance of the nonseparable components, but not eliminate them.

Several groups have investigated the use of chemical characterization for PVT calculations for residues. Rather than considering boiling point, direct measurement of molecular weight is used, and data from NMR and IR are used to characterize other chemical characteristics. One example is from Prausnitz [1988]:

Data required:	Elemental analysis (C,H,O,N,S) Molecular weight ^{1}H–NMR spectrum IR spectrum
Equation of State:	Modified perturbed–hard–chain (a three–parameter equation)
Approach:	Pure model compounds were used to develop correlations between the

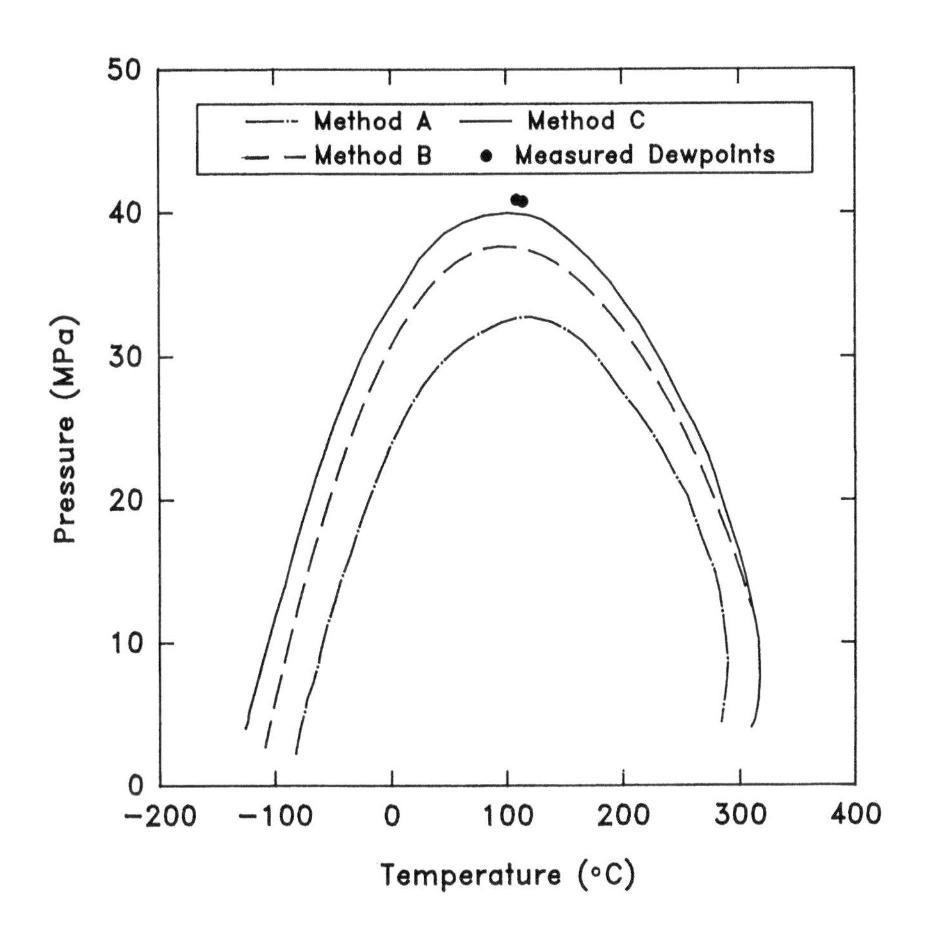

Figure 2–7

Phase envelope and measured dew points of a North Sea crude

Method A	Residue treated as single component
Method B	Residue treated a a number of pseudocomponents
Method C	Characterize all fractions over C_7 by paraffin, aromatic and naphthenic content

(Data from Pedersen *et al.*, 1984)

	EOS parameters and the chemical data. These correlations were then used to calculate parameters for heavy fossil fuel fractions.
Performance:	Tested against vapor pressure and liquid density for crude oil (up to 427—455° C cut) and coal liquid fractions (up to 325—425° C cut). Average absolute deviations for 12 samples were ca. 20% for vapor pressures and 5% for liquid density.

At present, the chemical characterization approach has several drawbacks: it requires significant amounts of data, the equations of state (e.g. perturbed hard chain or UNIFAC) are not normally used for lighter distillates, and the accuracy is poor. Prausnitz [1988] recommends that such methods only be used as a last resort. If a datum for vapor pressure is available for tuning parameters, then the predictions from these methods are much more accurate. In the context of most applications, however, a datum can be used to tune the extrapolation methods (Section 2.7.2) just as effectively, giving a thermodynamic representation of the residue that is compatible with lighter fractions.

The chemical characterization approach suffers from two other drawbacks. First, a heavy reliance is placed on model compounds for developing correlations between chemical structure and parameters for equations of state, but the model compounds which are available are relevant only to residues that are either highly paraffinic or mainly unsubstituted aromatics. Polyfunctional molecules, as discussed in Chapter 1, cannot be represented effectively using a data base from n—alkanes, isoparaffins, and low—molecular weight polyaromatics and heteroaromatics. One reason for the reliance on model compounds is the paucity of phase behavior data for well—characterized residue fractions. The second limitation is that some of the equations of state were developed for alkane—like materials, and deviations are treated as perturbations thereon. The cross—linked polynuclear characteristics of residue materials cannot be represented as a perturbation of a linear structure.

2.8 Hydrogen Solubility

Hydrogen is commonly used in upgrading processes to enhance product formation; therefore, the behavior of mixtures contianing this gas are of primary interest in process design. Mixtures of hydrogen and hydrocarbons require some special treatment in order to use cubic equations of state. Gray *et al.* [1985] studied the fit of the Soave—Redlich—Kwong and Peng Robinson equations of state to data ranging from cryogenic to high temperature conditions. The binary interaction parameter, δ_{ij}, was dependent on the critical temperature of the hydrocarbon (Figure 2—8). The model correlation was given by:

$$\delta_{ij} = A + \frac{BX^3}{1 + X^3} \tag{2.41}$$

$$X = \frac{T_{cj} - 50}{1000 - T_{cj}} \tag{2.42}$$

where T_{cj} was the critical temperature of the j—th component. The pure component parameters and mixing rules followed the normal Peng—Robinson method. Other specific correlations are available for hydrogen solubility in alicyclic and aromatic solvents [Shaw, 1987; Prausnitz, 1988], but these correlations lack the ease of application of the δ_{ij} approach for cubic equations of state.

2.9 Precipitation of Solids

Precipitation of solids can occur under a variety of process conditions, whenever the solvent characteristics of the liquid phase are no longer able to maintain polar or high—molecular weight material in solution. Some examples are:

Asphaltenes	precipitate when n—alkanes are used to dilute crude bitumen and residue fractions
Waxes	crystallize from solution in response to a drop in temperature or an increase in the aromatic content of the solvent
Sludge in fuel	formation of solids due to oxidation or other chemical reactions, leading to formation of insoluble polar material. This type of reaction occurs upon long—term storage.
Sludge in reactor	cracking and hydrogenation reactions can

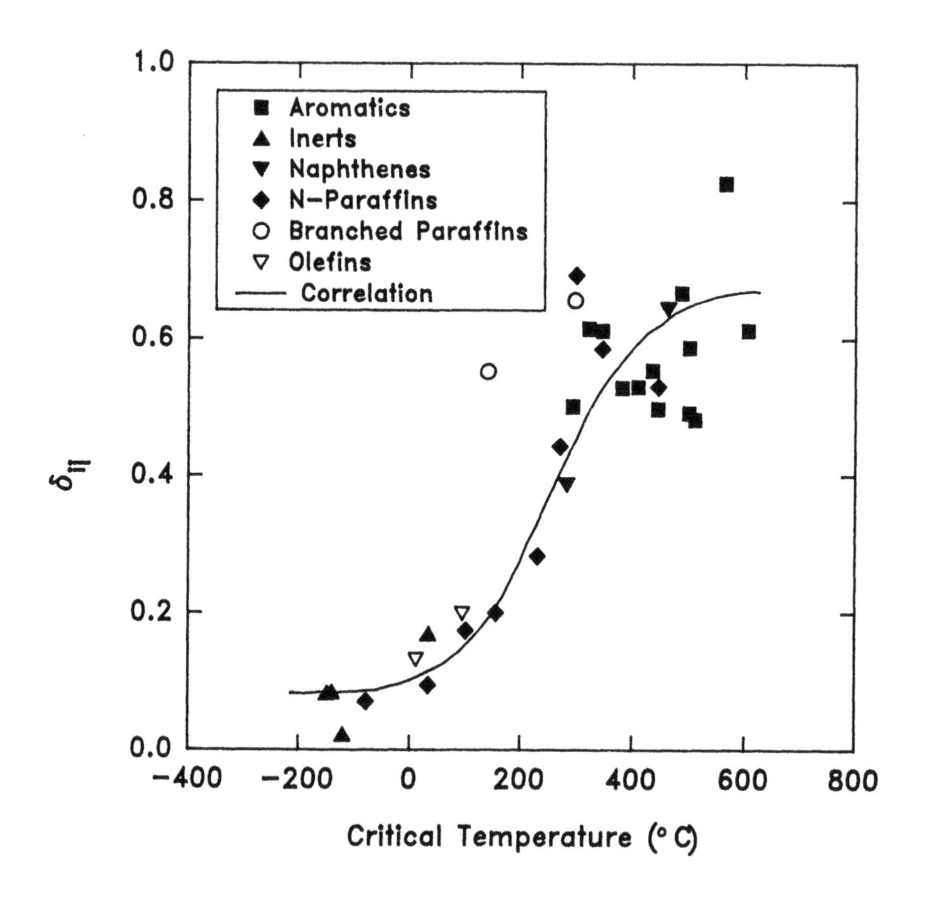

Figure 2–8
Binary interaction parameter, δ_{ij}, for hydrogen–hydrocarbon mixtures using the Peng–Robinson equation (Data from Gray *et al.*, 1985)

	change the solvent characteristics of the liquid phase, so that asphaltenic or waxy solids form either in the reactor or downstream.
Coke	at high temperatures some oil components undergo addition and polymerization reactions, forming an insoluble carbon–rich deposit known as coke.

All of these precipitation phenomena can in principle be analyzed as a thermodynamic phase equilibrium problem; the main limitation to making predictions is the complexity of both the solids and the liquid solutions. The work of Mitchell and Speight [1973] showed quantitatively how solvent strength affected solubility of a single asphaltene material. They precipitated asphaltenes from Athabasca bitumen using a range of nonpolar solvents and solvent blends. The amount of precipitate correlated linearly with the solubility parameter of the solvent, where this solubility parameter was defined by either of two equations:

$$\delta_1 = \sigma V^{-1/3} \tag{2.43}$$

$$\delta_2 = [(\Delta H_v - RT)/V]^{1/2} \tag{2.44}$$

Both forms were developed by Hildebrand and coworkers. Here σ is the surface tension, V is the molar volume, and ΔH_V is the enthalpy of vaporization. The precipitation data are shown in Figure 2–9 for both δ_1 and δ_2 values. The linear relationship was valid for pure paraffin, isoparaffin, olefin and cycloalkane solvents, as well as blends of benzene and n–pentane. No asphaltenes precipitated from polar solvents or aromatic solvents.

Figure 2–9 shows that solvent effects can be used to predict how changes in solubility parameter can be affect the amount of precipitate. Similarly, the effects of temperature and pressure on precipitation can be developed from this type of linear equation. These results do not, however, explain why the solids precipitate, or how a different oil or processed product fraction would behave. Figure 2–9, and other work to date, do not include a characterization of the solid so that the results can be applied only qualitatively to other solid–liquid equilibrium relationships.

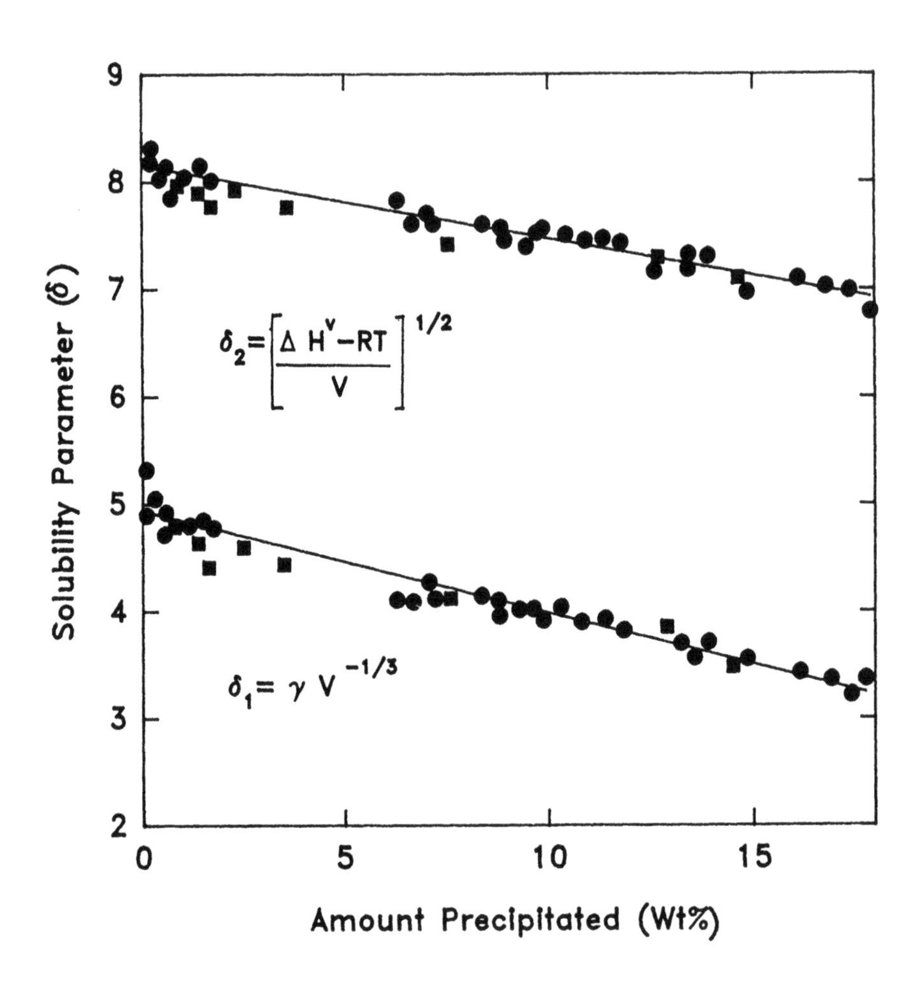

Figure 2–9

Relation of amount of asphaltene precipitated to solubility parameters δ_1 and δ_2 (• pure solvents; ■ mixed benzene + n–pentane)(Data from Mitchell and Speight, 1973)

2.10 Thermal Properties of Petroleum Fractions

Enthalpy and entropy departures can be calculated from the equation of state, enthalpy by equation (2.1) and entropy as follows:

$$\frac{S - S_0^{ID}}{RT} = \ln z - \ln P/P^\circ + \int_{\infty}^{V} [\frac{1}{R}(\partial P/\partial T)_v - 1/V]dV \tag{2.45}$$

The ideal gas properties are calculated from the heat capacity of the ideal gas [Kesler and Lee, 1976]:

$$C_p^{ID} = a + bT + cT^2 \tag{2.46}$$

$$a = -0.32646 + 0.02678 \cdot K_W - CF(0.084773 - 0.080809 \cdot SG) \tag{2.46a}$$

$$b = -[1.3892 - 1.2122 \cdot K_W + 0.0383 \cdot K_W^2 - CF(2.1773 - 2.0826 \cdot SG)] \times 10^{-4} \tag{2.46b}$$

$$c = -[1.5393 + CF(0.78649 - 0.70423 \cdot SG)] \times 10^{-7} \tag{2.46c}$$

where

$$CF = [(12.8/K_W - 1)(10.0/K_W - 1) \times 100]^2 \text{ for } 10 \leq K_W \leq 12.8$$

$$CF = 0 \text{ otherwise} \tag{2.46d}$$

where T is temperature (R) and SG is the specific gravity at 15°C. The ideal enthalpy and entropy are then calculated by integrating the heat capacity:

$$H^{ID} = \int_{T^\circ}^{T} C_p^* \, dT \tag{2.47}$$

$$S^{ID} = \int_{T^\circ}^{T} C_p^*/T \, dT \tag{2.48}$$

Equations (2.1) and (2.45) are then used to calculate the enthalpies of real vapor or liquid phases via an equation of state. An alternative method is to use empirical correlations to estimate enthalpy as a function of reduced temperature and reduced pressure [Kesler and Lee, 1976].

2.11 Surface Tension

The interfacial tension between vapor and liquid phases is important in distillation and reactor design. In both cases, the surface tension influences bubble size and mass transfer. In the case of reactor design, the liquid phase is at 10—20 MPa and saturated with hydrogen and other light components. Data for surface tensions under high pressures are scanty. Hwang *et al.* [1982] measured surface tensions of coal liquids with hydrogen and methane, at pressures up to 17 MPa and temperatures to 260° C. Surface tension was estimated as follows; the surface tension was estimated for each fraction at its normal boiling point:

$$\sigma_i = 673.7/K_{w,i} \, [1 - T_{b,i}/T_{c,i}]^{1.232} \tag{2.49}$$

where K_W is the Watson characterization factor and σ is in dynes/cm. The subscript i indicates the properties of a fraction or pseudocomponent. The parachor for each fraction was assumed constant with temperature, and was calculated from the molecular weight, M, and the density of the fraction as follows:

$$P_i^* = \frac{M_i \, \sigma_i^{1/4}}{\rho_{L,i} - \rho_{G,i}} \tag{2.50}$$

The mixture property was calculated as follows:

$$\sigma_m = [\, \Sigma \, P_i^* \, (x_i\rho_L/M_L - y_i\rho_G/M_G)]^4 \tag{2.51}$$

In this case, the densities and molecular weights are mixture properties, while x_i and y_i are component mol fractions in the liquid and vapor phases respectively.

The correlations agreed with the experimental data to within ± 16% on average, but the experiments and the use of the correlations were both open to question. In applying equations (2.49)–(2.51), Hwang *et al.* [1982] assumed that the vapor phase was pure gas (i.e. hydrogen or methane), and that no gas dissolved in the liquid. The correlation, applied on this basis, tended to overpredict the surface tension which would in fact be lowered by significant dissolution of gas in the liquid during the experiment. Nevertheless, prediction of surface tension ca. 17% should be sufficient for design calculations.

2.12 Transport Properties

2.12.1 Viscosity

Viscosity is the most important transport property for engineering design, because of its use in design calculations and its use in estimation of other transport properties. Correlation and prediction of viscosity of heavy oils as a function of temperature, pressure, and gas solubility has received much attention because of the importance of viscosity for production processes (see, for example, Mehrotra *et al.* [1989] and references therein). These correlations, however, have not been verified at temperatures > 200°C nor do they apply to cracked distillate products. In the context of upgrading processes, a useful estimation method should apply both to a feed residue and a product gas oil.

Bitumens and residue fractions are often solids or semisolids at room temperature, with viscosities of 10^5 mPa·s or more at 20° C. Their viscosity drops drastically with increasing temperature, so that they become free–flowing liquids at 100–150°C. The high viscosity of these fractions can be attributed to the high–molecular weight components, which become entangled and aggregated at low temperature (see Figure 1–6 illustrating asphaltene micelles). Despite the fact that the high–molecular weight components give high viscosity, heavy oils, bitumens and residues are Newtonian fluids.

The sensitivity of viscosity to temperature requires that any estimation method account both for composition dependence and temperature dependence. The correlation suggested by Twu [1985] follows the assay procedures for viscosity: viscosity is estimated at two different temperatures, then the viscosity–temperature relationship is interpolated or extrapolated. Twu's correlation for kinematic viscosity, ν, uses the principle of corresponding states with n–alkanes as a reference series (as in section 2.5 for critical properties).

For petroleum fractions at 210° F:

$$\ln(\nu_2 + 450/T_b) = \ln(\nu_2^\circ + 450/T_b)[(1+2f_2)/(1-2f_2)]^2 \quad (2.52)$$

$$f_2 = |x| \cdot \Delta SG - 21.1141 \cdot \Delta SG^2/T_b^{1/2} \quad (2.52a)$$

$$|x| = |1.99873 - 56.7394/T_b^{1/2}| \quad (2.52b)$$

At 100° F:

$$\ln(\nu_1 + 450/T_b) = \ln(\nu_1^\circ + 450/T_b)[(1+2f_1)/(1-2f_1)]^2 \quad (2.53)$$

$$f_1 = 1.33932 \cdot |x| \cdot \Delta SG - 21.1141 \cdot \Delta SG^2/T_b^{1/2} \quad (2.53a)$$

For petroleum fractions:

$$\Delta SG = SG - SG^\circ \quad (2.54)$$

The superscript "o" in equations indicates the properties of the corresponding n—alkane, ν is the kinematic viscosity in cSt, T_b is the normal boiling point (R), and SG is the specific gravity at 15° C. The viscosity of the n—alkanes were correlated as follows:

$$\ln(\nu_2^\circ + 1.5) = 4.73227 - 27.0975 \cdot \alpha + 49.4491 \cdot \alpha^2 - 50.4706 \cdot \alpha^4 \quad (2.55)$$

$$\ln(\nu_1^\circ) = 0.801621 + 1.37179 \ln(\nu_2^\circ) \quad (2.56)$$

where is α is a function of reduced boiling point (equations 2.24 and 2.28). The value of SG° in equation (2.54) is obtained from equation (2.26), which is also a function of α.

The temperature dependence can then be estimated using the Walther equation, which successfully represents the temperature dependence of both distillates and residues:

$$\ln(\ln Z) = A + B \cdot \ln T \quad (2.57)$$

$$Z = \nu + 0.7 + \exp(-1.47 - 1.84\cdot\nu - 0.51\cdot\nu^2) \quad (2.57a)$$

Solving for ν gives:

$$\nu = Z - 0.7 - \exp[-0.7487 - 3.295(Z-0.7) + 0.6119(Z-0.7)^2 - 0.3193(Z-0.7)^3] \quad (2.58)$$

At viscosities over 2 cSt, the exponential term in equation (2.57a) is negligible, and

$$Z = \nu + 0.7 \quad (2.59)$$

Example calculation:

Twu [1985] gives the example of a petroleum fraction with a boiling point of 400°C (1210.17 R) and a specific gravity of 0.8964. From equations (2.22), (2.55) and (2.56) the propeties of the n–alkane of the same boiling point are obtained

$$SG^\circ = 0.8044$$
$$\nu_2^\circ = 2.94 \text{ cSt}$$
$$\nu_1^\circ = 9.827 \text{ cSt}$$

From equation (2.54), $\Delta SG = 0.0920$, and the viscosities at 100 and 210°F are obtained from equations (2.52) and (2.53).

$$\nu_2 = 4.155 \text{ cSt}$$
$$\nu_1 = 24.26 \text{ cSt}$$

Equations (2.57) to (2.59) are then used to calculate the viscosity at other temperatures. Equation (2.59) gives $Z_1 = 24.96$ at 559.67 R, while $Z_2 = 4.855$ at 669.67 R. Solving for the parameter B in equation (2.57) gives

$$B = -3.963$$

The equation for temperature dependence is now given by:

$$\ln[\ln(\nu + 0.7)] = 26.2438 - 3.963\cdot\ln T$$

This equation is valid for $\nu \geq 2$ cSt. For example, $\nu = 9.196$ cSt at 606.67 R, compared to the experimental value of 9.4 cSt.

This viscosity correlation gave an average error of 8.53% in estimating kinematic viscosity for 344 petroleum fractions with boiling points up to 727° C (1800 R) and API gravity down to —30. Various modifications of equation (2.57) have been widely successful for correlating viscosity of heavy oils and bitumens over a wide range [Mehrotra *et al.*, 1989; Puttagunta and Miadonye, 1991]. The estimates of viscosity using this method should be satisfactory for design of upgrading processes, since the temperature is high enough that the residues are free—flowing liquids. The correlation of Twu [1985] is not recommended for temperatures below 100° C (i.e. for reservoir engineering calculations) where the viscosity changes dramatically with temperature. Methods based on equation (2.57), or similar expressions, and experimental data for the specific oil should be used instead.

The method of Twu [1985] is quite suitable for estimating the viscosity of either real boiling fractions or pseudcomponents. The problem which then arises is how to calculate the viscosity of a mixture of these components. A number of mixing rules have been proposed. For defined mixtures, Tsonopoulos *et al.* [1987] suggest:

$$\mu_m = [\sum_i x_i \mu_i^{1/3}]^3 \tag{2.60}$$

where μ is the dynamic viscosity, $\nu \cdot \rho$. Mehrotra *et al.* [1989] suggested the following weighted mean for viscosity of mixtures of Cold Lake fractions:

$$M_m^{1/2} \ln(\mu_m + 0.7) = \sum_i x_i M_i^{1/2} \ln(\mu_i + 0.7) \tag{2.61}$$

2.11.2 Thermal conductivity

Several methods based on corresponding states have been proposed for estimating the thermal conductivity of pure components [Reid *et al.*, 1987], but these methods are difficult to apply to petroleum fractions. In light of the poor quality of the available data, Tsonopoulos *et al.* [1987] recommended using the API method for engineering design [API, 1983]:

$$k_T = 0.07727 - 4.558 \times 10^{-5} \cdot T \tag{2.62}$$

where T is in °F and thermal conductivity is in BTU/(h·ft·°F). For liquid mixtures, the thermal conductivity is given by

$$k_{T,m} = \sum_i \sum_j \phi_i \phi_j k_{T,ij} \tag{2.63}$$

$$k_{T,ij} = 2\,(1/k_{T,i} + 1/k_{T,j})^{-1} \tag{2.63a}$$

$$\phi_i = x_i v_i / \Sigma x_i v_i \tag{2.63b}$$

Here ϕ_i is a volumetric fraction calculated from the mol fraction of component i, x_i, and the molar volume, v_i.

2.11.3 Diffusivity

The diffisivity of components in petroleum and residue fractions is mainly of interest in mass transfer and catalysis calculations. A widely used correlation for diffusivity in liquids at infinite dilution is the Wilke–Chang equation [Reid *et al.*, 1987]:

$$D^{\circ}_{AB} = \frac{7.4 \times 10^{-8} (\psi M_B)^{1/2} T}{\mu_B V_A^{0.6}} \tag{2.64}$$

where D°_{AB} is the diffusivity of solute A in solvent B at infinite dilution (cm^2/s), μ_B and M_B are the solvent viscosity (cP) and molecular weight respectively, temperature T is in K, and V_A is the molar volume of the solute at its normal boiling point, cm^3/mol. ψ is an association factor, which has a value of 1.0 for oils. Alternative methods are more accurate if the parachors of the solute and solvent are known accurately [Reid *et al.*, 1987]. Given the uncertainty in estimating the surface tension for petroleum fractions (section 2.10), these methods are unlikely to be better than Wilke–Chang.

Notation

a, A	empirical coefficients
b, B, c	empirical coefficients
C	carbon number

C_p	heat capacity, BTU/(lb—R)
CF	empirical coefficient
D	empirical coefficient
D°_{AB}	diffusivity of A in B at infinite dilution, cm^2/s
f, F	empirical coefficients
H	enthalpy, kJ/kmol or BTU/lb
k_T	thermal conductivity, BTU/(hr—ft—R)
K_w	Watson characterization factor
M, MW	molecular weight, Da
n	number of components
P	pressure, kPa, psia or atm
P^*	parachor
q	arbitrary independent variable
R	gas constant
S	entropy, kJ/(kmol—K) or BTU/(lb—R)
SG	specific gravity at 15°C, relative to water
T	temperature in °C, °F, K or R
v, V	molar volume, cm^3/mol or m^3/kmol
v_i	volume fraction
x_i	mol fraction
x	empirical coefficient in equations 2.33a, 2.52a, 2.53a
X	empirical coefficient
z, Z	compressibility

α	empirical coefficient
δ_{ij}	binary interaction parameter
δ_1, δ_2	Hildebrand solubility parameters
θ	empirical coefficient
κ	empirical coefficient
μ	dynamic viscosity, mPa·s or cP
ν	kinematic viscosity, cSt
σ	surface tension, dynes/cm
ϕ	volumetric fraction in equation 2.63
ψ	empirical coefficient
ω	acentric factor

Subscripts

b	boiling condition

c	critical
G	vapor phase
i, j	counters
L	liquid phase
m	molecular weight or mixture
p	pressure
r	reduced conditions
sat	saturation conditions
T	temperature
v	volume

Superscripts

ID	ideal gas
o	reference compound (n—alkane) or reference temperature or pressure

Further Reading

American Petroleum Institute Technical Data Book — Petroleum Refining, API, Washington DC, 1983.

Annual Book of ASTM Standards, Volume 05, Petroleum Products and Lubricants, 1990.

Lu, B.C.—Y.; Fu, C.—T. "Phase equilibria and PVT properties", in *AOSTRA Technical Handbook on Oil Sands, Bitumens, and Heavy Oils*, L.C. Hepler and C. Hsi, eds., AOSTRA, Edmonton, 1989, pp 129.

Pedersen, K.S.; Thomassen, P.; Fredenslund, A. "Thermodynamics of petroleum mixtures containing heavy hydrocarbons. 1. Phase envelope calculations by use of the Soave—Redlich Kwong equation of state", *Ind. Eng. Chem. Process Des. Dev.* 1984, 23, 163—170.

Walas, S.M. *Phase Equilibrium in Chemical Engineering*, Butterworth, Stoneham, MA, 1985.

Problems

2.1 The following TBP data are available from distillation of a hydrotreated coker gas oil:

Cut #	Max. B.P. of Cut, °C	Wt% in Cut	Density of Cut, g/cm^3
1	239	7.60	0.831
2	292	12.91	0.873
3	331	13.10	0.904
4	363	13.62	0.925
5	383	13.80	0.943
6	440	14.07	0.955
7	481	16.29	0.970
8	>481	8.60	nd

a) What would the 50% point of the oil be? (i.e. the temperature where 50% would be distilled off)

b) Calculate the weight average boiling point, the volume average boiling point, and the mean average boiling point.

2.2 For fraction # 5 in the oil from Problem 2.1, calculate the critical temperature, critical pressure, molecular weight, and acentric factor.

2.3 Estimate the equilibrium flash vaporization curve for the oil in Problem 2.1 at 10 MPa. Repeat the calculation for a mixture of the oil with hydrogen gas, at a ratio of 2000 SCF/bbl. (SCF= standard cubic feet).

CHAPTER 3

Chemistry of Upgrading and Hydrotreating Reactions

3.1 Overall Objectives of Primary Upgrading

Given the composition of the residues, and the specifications for synthetic crude oils and petroleum fractions, the general chemical objectives of upgrading processes are easily defined:

1. Convert high–molecular weight residue components to distillates, i.e. compounds with a boiling point below 525° C. This conversion requires breakage of the carbon–carbon and carbon–sulfur bonds in the residue fraction.
2. Improve the H/C ratio of the distillate products, moving from 1.5 in the feed toward 1.8 mol/mol as suitable for transportation fuels. Note that a high H/C ratio is most desirable in the middle distillates for diesel and jet fuels. These fractions must be highly aliphatic in order to give appropriate ignition in diesel engines (i.e. the cetane number) and to minimize smoke production. In the gasoline fraction, on the other hand, alkylated aromatics with an H/C ratio of 1–1.5 are desirable because they boost the octane rating without a requirement for additives. Nevertheless, the product blend will have a higher H/C ratio than the residue feed.
3. Remove the heteroatoms down to acceptable levels. The main

heteroatoms of interest are the sulfur and nitrogen in the distillate products, which must meet the levels listed in Chapter 1.

Achieving these process goals requires that the the residue molecules undergo a number of different reactions, some thermal and some catalytic. The terminology of processes for converting and treating residues is confused, but the following terms are defined for consistent use:

Primary Upgrading	The first reaction step for conversion of residue (from bitumen, heavy oil, or petroleum) into distillate products. This conversion requires significant breakage of carbon—carbon bonds. Processes of this type are discussed in Chapters 5 through 7.
Hydrotreating	Hydrogenation of distillates or residues to further improve their properties. Distillate fractions may be subjected to more than one stage of hydrotreating, depending on product specifications and reactor performance. Hydrotreating, and other secondary treatment processes, are covered in Chapter 8. Synonymous terms include hydroprocessing and hydrofining.

The differentiation between primary and secondary processing is useful because the process designs are substantially different, but these processes still share a common underlying chemistry. The difference between primary and secondary stages is one of severity and selectivity. Thermal breakage of C—C bonds predominates over catalytic reactions at the severe conditions of primary upgrading, while secondary hydrotreating at milder conditions gives selective catalysis of desired reactions. The common reaction chemistry of the two types of processes will be considered together in this chapter. The differences in operating conditions and performance of the various processes will be discussed in subsequent chapters.

3.2 Thermal Reactions

3.2.1 Carbon—carbon bond breakage

Thermal reactions occur spontaneously in organic mixtures at a high enough temperature, and are noncatalytic. The primary objective of any upgrading process is to break chemical bonds, giving a lighter product. The main bonds of interest during primary upgrading are carbon—carbon, carbon—sulfur and carbon—hydrogen. The only feasible means to achieve carbon—carbon bond breakage in residues is via thermal reactions. Bond energies give an indication of the difficulty of breakage:

Table 3—1
Bond Dissociation Energies
(Data of Benson, 1976)

Chemical Bond	Energy, kcal/mol
C———C (aliphatic)	85
C———H (n—alkanes)	98
C———H (aromatic)	110.5
C———S[1]	77
C———N (in amine)	84
C———O (in methoxy)	82

1. Estimated from methyl sulfide and methyl radical formation from dimethyl sulfide, using additivity data of Benson [1976].

The carbon—carbon bonds in aromatic compounds are much stronger because of the resonance stabilization, which amounts to an extra 6 kcal/mol of C—C bonds in benzene. Polyaromatics have lower stabilization energies, for example, naphthalene has an average stabilization energy of 5.5 kcal/mol of C—C bonds, based on 61 kcal/mol of naphthalene and 11 C—C bonds. The resonance stabilization of any aromatic bonds renders them unbreakable at normal process temperatures ($< 600^\circ C$) until the aromatic character is destroyed by hydrogenation. In the absence of a catalyst, the breakage of C—C bonds requires a temperature on the order of $420^\circ C$ to achieve useful rates.

3.2.2 Limits on catalytic C—C bond breakage in residues

Breakage of C—C bonds below ca. $400^\circ C$ requires very active catalysts which, to date, have not been effective when applied to residues. Catalysts such as zeolites will promote breakage of C—C bonds at hydroconversion temperatures (over $400^\circ C$), but are fouled so rapidly in primary upgrading that they have no commercial significance. Any catalyst for breaking C—C bonds relies on carbenium ion intermediates, promoted by the acid site on the catalyst. The stronger the acid site, the more facile the reaction. Superacids, for example, can activate C—C bonds at temperatures as low as 25° C. The basic nitrogen compounds in bitumen, however, adsorb to acid sites and render then inactive [Mills *et al.*, 1950].

Primary upgrading processes, therefore, must operate at temperatures above 420° C in order to break C—C bonds, and the contribution of any catalyst to this process is limited. The higher the temperature, the shorter the time required to achieve useful conversion levels of over 50% of the feed residue. The possible reactions of a hydrocarbon mixture do not change with temperature, but the relative rate of different reactions is sensitive. The extent of reaction, therefore, and the yield of specific products does change with process severity.

3.2.3 Cracking of paraffins

The modern commercial processes for the conversion of high—molecular weight hydrocarbons to liquid products use moderate temperatures (400—450° C) and high pressures ($\geq$ 10 MPa). The breakage of carbon—carbon bonds is mainly due to free—radical chain reactions. Free radicals are highly reactive intermediates which have an unpaired electron. Their transitory existence begins with the breakage of a bond. The stability of alkyl free radicals is in the order 3° > 2° > 1°, and reaction pathways favor the formation of the most stable radical species. Hence, formation of t—butyl radical (a tertiary radical) is energetically much more favorable than the ethyl radical, which is primary. Radicals can also be stabilized by delocalization or resonance, as in the case of the benzyl radical. Free radicals are present at low, but measurable, concentrations during thermal reactions. Their importance is due to the occurence of chain reactions of hydrocarbons as follows:

Initiation:

$$M \xrightarrow{k_1} 2\,R^{\bullet} \tag{3.2}$$

Propagation:

Chain transfer

$$R^{\bullet} + M \xrightarrow{k_2} RH + M^{\bullet} \tag{3.3}$$

β—scission

$$M^{\bullet} \xrightarrow{k_3} R^{\bullet} + \text{Olefin} \tag{3.4}$$

Termination:

$$\text{Radical} + \text{Radical} \xrightarrow{k_4} \text{Products} \tag{3.5}$$

Here M is the parent compound, and $R^{\bullet}$ is a smaller alkyl radical.

Reaction (3.4) is called β–scission because the C–C bond β to the radical center (i.e. two carbons away) breaks to form an olefin and a smaller alkyl radical. For example, the n–nonane radical would decompose via β–scission as follows:

Ethylene formation

$$\longrightarrow \quad + \ C_2H_4 \tag{3.6}$$

Propylene formation

$$\longrightarrow \quad + \tag{3.7}$$

A chain reaction occurs when reactions (3.3) and (3.4) are much more frequent than (3.2) and (3.5). The free radicals form a steady state population that promotes the reaction but changes little in concentration. Kossiakoff and Rice [1943] developed this mechanism for cracking reactions, and it forms the basis for our understanding of high–temperature reactions of hydrocarbons. At low pressure (ca. 101.3 kPa) these reactions favor the formation of light olefins such as ethylene and proylene, and indeed low pressure pyrolysis (thermal breakage) at 700 – 800°C is used to manufacture these petrochemicals from light hydrocarbons, naphthas, and gas oils.

At pressures of >10 MPa the product distribution no longer follows the Rice–Kossiakoff mechanism for low pressure pyrolysis, although the reaction still proceeds via a free–radical mechanism. At elevated pressures, hydrogen abstraction (reaction (3.3)) and radical addition reactions (reaction (3.5)) become more favorable. β–scission is less important at high pressure, and the cracking of alkanes can be described in terms of a single–step mechanism [Fabuss *et al.*, 1964; Ford, 1986; Kissin, 1987]. Rather than the multistep cracking via reaction (3.4), a radical is rapidly stabilized without the formation of olefins. At intermediate pressures of 3 to 7 MPa, a two–step mechanism has been reported for thermal cracking of alkanes [Mushrush and Hazlett, 1984; Zhou *et al.*, 1986, 1987].

An example of this shift in mechanism was observed in studies by Khorasheh [1992]. A series of thermal cracking experiments were performed using n–hexadecane as a model compound and benzene as solvent. Experiments were conducted in a plug–flow reactor at 14 MPa and 673 to 713 K using feeds containing 0.01 and 0.05 mol fraction n–hexadecane. The product distributions in terms of mol/100 mol of

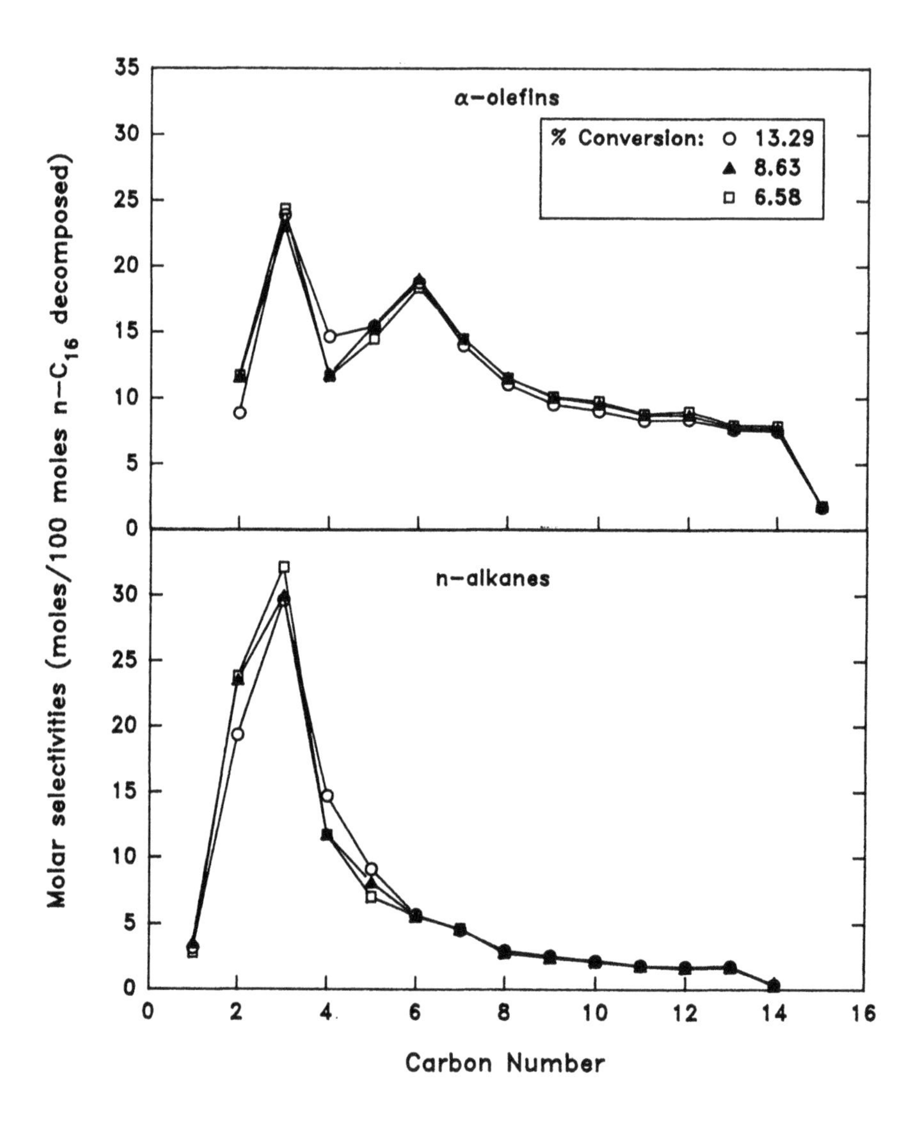

Figure 3–1

Distribution of products from thermal cracking of n–hexadecane in benzene. Experiments were carried out at 438° C with 0.03 mol fraction n–hexadecane in benzene (Data from Khorasheh, 1992)

Reactions of parent radicals:

$$n\text{—}C^{\bullet}_{16} \longrightarrow R^{\bullet}_1 + \text{olefin} \quad (a)$$

Reactions of first generation radicals:

$$R^{\bullet}_1 + n\text{—}C_{16} \longrightarrow R_1H + n\text{—}C^{\bullet}_{16} \quad (b)$$

$$R^{\bullet}_1 \longrightarrow \text{olefin} + R^{\bullet}_2 \quad (c)$$

Reactions of second generation radicals:

$$R^{\bullet}_2 + n\text{—}C_{16} \longrightarrow R_2H + n\text{—}C^{\bullet}_{16} \quad (d)$$

Termination Reactions:

$$R^{\bullet}_1 + R^{\bullet}_2 \longrightarrow R_1\text{—}R_2 \quad (e)$$

$$n\text{—}C^{\bullet}_{16} + R^{\bullet}_2 \longrightarrow R_3 \quad (f)$$

$$n\text{—}C^{\bullet}_{16} + n\text{—}C^{\bullet}_{16} \longrightarrow C_{32} \quad (g)$$

Figure 3—2
Propagation and termination reactions for the reactions of n—alkanes (n—C_{16} for example)

n—hexadecane cracked are presented in Figure 3—1 for representative conditions. The product distributions can described in terms of a modified two—step cracking mechanism (Figure 3—2), where alkyl radicals generated from β—scission of the parent hexadecyl radical either undergo a second decomposition step by β—scission, or participate in hydrogen abstraction from n—hexadecane to give the corresponding alkane. The second cracking step was favored as the temperature was increased, giving lower yields of n—alkanes and increasing yields of α—olefins, while hydrogen abstraction was favored at lower temperatures.

As the mol fraction of n–hexadecane in the feed was increased, the product distribution shifted towards a single–step mechanism (hydrogen abstraction favored over the second cracking step), because the hydrogen abstraction reactions were bimolecular.

The thermal chemistry of paraffins is summarized in the following table:

Table 3–2
Thermal Cracking of Paraffins

H_2	Pressure	Temp, °C	Products
No	101.3 kPa	>700	Light olefins, coke
No	3–7 MPa	430	Alkanes and olefins
No	14 MPa	430	Alkanes and olefins
Yes	14 MPa	430	Alkanes, less olefins

Although alkanes are not normally a significant constituent of residues, the thermal chemistry of paraffins provides the framework for interpreting the thermal reactions of any of the compound types. The paraffinic side chains in the residue would tend to react according to the mechanisms described above.

3.2.4 Further reactions of olefins

The reactions of olefins depend on two major factors: the presence of hydrogen and a catalyst to promote hydrogenation, as in reaction (3.8), and the total concentration of olefins, which in turn depends on the level of conversion.

$$\text{olefin} + H_2 \longrightarrow \text{alkane} \qquad (3.8)$$

At high pressures, the olefins can undergo addition reactions with radicals:

$$\text{olefin} + R^{\cdot} \longrightarrow \text{R-substituted radical} \qquad (3.9)$$

Reaction (3.9) is essentially the reverse of the β–scission reaction. When olefins are present at sufficient concentration, i.e. high conversion in the absence of a catalyst or hydrogen gas, then they participate in free-radical reactions similar to alkanes:

Hydrogen abstraction

+ R· ⟶ + RH (3.10)

β–scission

⟶

\+ (3.11)

Addition

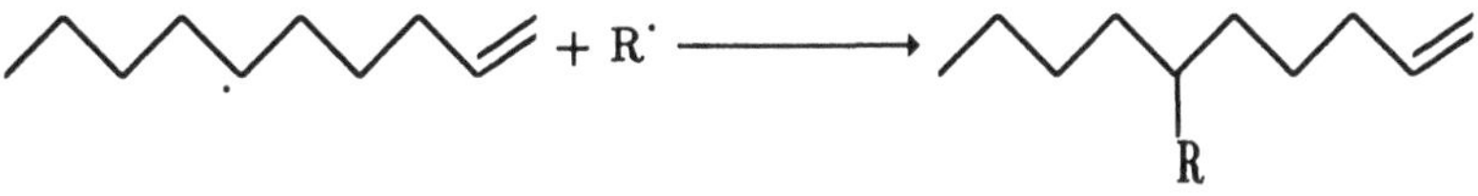

+ R· ⟶

R

(3.12)

The diolefin products of β–scission, such as 1,4–butadiene in reaction (3.11), tend to polymerize in the product liquid and on catalysts, and therefore require special treatment if they are present. Only hydrogen deficient, noncatalytic conditions favor olefin formation, and diolefin yields are significant only at high conversion levels.

3.2.5 Reactions of naphthenes

The thermal reactions of naphthenic ring compounds are similar to the paraffins, with additional reaction pathways of ring opening and dehydrogenation. Ring opening is a variant of β–scission wherein the product of the bond breaking reaction is a single molecule. Decalin provides an example of this type of reaction:

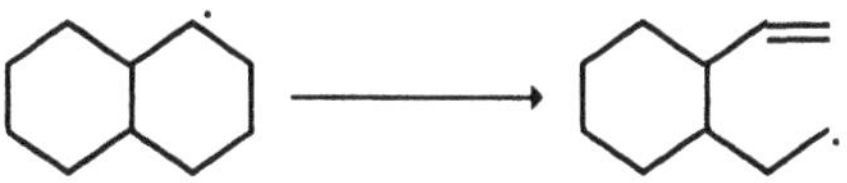

⟶ (3.13)

Ring opening is less favorable than the equivalent reaction in paraffins, making the naphthenes less reactive than paraffins. The decomposition of tridecylcyclohexane provides an example of this stability [Savage and

Klein, 1988]. A variety of cycloalkanes and alkenes were observed, corresponding to breakage at different points along the alkyl chain. The most favorable points for breakage were adjacent to the cyclohexyl ring, because the free radical intermediates were more stable:

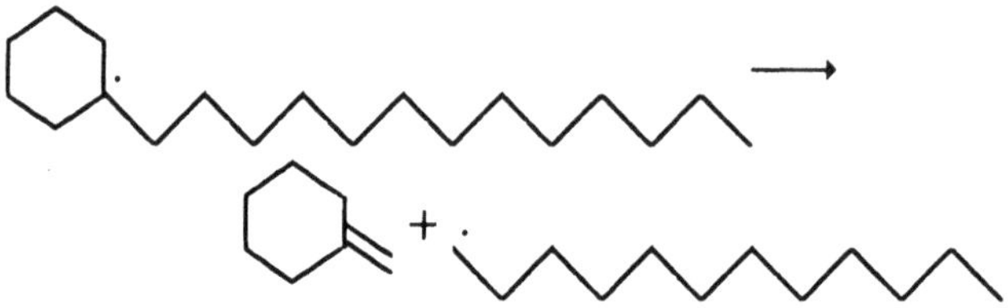

(3.14)

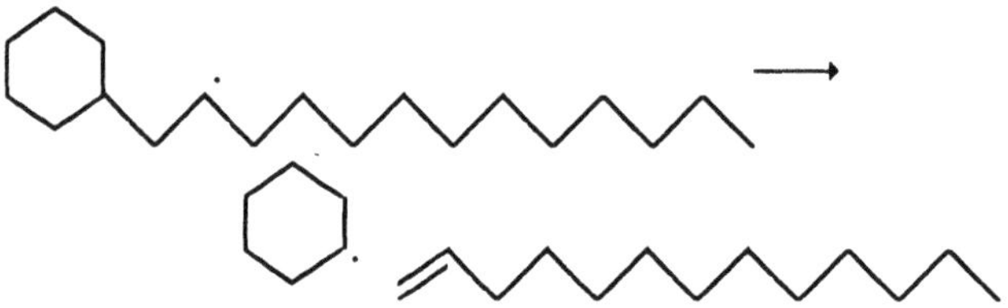

(3.15)

Reaction (3.14) has a tertiary radical, making dodecane and methylenecyclohexane the second most abundant pair of products. Reaction (3.15) was the most favored reaction because the β–scission reaction gives a secondary cyclohexyl radical, rather than the usual primary radical. No paraffins or olefins above C_{13} were detected, indicating that opening of the cyclohexyl ring was very unfavorable.

The residue fraction can contain large amounts of naphthenic structures, linked to other groups by paraffinic bridges and substituted with side chains (see Figure 1.7). The above reaction chemistry shows that cracking of these molecules will release the naphthenic groups and their side chains, more easily than breaking up the naphthenic aggregates.

In the absence of hydrogen, the naphthenic groups can dehydrogenate to form aromatics. In the case of tridecylcyclohexane, for example, toluene was formed as a significant product at 450° C:

(3.16)

No toluene was detected at lower temperatures, and no other aromatics were observed. Clearly the reverse reaction is favored under hydrogen–rich conditions in the presence of a catalyst, so that hydrocracking can form naphthenes from aromatic compounds.

3.2.6 Reactions of alkylaromatics and aromatics

The chemistry of the alkyaromatics is similar to the alkylnaphthenes; cracking removes the side chains to give a distribution of aklylaromatics with shorter chains, alkanes, and olefins. Aromatic groups such as benzene are not cracked by thermal reactions at normal operating temperatures. For example, pyrolysis of n–pentadecylbenzene gave toluene and styrene as the main aromatic products, with lower yields of phenyl alkanes and phenyl olefins with chains of 2–12 carbons [Savage and Klein, 1987]. Toluene was the favored product because of the high stability of the corresponding benzyl radical.

Side chains on condensed aromatic groups such as pyrene may undergo an unusual autocatalytic reaction, wherein the side chain is rapidly removed α to the aromatic cluster. For example, pyrolysis of 1–dodecylpyrene at high conversion gave mainly pyrene and dodecane as products [Savage *et al.*, 1989]. The nature of the autocatalysis and the required characteristics of the aromatic cluster have not yet been defined, but this pathway is consistent with the low yields of olefins during pyrolysis of asphaltenes.

Alkylaromatic radicals can also participate in condensation reactions, which give rise to fused–ring aromatics and eventually coke. The exact sequence of steps in this process is not well defined, but one starting point is likely a radical reaction [Poutsma, 1990]:

$$C_6H_5^{\bullet} + C_6H_6 \longrightarrow C_6H_5{-}C_6H_5 + H^{\bullet} \quad (3.17)$$

(biphenyl)

Another possible pathway to form linkages between aromatics is recombination of radicals:

$$C_6H_5CH_2^{\bullet} + C_6H_5^{\bullet} \longrightarrow C_6H_5CH_2C_6H_5 \quad (3.18)$$

(benzyl + phenyl radicals) (diphenylmethane)

Biphenyl was observed as a significant product in cracking of n–hexadecane in benzene, while diphenylmethane was significant when n—hexadecane was cracked in toluene [Khorasheh, 1992]. Cracking of the n—alkane accelerated the formation of biphenyl, which normally would only occur at temperatures over 550° C. Formation of such products would be most favorable at a high concentration of aromatics.

Higher aromatics (naphthalene, anthracene etc.) undergo addition reactions like (3.17) more easily than benzene and at lower temperatures [Poutsma, 1990]. These bridged aromatic compounds would then undergo further addition reactions to form larger and larger networks of aromatic rings, leading to coke formation.

3.2.7 Reactions of bridged aromatics

Bridged structures such as biphenyl, diphenylmethane, and longer bridges occur naturally in residues [Strausz, 1989], as well as being formed by addition reactions. The strength of the bond between aromatic carbon and the α—aliphatic carbon is stronger than aliphatic C—C bonds, which would be expected to make cracking more difficult. For example, the biphenyl linkage has a bond dissociation energy of 476 kJ/mol, compared to 375 kJ/mol for diphenyl methane and 309 kJ/mol for C—C bonds in longer bridges [Poutsma, 1990].

Freund *et al.* [1991] studied the cleavage of long bridges between aromatic clusters using the model compound dipyrenyl eicosane (i.e. a pyrene group on either end of a twenty—carbon n—alkyl bridge. Reactions were conducted in the liquid phase at low pressure, and at temperatures ranging from 400—500° C. The products were pyrene and alkyl pyrene, with pyrene, methyl pyrene, C_{19}—pyrene and C_{20}—pyrene as the most abundant products, consistent with the pattern observed in side—chain cracking (reactions 3.14—3.16).

Under suitable conditions, the strong bonds in short bridges such as biphenyl or diphenyl methane can be broken. Malhotra and McMillen [1990, 1993] have proposed that transfer of hydrogen by an aromatic solvent molecule can account for the cracking of strong bonds. The reaction scheme is as follows:

$$\text{Solvent} + \text{H}^{\bullet} \longrightarrow \text{Solvent—H}^{\bullet} \qquad (3.19a)$$

$$\text{Solvent—H}^{\bullet} + \text{(biphenyl)} \longrightarrow$$

$$\text{Solvent} + \text{(H-adduct radical of biphenyl)} \qquad (3.19b)$$

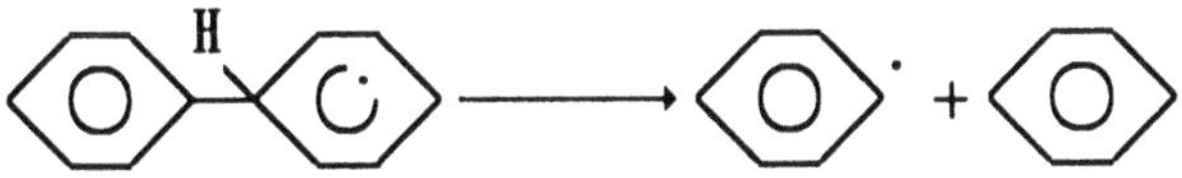

(3.19c)

Reaction (3.19c) is the reverse of reaction (3.17). The aromatic group is represented as a benzene ring in (3.19) for convenience, but this radical hydrogen transfer mechanism is actually observed for naphthyl and larger aromatics. The solvent molecule is an aromatic compound such as anthracene or pyrene.

The cleavage of bonds by hydrogen addition, or *hydrogenolysis*, is favored by the presence of hydrogen and a catalyst. A possible role for the catalyst is to provide hydrogen atoms which would activate solvent liquid–phase compounds (3.19a) for hydrogenolysis [Malhotra and McMillen, 1993; Sanford, 1993]. In a reactor, therefore, the presence of hydrogen and catalyst tends to inhibit addition reactions which minimizes the formation of bridged compounds and, ultimately, minimizes the formation of coke. Sanford [1993] has proposed that hydrogenolysis reactions are responsible for both suppression of coke and for breaking up large aromatic ring groups, so that hydrogenolysis reactions like (3.19) may be predominant at high levels of residue conversion.

3.2.8 Example: Cracking of a hypothetical residue molecule

Consider how a typical bridged structure, such as would be found in residues, would behave under thermal cracking conditions:

X =

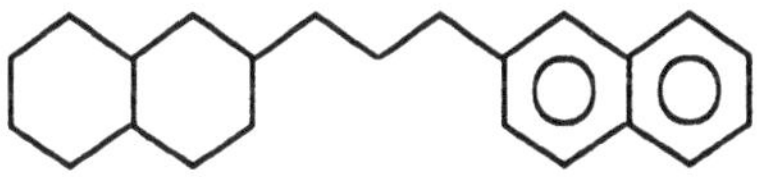

This compound would be activated by hydrogen abstraction (Reaction 3.3) whereby a hydrogen atom is removed by a free radical.

$$R^{\cdot} + X \longrightarrow X^{\cdot} + RH$$

The aliphatic C—H bonds are weaker than aromatic CH, therefore these atoms are most likely to be abstracted. The potential thermal reactions depend on which hydrogens are abstracted.

Case 1 – Abstraction of Bridge H

Abstraction of a hydrogen from the α–carbon is favored because this benzylic C—H bond is relatively weak due to stabilization of the radical by the aromatic ring. Reaction by β–scission gives:

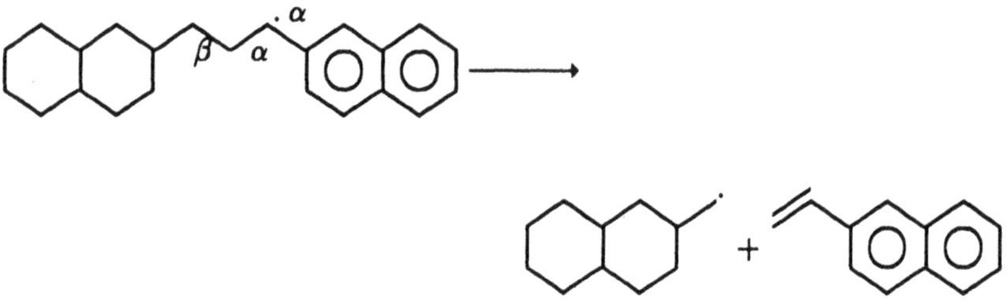

The radical would then go on to other reactions. Note that an aromatic C—C bond was β to the radical, but due to its strength this bond would not break.

Case 2 – Abstraction of Naphthenic H

Abstraction of a naphthenic H gives a radical with two ring C—C bonds in the β–position.

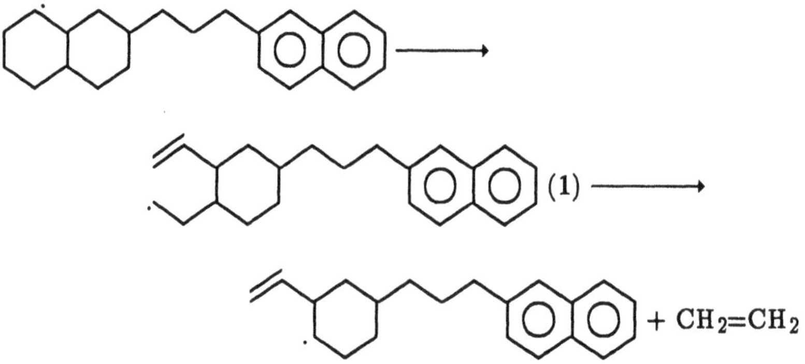

(1)

+ $CH_2{=}CH_2$

β–scission of the naphthenic ring gives an olefin and a radical within the same molecule (1). The most likely reaction of (1) is recombination of the radical with the olefin to reform a five– or six–membered ring. An alternative reaction would be further β–scission to give ethylene, as shown above.

Case 3 — Abstraction of Aromatic H

The aromatic C—H bond is very strong, therefore abstraction in this position is an unfavorable reaction:

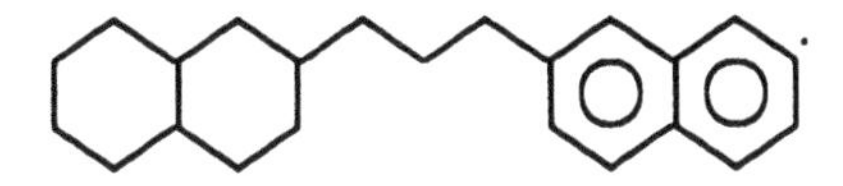

In this case the bonds in the β position are unreactive aromatic C—C bonds, and no scission could occur. This radical could only participate in a hydrogen abstraction reaction.

Case 4 — Hydrogenolysis

In the presence of hydrogen, the radical hydrogen transfer reaction can provide and alternative pathway for cracking, following the scheme of Reaction (3.19).

This simple example illustrates why side chains and bridges tend to crack most easily, giving naphtha and light paraffin products. Cracked aromatic and naphthenic products tend to retain some alkyl substituents.

3.2.9 Reactions of mixtures

To a first approximation, the cracking of a mixture such as bitumen will be determined by the reactivity of its constituents. The various compounds will interact during the degradation process, because the free radical intermediates are "shared" between the different species according to concentration and stability. In broad terms we can define the hierarchy of cracking reactions as follows, from most reactive to least reactive:

Paraffins > Linear olefins > Naphthenes >
Cyclic olefins > Aromatics

Crackability tends to increase with molecular weight (or boiling range). This observation may seem contradictory, but larger molecules have more bonds which can easily rupture, increasing the probability of breakage.

3.2.10 Sulfur compounds

Thiophenic sulfur compounds are unaffected by thermal reactions, as with carbon—based aromatic compounds. The thiols, thioethers, and disulfides, however, are quite reactive under thermal process conditions. Estimates for this type of sulfur range as high as 50% of the total sulfur in bitumens and asphalts [Green *et al.*, 1993]. The thermal reactions of sulfur are favorable because the C—S bond is weaker than other aliphatic bonds (Table 3—1), for example:

$$R{-}S{-}R' + 2H_2 \longrightarrow H_2S + RH + R'H \qquad (3.20)$$

Thermal processing of bitumen is always accompanied by the evolution of hydrogen sulfide, even at temperatures as low as 250° C. The facile rupture of the sulfide bonds has been postulated as a major mechanism for cracking of the high molecular weight components of bitumen. The thermal reactions of sulfur compounds can give conversion levels of order 30—50% without any catalyst at all, with an attendant evolution of H_2S.

The nitrogen compounds are almost entirely present as heteroaromatic species, and are almost unaffected by thermal reactions. Oxygen species in bitumens and residues have not been studied as intensively; carboxylic acids and ketones are relatively reactive, while ethers and furans follow the behavior of the analogous sulfur compounds.

3.3 Catalytic Reactions

3.3.1 Catalyst composition

The formation of distillate products from bitumen residues is dominated by thermal reactions; the main role of additives or catalysts in primary upgrading is to enhance the uptake of hydrogen and prevent condensation and coking reactions [Miki *et al.*, 1983; Le Page and Davidson, 1986]. A variety of transition metals are active for hydrogenation reactions, including iron, nickel, cobalt, molybdenum, tin, and tungsten [Weisser and Landa, 1973]. These metals have the virtue of being active in the sulfide state, as well as in the metallic form. The high sulfur concentrations in upgrading processes guarantee that metal surfaces are in the sulfide form, so that more active hydrogenation catalysts such as platinum cannot be used. The activity of a metal sulfide catalyst depends on three major factors: the metal used, the surface area of the metal, and the presence of any alloyed metals. The normal practice for obtaining very high surface areas is to support the metal sulfide as very small crystallites on a high—surface area support such as γ—alumina. Alternatives include adding the metal as a powdered sulfide or oxide, impregnating metal salts on a support, or using dissolved organo—metallic compounds.

Catalysts or additives for treating residues are selected based on activity and cost. High activity hydrogenation catalysts for upgrading are mixtures of Co and Mo or Ni and Mo, supported on alumina and formed into pellets or beads. Under hydroprocessing conditions, the active metals are in the sulfide form, with Co or Ni acting as a promoter on crystallites of MoS_2 [Ho, 1988; Prins *et al.*, 1989]. These catalysts are gradually deactivated in service due to deposition of vanadium and nickel sulfides from the feed oil, and due to coke formation.

Iron compounds are commonly used as powdered or impregnated additives, due to their very low cost. For example, iron oxide will form the sulfide in situ, and help to promote hydrogen transfer reactions. Iron sulfate is also an effective additive. Processes which use iron do not normally recyle the catalyst.

A range of catalysts are used for demetallation, i.e. removal of vanadium and nickel. The choice of catalyst tends to be more dependent on the pore characteristics of the support than on the activity of the catalyst itself, since a controlling factor in the life of the catalyst is ability to continue to function with high loadings of vanadium.

3.3.2 Reactions of aromatic compounds

In the presence of a catalyst and hydrogen gas, aromatic groups are hydrogenated to give hydroaromatics and naphthenes. The more rings in an aromatic cluster, the more thermodynamically favorable the hydrogenation reaction. Benzene rings are the least reactive, which can be interpreted as being due to the unusually high stability of benzene due to resonance.

Table 3—3
Resonance Stabilization of Aromatic Compounds

Compound	Resonance Energy		Average Resonance Energy per Bond
	kcal/mol	eV (1)	kcal/mol
Benzene	36	0.838	6
Naphthalene	61	1.34	5.5
Anthracene	84	1.6	5.25
Phenanthrene	92	1.75	5.75
Pyrene		2.13	
Furan, pyrrole and thiophene	22—28		4.4—5.6
Pyridine	23		3.8

1. Resonance energies in kcal/mol are from enthalpy of combustion, relative to equivalent polyenes [Morrison and Boyd, 1973]. Energies for stabilization in eV are from molecular orbital calculations and a resonance structure count method [Carey and Sundberg, 1980].

The resonance stabilization energy of aromatic structures decreases on a per—bond basis from benzene to higher fused—ring aromatics. The delocalization of the π—electrons due to resonance gives a very low electron density in benzene, and this property correlates with reactivity, as illustrated in Figure 3—3.

Polyaromatics such as naphthalene and phenanthrene are more easily hydrogenated than benzene, and in the presence of hydrogen, a catalyst, and a hydrogen acceptor these compounds can undergo cycles of hydrogenation and dehydrogenation:

$$+ 2H_2 \longrightarrow \qquad (3.21)$$

(1,2,3,4—tetrahydronaphthalene)

$$+ 4\ R^{\cdot} \longrightarrow \qquad + 4\ RH \qquad (3.22)$$

In the case of phenanthrene:

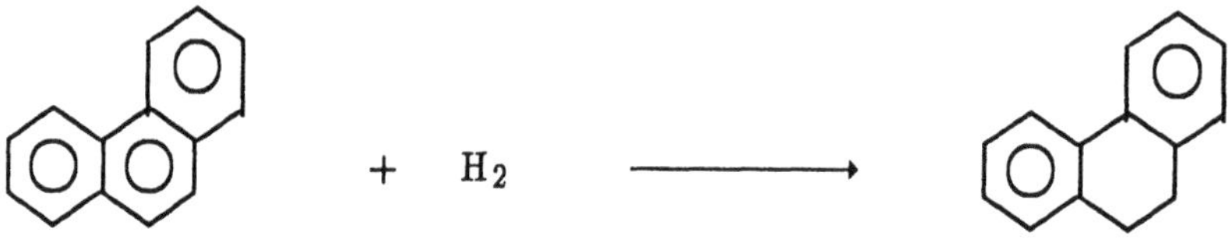

$$+ \quad H_2 \longrightarrow \qquad (3.23)$$

9,10—Dihydrophenanthrene

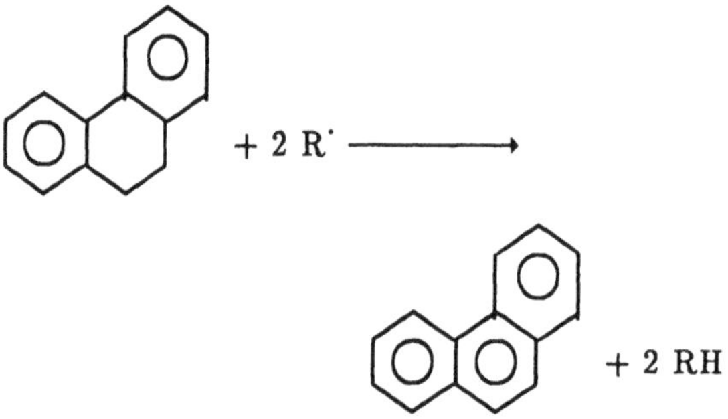

$$+ 2\ R^{\cdot} \longrightarrow \qquad + 2\ RH \qquad (3.24)$$

These reactions shuttle hydrogen from the gas phase to reactive radicals

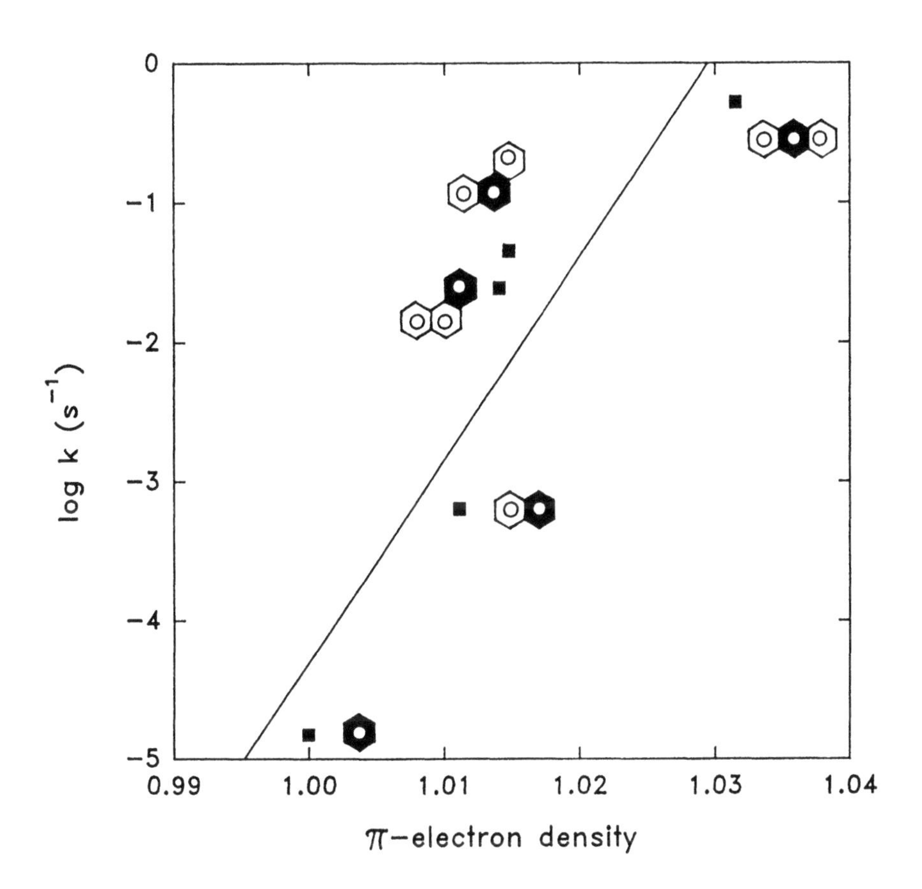

Figure 3–3
A linear free–energy relationship between the first order rate constant for hydrogenation of aromatic compounds (k) and the electron density (Data from Neurock *et al.*, 1990)

in the liquid phase. These hydrogen transfer reactions are thought to be very important in the liquefaction of coal, where the solvents tend to resemble naphthalene and phenanthrene.

Residues have high concentrations of hydroaromatics and naphthenes such as steranes which could undergo cycles of hydrogenation and dehydrogenation. In principle then, these reactions should be important in upgrading of bitumen. Studies of upgrading using deuterium as a tracer for hydrogen transfer [Steer *et al.*, 1992] and studies of hydrogen with model compounds for heavy oil [Khorasheh, 1992], however, indicate that little shuttling of hydrogen occurs due to free radical reactions such as equations (3.22) and (3.23) at temperatures ca. 450° C.

The catalytic hydrogen exchange reactions (equations (3.20) and (3.22)) are reversible; the direction of the net reaction depends on the specific aromatic structure undergoing reaction, the hydrogen fugacity, and the temperature. In general, at a given temperature some components of the oil will dehydrogenate to liberate hydrogen at low pressure, while at high pressure hydrogen will be consumed in net conversion of aromatics.

Quantitative estimates for these reactions cannot be made due to the lack of thermochemical data for appropriate aromatic compounds. The major emphasis in measuring thermochemical properties of multiring aromatics has been for hydrogenation at temperatures below 350° C to form naphthenics. The equilibrium constants for hydrogenation reactions have been measured for alkyl benzenes, naphthalene, and phenanthrene, in some cases up to 450° C [Girgisz and Gates, 1991].

Figure 3—4 illustrates an example equilibrium calculation from the available data for two— and three—ring compounds from Frye and Weitkamp [1969]. These results show the effect of hydrogen pressure on equilibrium conversion of naphthenics at 430° C. At pressures *ca.* 8—9 MPa, significant dehydrogenation of three—ring naphthenes to substituted phenanthrenes would be expected at equilibrium, as well as some conversion of two—ring naphthenes. The effects of substituent groups on naphthalene and phenanthrene equilibrium have not been determined. Biomarkers such as terpanes and steranes, which are abundant in Alberta bitumens [Strausz, 1989], could dehydrogenate as indicated in Figure 3—4 and form more aromatic products. Such biomarkers are altered by refinery processing [Peters *et al.*, 1992] but their hydrogenation/dehydrogenation equilibria have not been measured. Similarly, no equilibrium data are available for pyrene, anthracene, chrysene, or their alkylated derivatives.

Although the breakage of C—C bonds is a thermal process, the presence of a catalyst serves to enhance the exchange of hydrogen with

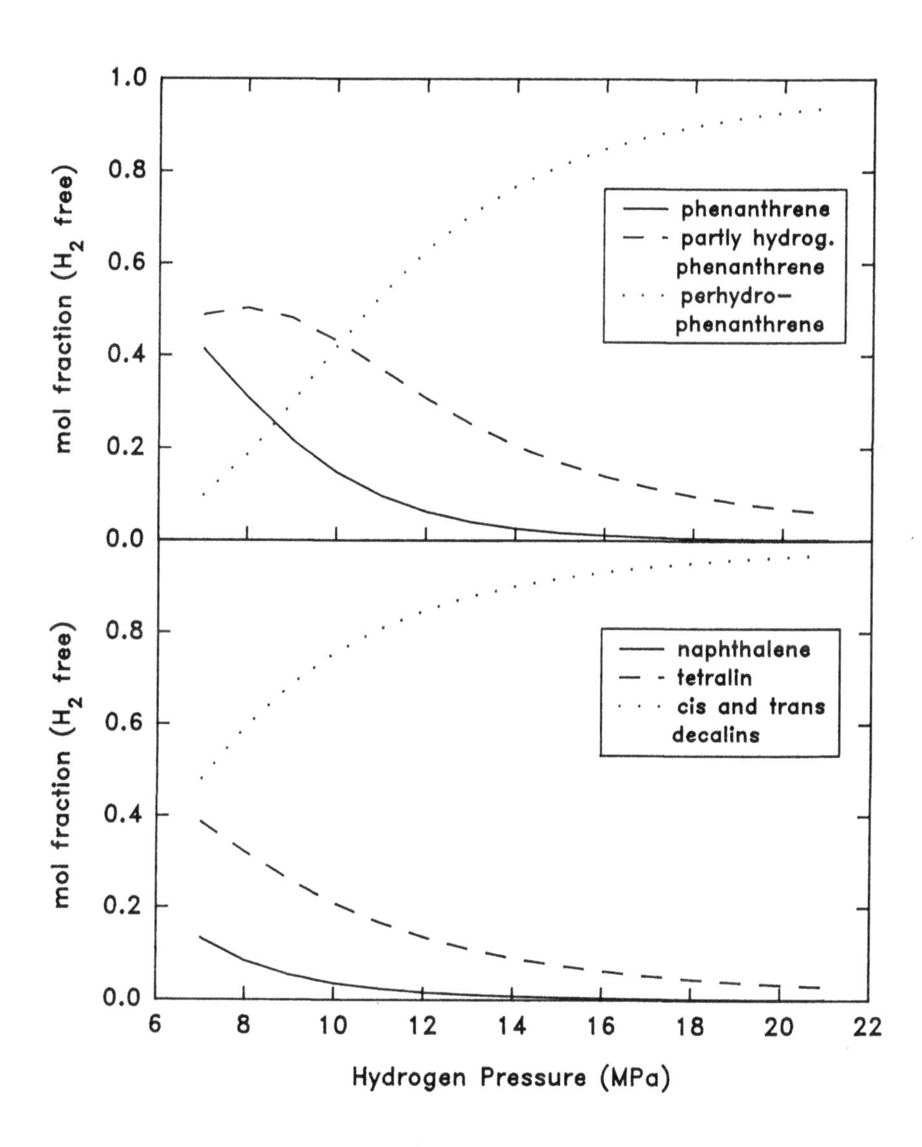

Figure 3–4
Equilibrium compositions of hydrogenated and aromatic compounds from phenanthrene (upper panel) and naphthalene (lower panel) as a function of hydrogen pressure at 430°C (Thermophysical data are from Frye and Weitkamp, 1969)

the liquid hydrocarbons [Miki *et al.*, 1983]. This supply of hydrogen directs the products toward saturated alkanes and helps to control coke formation. For example, conversion of olefins likely helps to suppress coke formation:

$$+ H_2 \rightleftharpoons \qquad (3.25)$$

(1—nonene)

Even at identical temperature and pressure, different feeds may hydrogenate or dehydrogenate depending on composition [Steer *et al.*, 1992]. Nevertheless, the qualitative behavior of different residues will be similar, as illustrated in Figure 3—5. Increasing temperature or decreasing hydrogen pressure crosses the boundary into the coking regime. At low temperatures, thermal reactions occur too slowly to be of interest. The catalyst activity, in general, augments hydrogen pressure in suppressing coke formation.

3.3.3 Reactions of sulfur, nitrogen, and oxygen compounds

The sulfur compounds in the bitumen exhibit very different reactivities. As mentioned above, the aliphatic sulfur species are readily converted by thermal reactions. Conversion of thiophene compounds requires catalytic hydrodesulfurization. The action of the catalyst gives rise to two parallel reaction pathways:

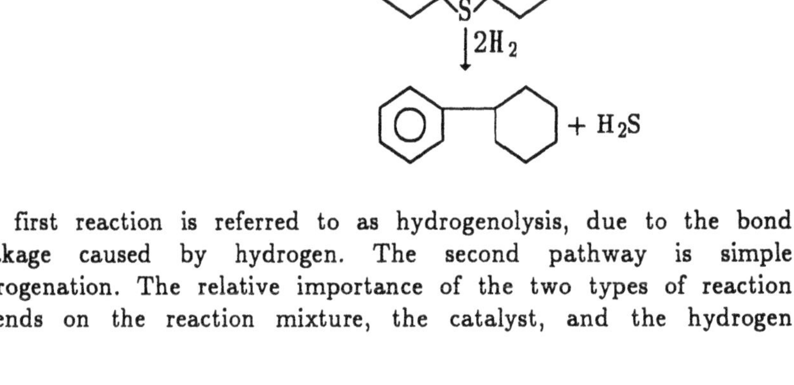

S $\xrightarrow{2H_2}$ + H_2S (3.26)

$3H_2$

S

$2H_2$

$+ H_2S$

The first reaction is referred to as hydrogenolysis, due to the bond breakage caused by hydrogen. The second pathway is simple hydrogenation. The relative importance of the two types of reaction depends on the reaction mixture, the catalyst, and the hydrogen

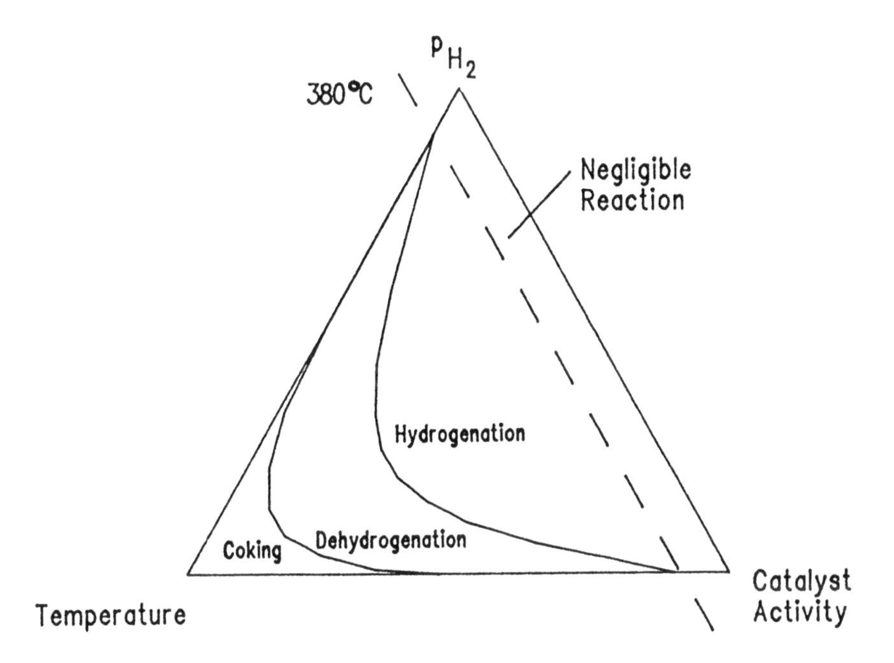

Figure 3–5
Regions of operation for thermal and catalytic processes with hydrogen pressure, temperature, and catalyst activity

pressure. Ni/Mo and Co/Mo catalysts are selective for sulfur heterocycles, in preference to aromatic hydrocarbons. This selectivity is thought to be due to the favorable energetics of coordination of the sulfur atom with active sites on the catalyst surface. One representation of the reaction mechanism is from Prins *et al.* [1989], where □ represents the active site:

```
  ,H            H,
 S                S
 |        □       |
-Mo-S----Mo----S-Mo-  +  (thiophene)  ------>

  ,H   (thiophene)   H,                      (ring)       H,
 S          S          S          S           S             S
 |          |          |          ||          |             |
-Mo-S------Mo------S--Mo-  ----> -Mo-S-------Mo-------S----Mo-

          (C4 chain)                     ,H  H, ,H  H,
 S          S          S               S      S      S
 ||         ||         ||    +2H2      |      ||     |
-Mo-S------Mo------S--Mo-  ------>   -Mo-S---Mo---S-Mo-  ------>

                                       ,H            H,
                                      S                S
                                      |        □       |
                                     -Mo-S----Mo----S-Mo-  + H2S      (3.27)
```

In this catalytic cycle, the active site on the molybdenum is regenerated by hydrogenation of the attached sulfur groups to give hydrogen sulfide. This mechanism suggests that sulfur in the catalyst should exchange with gas–phase H_2S, which is indeed observed in tracer studies with radioactive H_2S.

Nitrogen compounds are much less reactive, in part because only the hydrogenation pathway is observed in catalytic hydrodenitrogenation, as illustrated for pyrrole:

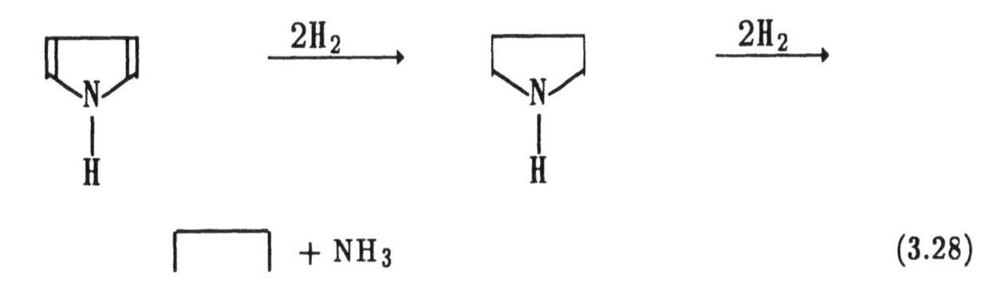

Nitrogen compounds build up in the residue fraction during upgrading, due to their low reactivity. For example, Gray *et al.* [1989] reported that a pitch after hydrocracking had a similar sulfur content to the feed oil, indicating that sulfur removal was in balance with cracking reactions. Nitrogen content, on the other hand, doubled due to an accumulation of both pyrrolic and basic compounds (Table 3—4).

Several interactions between the simultaneous catalytic reactions have been defined unambiguously, largely through the study of model compounds. Hydrogenation and hydrogenolysis (i.e. direct removal of sulfur) sites appear to be distinct on Ni/Mo and Co/Mo catalysts. The hydrogenation sites are more acidic, and hence tend to be poisoned by adsorption of basic nitrogen compounds, hence simultaneous HDS of dibenzothiophene and HDN of quinoline showed that the quinoline inhibited the hydrogenation of aromatic rings [Sundaram *et al.*, 1988].

Table 3—4
Nitrogen Types in Heavy Residues from Athabasca Bitumen
(Data from Gray *et al.*, 1989)

	Athabasca Bitumen (427° C+)	Hydrocracker Pitch (427° C+)
Total S, wt%	5.54	5.13
Total N, wt%	0.53	0.99
Pyrrolic N, mol/kg[1]	0.12	0.22
Basic N, mol/kg[2]	0.09	0.16

1. Determination by IR spectroscopy of 3500—3420 cm^{-1} band
2. Determined by titration.

Lavopa and Satterfield [1988] showed that this inhibition was dependent on temperature; at 260° C the presence of quinoline enhanced HDS of dibenzothiophene, whereas at 360° C the quinoline was inhibitory. The hydrogen sulfide released by HDS, and the water released by HDO together promote the HDN of basic nitrogen compounds [Satterfield *et al.*, 1985], probably by increasing the acidity of the catalyst surface. The effect of hydrogen sulfide and water is synergistic; two combined have a larger effect than either compound by itself.

3.3.4 Demetallization reactions

Recalling the discussion in Chapter 1, the Ni and V are present in residue fractions as organometallic complexes, but only the poryphyrin types are chemically well defined. Consequently, only porphyrin types have been well studied in terms of upgrading chemistry, but the behavior of nonporphyrin metals is thought to be quite similar. Metal removal occurs during both thermal and catalytic processing, but removal is more complete when a catalyst provides much more effective hydrogen transfer. The catalytic reaction is called hydrodemetallization. Figure 3—6 gives a probable sequence of events. Hydrogenation of the porphyrin forms a chlorin structure, which loses the vanadium to the surface of the catalyst. The porphyrin backbone then appears as pyrrole groups in the product oil. The vanadium (shown here coordinated to oxygen in the porphyrin) is coordinated to the sulfide surface, and the oxygen is released upon hydrogenation. The vanadium gradually develops as a bulk sulfide phase with the composition V_3S_4, which is nonstoichiometric. The sulfide phase forms rod—shaped deposits on the surface of the catalyst.

Once the original catalyst surface is covered by vanadium sulfide, then the reaction continues as an autocatalytic process. Indeed, once the vanadium sulfide layer is established, supported—metal catalysts are indistinguishable from γ—alumina support alone.

3.3.5 Catalytic C—C bond breakage in distillates

Once nitrogen compounds have been reduced to a low level, then distillates can be catalytically cracked over acidic catalysts. Two main classes of catalysts are used:

1. Zeolite catalysts, mainly used in fluid—catalytic cracking reactors.

2. Bifunctional catalysts, consisting of metals such as nickel supported on silica/alumina or zeolites. These catalysts are normally used in distillate hydrocracking.

In both cases, the active sites for cracking are highly acidic. Alkanes adsorb to the catalyst, and undergo cracking and rearrangement via

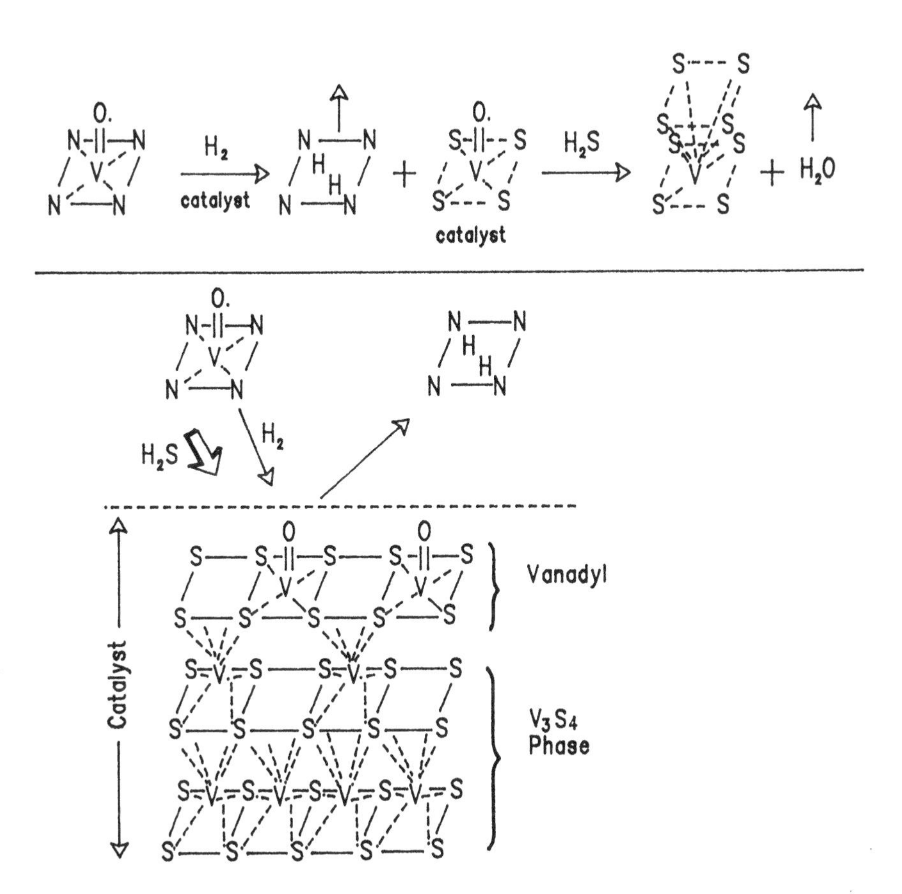

Figure 3–6
Mechanism for vanadium removal (From Asaoka *et al.*, 1987)

carbenium ion intermediates:

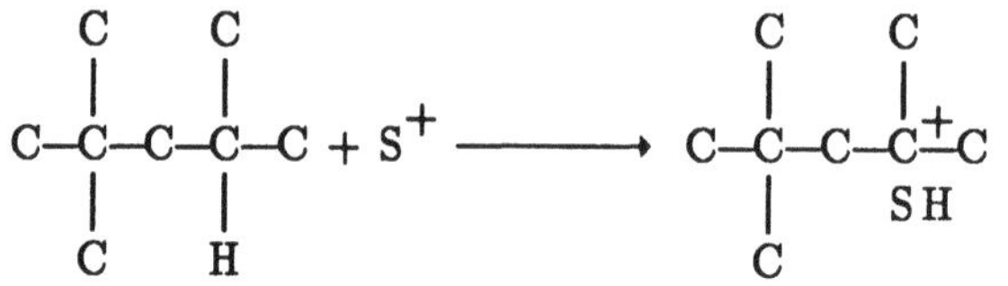

$$\begin{array}{ccccccccc} & & C & & & & C & & \\ & & | & & & & | & & \\ C & - & C & - & C & - & C & - & C \\ & & | & & & & | & & \\ & & C & & & & H & & \end{array} + S^{+} \longrightarrow \begin{array}{ccccccccc} & & C & & & & C & & \\ & & | & & & & | & & \\ C & - & C & - & C & - & \overset{+}{C} & - & C \\ & & | & & & & & & \\ & & C & & & & SH & & \end{array} \tag{3.29}$$

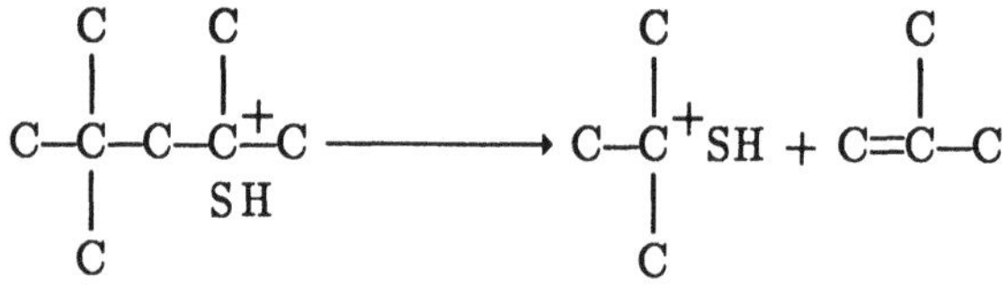

$$\begin{array}{ccccccccc} & & C & & & & C & & \\ & & | & & & & | & & \\ C & - & C & - & C & - & \overset{+}{C} & - & C \\ & & | & & & & & & \\ & & C & & & & SH & & \end{array} \longrightarrow \begin{array}{cccc} & & C & \\ & & | & \\ C & - & C^{+} & SH \\ & & | & \\ & & C & \end{array} + \begin{array}{ccccc} & & C & & \\ & & | & & \\ C & = & C & - & C \end{array} \tag{3.30}$$

$$\begin{array}{cccc} & & C & \\ & & | & \\ C & - & C^{+} & SH \\ & & | & \\ & & C & \end{array} \longrightarrow \begin{array}{ccccc} & & C & & \\ & & | & & \\ C & - & C & - & C \\ & & | & & \\ & & H & & \end{array} + S^{+} \tag{3.31}$$

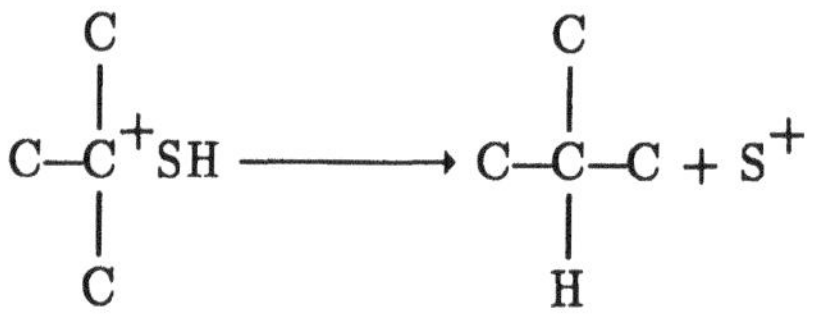

In this catalytic cycle, the acid site is represented by S^+, and the alkane is shown as a carbon skeleton. The stability of carbenium ions is $3^\circ > 2^\circ > 1^\circ$ (like free radicals), and therefore catalytic cracking favors isomerization to give branched structures which stabilize the positive charge on the carbon.

Active cracking catalysts are poisoned by nitrogen compounds that react with the acid sites. Because olefins are formed, these catalysts also tend to accumulate coke. The two main processes that use acid catalysts for cracking, fluidized–bed cracking and hydrocracking, use complimentary catalyst chemistry and process conditions. The high–activity zeolite catalysts tend to accumulate coke rapidly, so their active time–on–stream must be kept short. Consequently, they are used in riser or fluid–bed reactors where the catalyst can be quickly blended with the feed, then removed continuously for regeneration. These processes are described in detail in Chapter 7.

The opposite approach, hydrocracking, uses hydrogen pressure and metallic catalysts to control coke formation and to simultaneously stabilize unsaturated products. The catalysts are bifunctional to provide the appropriate blend of acidic cracking activity and hydrogenation activity. Additional information on catalysts for cracking and hydrocracking of distillates is available from Gates [1992].

3.4 Interactions Between Catalytic and Thermal Reactions

The reactions of a petroleum fraction involve a large number of simultaneous reactions of different species. This effect is compounded in hydroconversion processes because thermal and catalytic reactions are occurring simultaneously. Given that the addition of a catalyst does not enhance every single type of reaction, are there any interactions between catalytic and noncatalytic reactions?

3.4.1 Mechanisms of thermal and catalytic interactions

The extent of interactions appears to depend on the products. Light ends, for example, are almost completely unaffected by catalyst activity. Khorasheh *et al.* [1989] found that yields of methane, ethane, and propane were independent of catalytic activity. They compared catalytic and thermal reactions of gas oil in the same reactor with and without catalyst (Figure 3—7). Only the yields of butanes and hexanes were enhanced by the presence of a Ni/Mo on γ—alumina catalyst. On the other hand, addition of a catalyst in hydroconversion of residues at 430—450° C can boost residue conversion from ca. 60% to 65 to 70%, depending on activity [Gray *et al.*, 1992]. The role of catalyst in preventing coke formation is also a critical aspect of process performance.

The most obvious interaction is chemical; by hydrogenating feed compounds, the catalyst transforms the reactants that are involved in subsequent cracking. For example, hydrogenation of a phenanthrene ring by a catalyst creates a compound that is much more easily cracked. Hydrogenation of aromatic rings and heterocyclic compounds also reduces boiling point with actual breaking any carbon—carbon bonds, as illustrated in the following reaction:

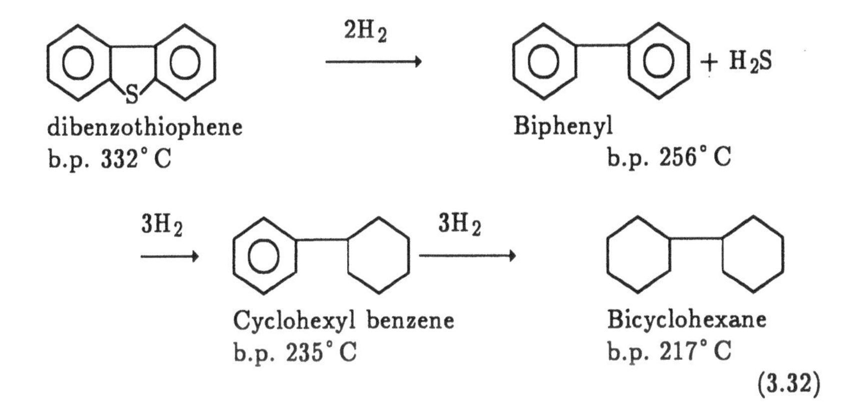

(3.32)

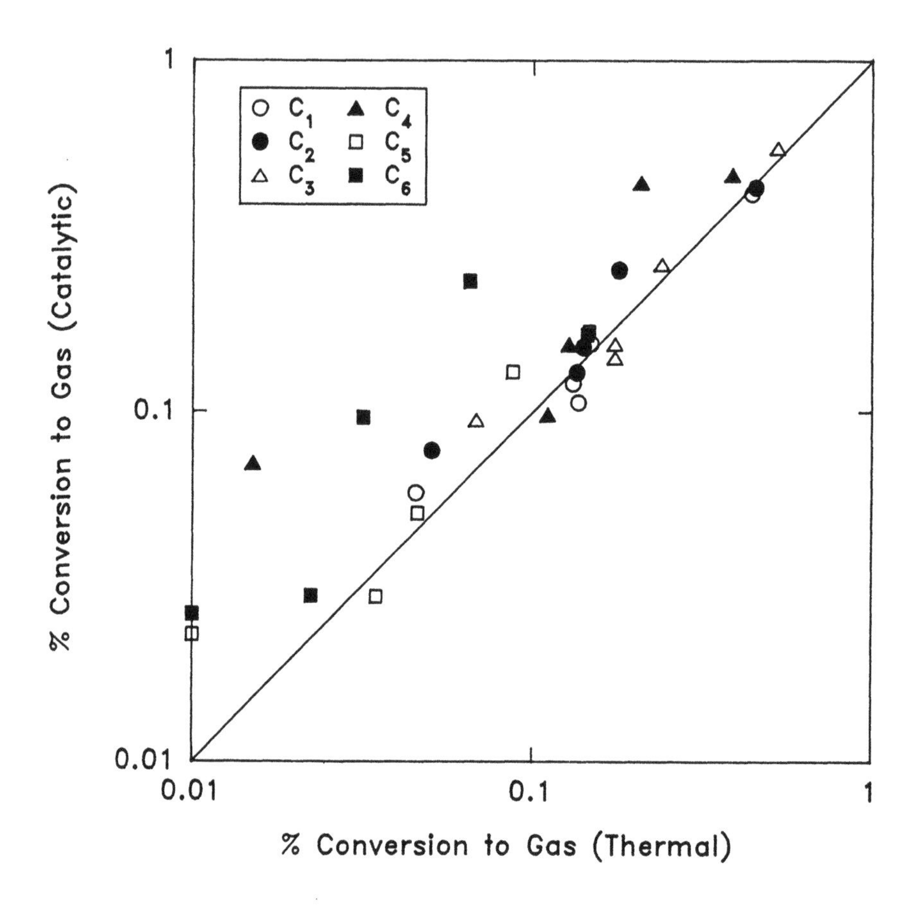

Figure 3–7
Gas yields from catalytic and thermal hydroprocessing of gas oil at 380–420° C, 13.9 MPa hydrogen pressure (Data from Khorasheh *et al.*, 1989)

Since cracking is always defined based on distillation yields, a decrease in boiling point due to catalytic hydrogenation and hydrogenolysis can give an apparent contribution of hydogenative catalysts to cracking conversions.

In the presence of even a weak hydrogenation catalyst, the olefins produced by cracking are hydrogenated. If these compounds are not hydrogenated, they can undergo addition reactions with free radicals to increase molecular weight [Khorasheh, 1992]. A decrease in such regressive reactions is indistinguishable from an increase cracking reactions on the basis of overall residue conversion.

At severe reactor conditions, the addition of catalyst can prevent undesirable reactions in the feed, such as dehydrogenation and coking, thereby sustaining a higher rate of thermal cracking of residue species. Malhotra and McMillen [1993] and Sanford [1993] have suggested that hydrogen radicals spillover or migrate from the catalyst surface and promote hydrogenolysis. Malhotra and McMillen [1993] suggest that solvent molecules serve as hydrogen carriers (Reaction (3.19)) while Sanford [1993] has suggested direct migration of hydrogen radicals. In both cases, the catalyst serves to generate hydrogen radicals which direct the thermal reactions toward low yield of coke and high yield of liquids.

3.4.2 Reaction scheme for residue conversion

Sanford [1993] proposed the following series of steps for the thermal conversion of residue in the presence of a catalyst:

1. Thermal cracking of side groups and side chains, leaving a naphtheno-aromatic core. This initial phase of cracking would account for 55—65% of the residue conversion.

2. In the presence of hydrogen, hydrogenolysis would gradually degrade the aromatic groups giving distillate and gas products. Catalyst would enhance these hydrogenolysis reactions by generating hydrogen radicals.

3. In an inert atmosphere, condensation reactions would predominate and solids would make up the major portion of the product.

This scheme is illustrated in Figure 3—8, where the compositions and conversion levels represent the behavior of Athabasca bitumen. Other residues would give coke formation at different levels of conversion. This reaction scheme effectively illustrates the complexity of the chemistry of residue conversion.

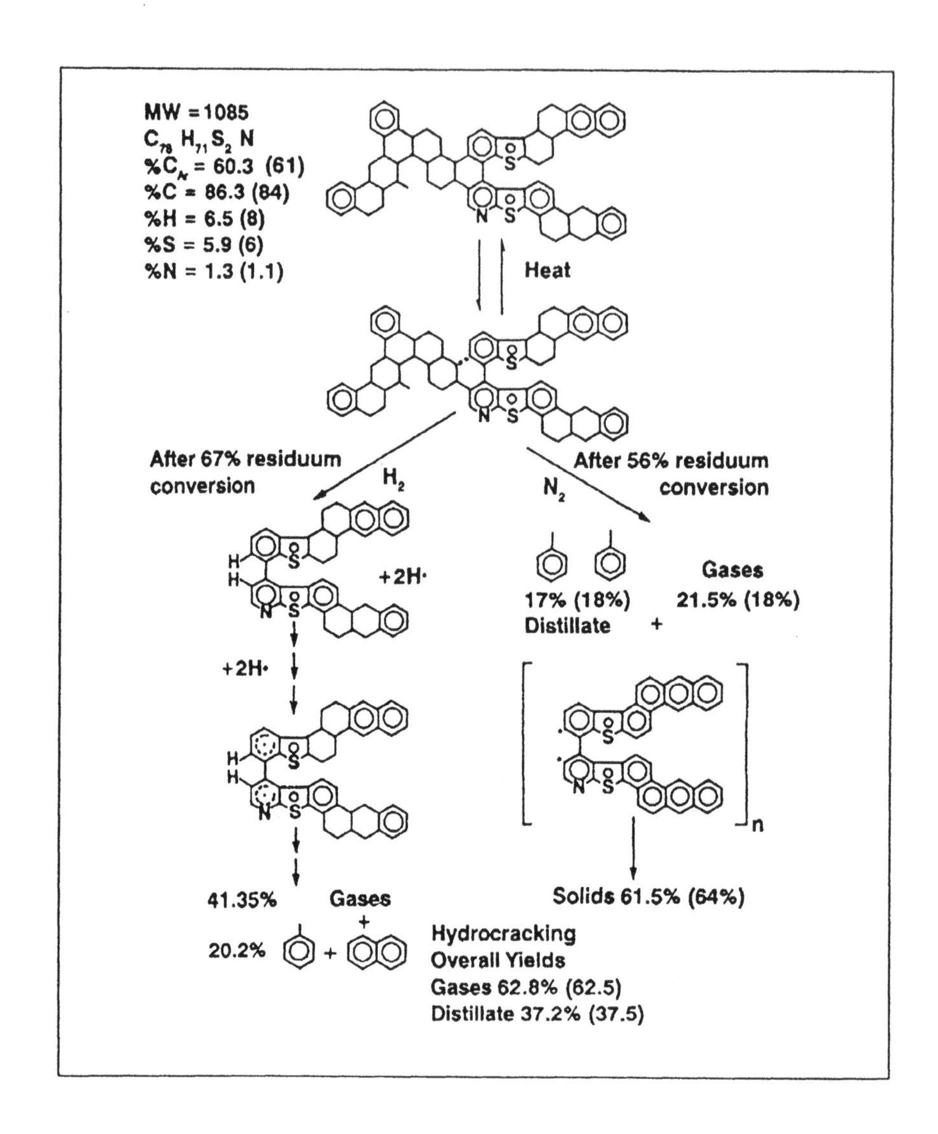

Figure 3–8

Possible conversion pathways for residue. Data in brackets represent measurements for Athabasca bitumen (*Reprinted from E.C. Sanford "Mechanism of coke prevention by hydrogen cracking during residuum hydrocracking", Prepr. Div. Petrol. Chem., 38(2), p. 413, 1993. Copyright by American Chemical Society*)

Notation

n—C_{16}	n—hexadecane
n—$C_{16}^{\cdot}$	n—hexadecyl radical
k	first—order reaction rate constant, s^{-1}
M	parent compound
$M^{\cdot}$	radical of parent compound
R	alkyl group
$R^{\cdot}$	alkyl radical
RH	hydrogenated alkyl radical
S	active site on catalyst

Further Reading

Gates, B.C. *Catalytic Chemistry*, John Wiley and Sons, New York, 1992.

Problems

3.1 List the primary products of thermal decomposition from the following compounds. Which pathway is most favored in each case?

a) n—dodecane
b) 2—methylundecane
c) 1,2—diphenylethane
d) 1,2—dicyclohexylethane

3.2 Tetralin (7,8,9,10—tetrahydronaphthalene) is commonly used as a hydrogen—transfer agent because of its ability to give up 4 moles of hydrogen atoms to form naphthalene. Write the series of elementary reactions for the overall reaction, showing the intermediate species.

3.3 List 8 classes of product molecules that would be obtained from thermal cracking of the hypothetical residue—fraction compound shown below. (Sketch the structures and name the chemical types; don't worry about naming substitutents)

CHAPTER

4

Fundamentals for Kinetic Analysis of Upgrading Reactions

Petroleum distillates and residues are complex mixtures containing a very large number of components that cannot be resolved and identified. Despite this complexity, we need kinetic analysis of the reactions of these materials in order to design large–scale reactors based on laboratory data. This chapter bridges the gap between the well-defined kinetics of simple chemical mixtures and the complexities of ill–defined petroleum fractions and residues. Detailed kinetics are presented for the major reaction types (thermal and catalytic), and then a series of examples show how these underlying kinetics relate to the observable changes due to chemical reactions in complex feeds.

4.1 Thermal Reactions

The thermal reactions of liquid–phase hydrocarbons occur via highly activated free–radical intermediates. These radicals are present only in very low concentrations, but they participate in a chain process to give significant conversion of the feed mixture. The detailed kinetics of this chain process are best illustrated by the example of cracking n–alkane in a solvent. For simplicity, consider the following single–step mechanism for high–pressure thermal cracking of n–alkanes:

Initiation: $$M \xrightarrow{k_1} 2\,R^{\bullet} \tag{4.1}$$

Propagation: $$R^{\bullet} + M \xrightarrow{k_2} RH + M^{\bullet} \tag{4.2}$$

$$M^{\bullet} \xrightarrow{k_3} R^{\bullet} + O \tag{4.3}$$

Termination: $$\text{Radical} + \text{Radical} \xrightarrow{k_4} \text{Products} \tag{4.4}$$

where M and $M^{\bullet}$ are the parent alkane and parent radicals, respectively, $R^{\bullet}$ and RH are lower alkyl radicals and the corresponding alkanes, respectively, and O represents olefins. Different termination reactions are possible as follows:

(i) $R^{\bullet} + R^{\bullet} \longrightarrow$ Products

(ii) $R^{\bullet} + M^{\bullet} \longrightarrow$ Products

(iii) $M^{\bullet} + M^{\bullet} \longrightarrow$ Products

(iv) all of the above.

The detailed kinetics will be derived for termination reaction (i). If cracking takes place in a batch reactor, then the rate of change of the radicals can be written:

$$\frac{d[R^{\bullet}]}{dt} = 2\,k_1[M] - k_2[M][R^{\bullet}] + k_3[M^{\bullet}] - 2\,k_4[R^{\bullet}]^2 \tag{4.5}$$

$$\frac{d[M^{\bullet}]}{dt} = k_2[M][R^{\bullet}] - k_3[M^{\bullet}] \tag{4.6}$$

The concentration of radicals is small, and will change very little through the course of the reaction. These intermediates will be in a *pseudo–steady–state*, where the concentration of radicals changes more slowly than the reactants, and the rate of change in the concentration of radical species is negligible:

$$\frac{d[R^{\bullet}]}{dt} = \frac{d[M^{\bullet}]}{dt} \cong 0 \tag{4.7}$$

Solving for the concentration of radicals gives:

$$[R^{\bullet}] = \left[\frac{k_1[M]}{k_4}\right]^{1/2} \tag{4.8}$$

$$[M^{\bullet}] = \frac{k_2}{k_3}\left[\frac{k_1}{k_4}\right]^{1/2} [M]^{3/2} \tag{4.9}$$

Writing the equation for the disappearance of the reactant:

$$-\frac{d[M]}{dt} = k_1[M] + k_2[R^{\bullet}][M] \tag{4.10}$$

The free radical reaction is a chain reaction, and once the radicals form they go through many cycles of the propagation steps. If we apply the *long–chain approximation*, then the rate of propagation >> rate of initiation, so that $k_2[R^{\bullet}][M] << k_1[M]$, giving:

$$-\frac{d[M]}{dt} = \frac{d[RH]}{dt} = \frac{d[O]}{dt} = k_2 (k_1/k_4)^{1/2} [M]^{3/2} \tag{4.11}$$

The overall reaction in this case gives fractional kinetics because of the combination of the elementary reaction steps. Termination reactions (ii) and (iii) give equations (4.12) and (4.13) respectively:

$$-\frac{d[M]}{dt} = \frac{d[RH]}{dt} = \frac{d[O]}{dt} = (k_1k_2k_3/k_4)^{1/2} [M] \tag{4.12}$$

$$-\frac{d[M]}{dt} = \frac{d[RH]}{dt} = \frac{d[O]}{dt} = k_3 (k_1/k_4)^{1/2} [M]^{1/2} \tag{4.13}$$

These analytical solutions for termination reactions (i) to (iii) indicate that the overall reaction order with respect to the parent alkane can vary between 1/2 and 3/2 order, depending on the nature of the termination step. The solution for case (iv) would vary between the two extremes, depending on the relative magnitude of the two propagation steps. If the rate of reaction (4.2) is much faster than the rate of reaction (4.3), as at very low temperature and high pressure, then the reaction order with respect to the parent alkane is 1/2. If the rate of reaction (4.3) is much faster than the rate of reaction (4.2), then the reaction order with respect to the parent alkane is 3/2. This situation occurs at high temperatures and low alkane concentrations, as in a dilute solution where the solvent can form radicals.

4.1.1 Example of thermal cracking: Kinetics for cracking of n–hexadecane

Solutions of n–hexadecane in benzene were reacted at 400–440°C and 14 MPa in a tubular reactor [Khorasheh, 1992]. At these conditions, the alkane was a solution in a supercritical solvent. The governing equation for the reactor was

$$\frac{dF_i}{dV} = r_i \tag{4.14}$$

where F_i is the molar flow rate of species i, V is reactor volume, and r_i is net rate of formation. The apparent rate constants were calculated for n–hexadecane using first–order kinetics ($r_i = -k[M]$). The Arrhenius plot of the first order rate constants for the overall conversion of n–hexadecane at three different rate constants is presented in Figure 4–1. The activation energy for the overall conversion of n–hexadecane was approximately 60 kcal/mol for all three mol fractions, in good agreement with the data of Ford [1986], (top line of figure) which represent rate constants for thermal cracking of pure n–hexadecane under gas–phase conditions. Under these conditions, hexadecane cracks following first–order kinetics.

The increase in the apparent first–order rate constants with increasing mol fraction of n–hexadecane in the feed suggests that the overall order of the reaction with respect to n–hexadecane concentration is greater than 1, following a termination reaction of type (i). The Arrhenius plot of the 3/2–order rate constants for the overall conversion of n–hexadecane is presented in Figure 4–2. Although 3/2–order kinetics seem adequate for 0.03 and 0.05 mol fraction data, the 3/2–order rate constants for 0.01 mol fraction are still lower than those for 0.03 and 0.05 mol fractions. For such a complex free–radical mechanism, especially with solvent interactions, a simple overall reaction

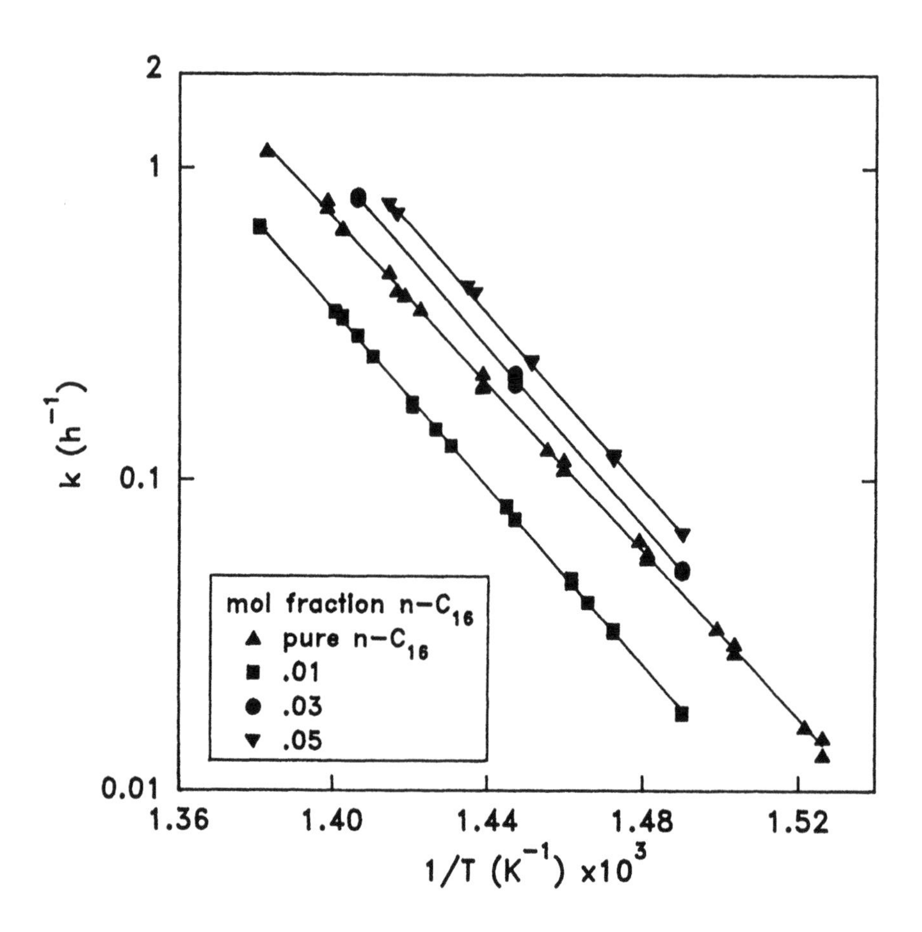

Figure 4–1
Arrhenius plot for cracking of n–hexadecane assuming first–order kinetics. Reaction at 13.9 MPa, 400–440° C in a tubular reactor in benzene solution, with mol fraction n–hexadecane from 1.0 to 0.01 (Data from Khorasheh, 1992).

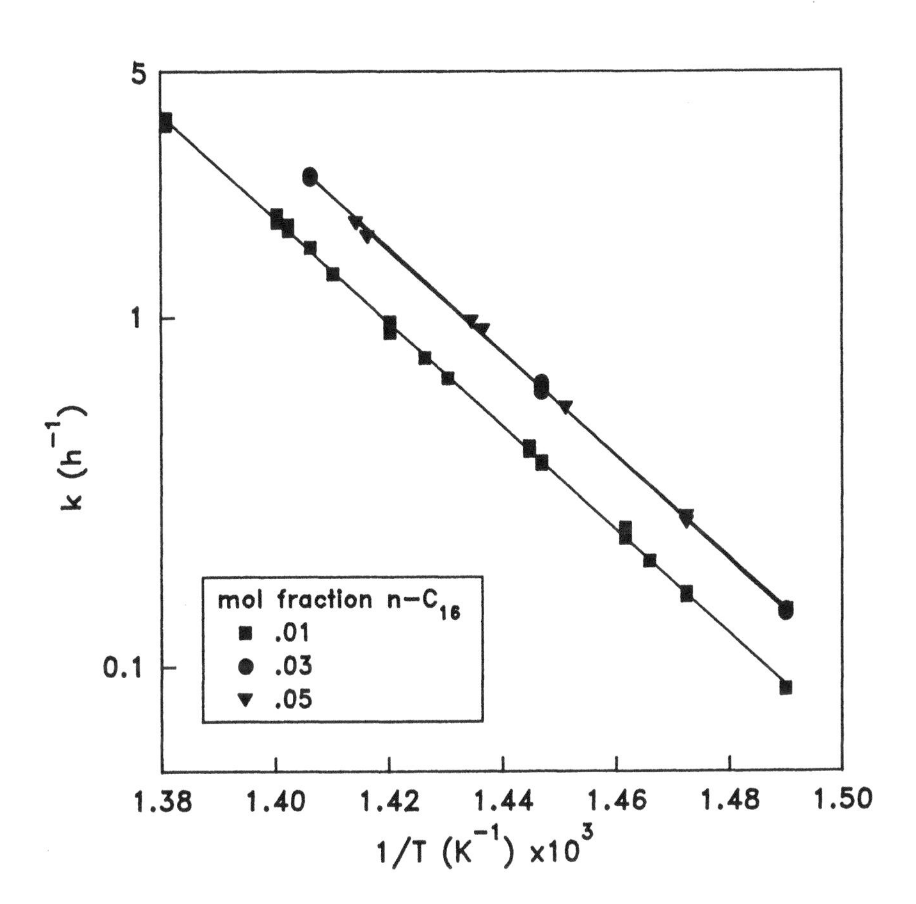

Figure 4–2
Arrhenius plot for cracking of n–hexadecane assuming 3/2–order kinetics. Reaction at 13.9 MPa, 400–440° C in a tubular reactor in benzene solution, mol fraction n–hexadecane from 1.0 to 0.01 (Data from Khorasheh, 1992).

order is not adequate to represent the observed kinetics over a range of temperatures and concentrations.

These results do not **prove** the mechanism, they merely show consistency between the kinetics and the mechanism. Additional evidence would either require identification of the radicals directly (e.g. by mass spectrometry) or by identifying termination products. In benzene, phenyl radicals were created by hydrogen abstraction:

$$R^{\bullet}\ (\text{or } M^{\bullet}) + C_6H_6 \longrightarrow RH + C_6H_5^{\bullet} \qquad (4.15)$$

These phenyl radicals will then react

$$C_6H_5^{\bullet} + n\text{–}C_{16} \longrightarrow C_6H_6 + n\text{–}C_{16}^{\bullet} \qquad (4.16)$$

$$C_6H_5^{\bullet} + C_6H_5^{\bullet} \longrightarrow C_6H_5\text{–}C_6H_5 \ \text{(biphenyl)} \qquad (4.17)$$

Biphenyl was indeed detected in the product mixture, supporting the proposed mechanism.

This example illustrates how free–radical reactions of hydrocarbons can give apparent reaction orders other than first order. It also demonstrates that a reaction can appear to be first order as long as the feed composition is constant, but that changes in inlet composition can give an unexpected change in the rate of reaction.

4.1.2 Cracking of mixtures

Thermal hydrocracking of hydrocarbon mixtures normally follows apparent first–order kinetics to a good first approximation. For example, Chung [1982] measured the kinetics of noncatalytic hydrocracking of a Lloydminster vacuum gas oil (s.g. = 0.9239; 62.8 mass % boiling between 343° C and 566° C) in a CSTR at 380–440° C, 13.9 MPa, LHSV 0.75 to 1.5 h^{-1}. Hydrocracking of the 343° C+ fraction gave conversions of 11 to 65%, and followed apparent first—order kinetics in a continuous—flow stirred tank reactor (CSTR):

$$-r_{HC} = k_1\, C_{ar} = 6.165 \times 10^{10} \exp(-170.6/RT)\, C_{ar} \tag{4.18}$$

where R has units kJ/mol—K and C_{ar} is the concentration of atmospheric residue.

To a first approximation, the cracking of a mixture such as a residue or bitumen will be determined by the reactivity of its constituents. The various compounds will interact during the degradation process, because the free radical intermediates are "shared" between the different species accrding to concentration and stability. Interactions between two components in a mixture will depend on the ability of a given radical to abstract hydrogen from component A versus component B [LaMarca *et al.*, 1990]. Easier abstraction from component A, for example, will enhance its cracking rate. In broad terms we can define the hierarchy of hydrocarbon cracking reactions as follows, from most reactive to least reactive:

Paraffins > Linear olefins > Naphthenes >
Cyclic olefins > Aromatics

Crackability tends to increase with molecular weight (or boiling range). This observation may seem contradictory, but larger molecules have more bonds which can rupture, increasing the probability of breakage. Figure 4—3 illustrates this trend for distillate fractions, and shows the increase in cracking as a function of both temperature and molecular weight. Stangeland [1974] proposed an empirical equation for catalytic hydrocracking for rate of reaction, k, as a function of boiling point T_b:

$$k\,(T_b) = k_o \left[T_b/1000 + B[(T_b/1000)^3 - T_b/1000] \right] \tag{4.19}$$

The boiling point in this equation must be in °F; the basis temperature of 1000°F corresponds to the highest temperature of heavy gas oil. The base reactivity is k_o, the limiting reactivity of the vacuum gas oil fraction. The factor B is an empirical constant in the range $0 < B < 1$. A value of 0 gives a linear relationship between rate and boiling point, while values greater than one give rapidly increasing reactivity with boiling point (Figure 4—4). Actual observations indicate a value of about

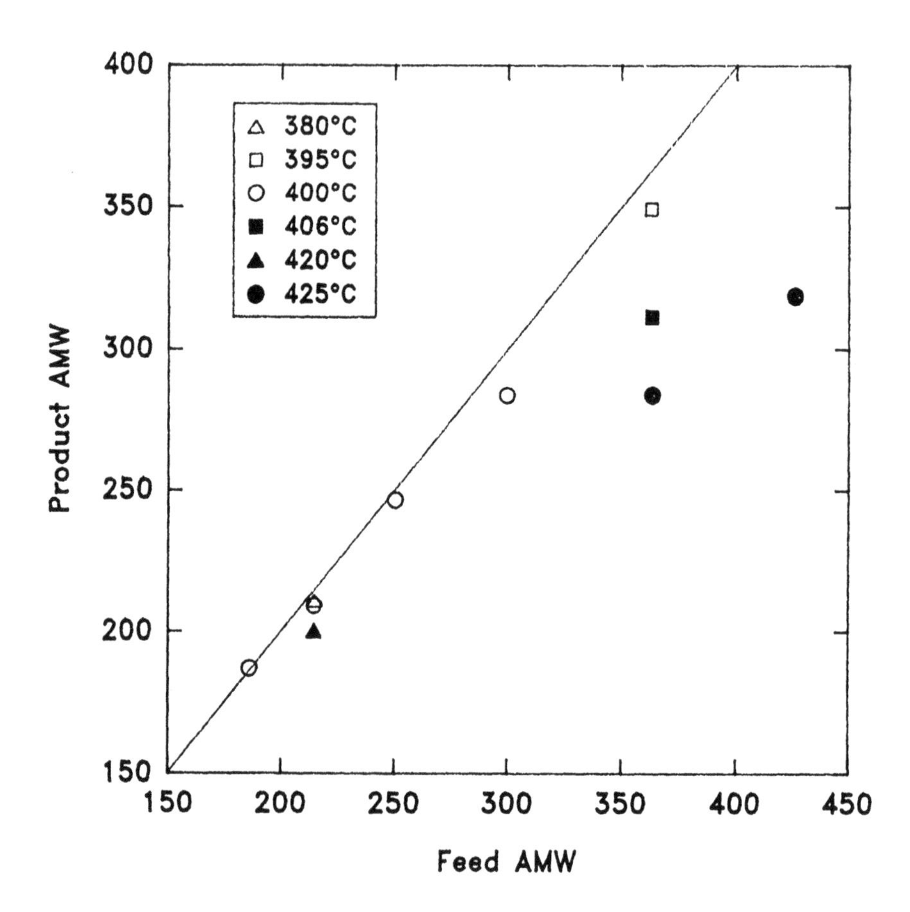

Figure 4–3

Hydrocracking of gas oil fractions as a function of molecular weight (Data from Trytten and Gray, 1990).

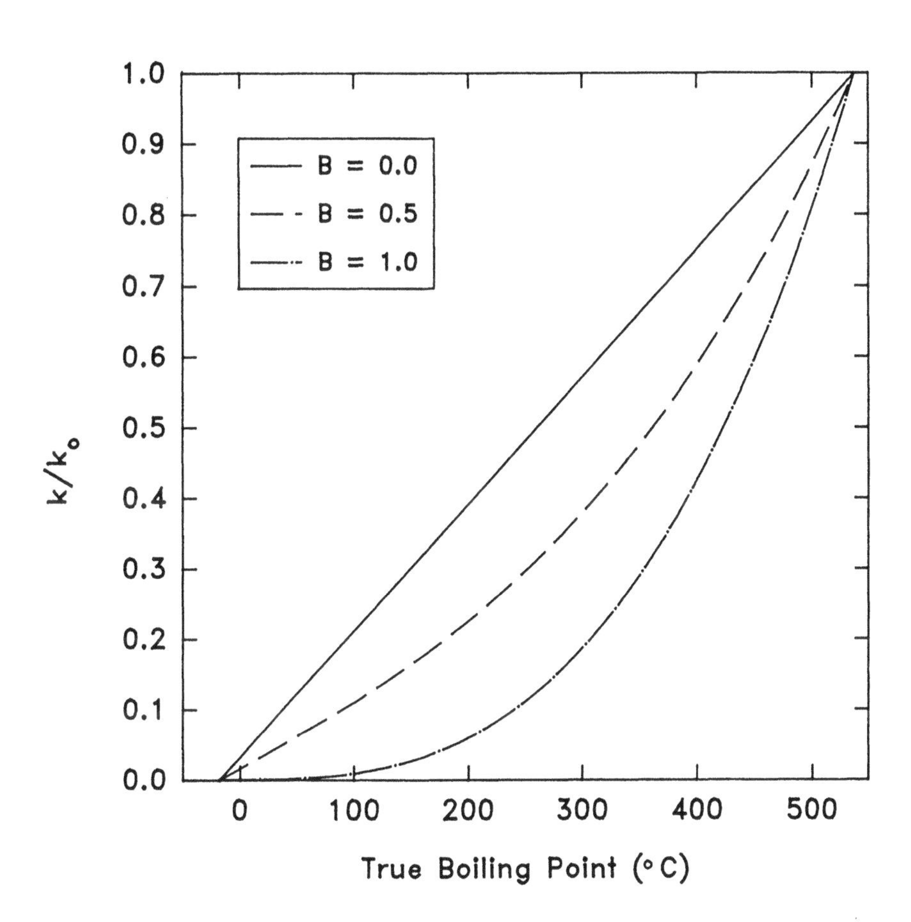

Figure 4–4
Rate constant for hydrocracking of gas oil fractions (Curves calculated from equation (4.19))

0.5, so that light fractions in the naphtha range have almost no cracking reactivity at temperatures in the 400—440° C range (compared to the heavier material).

This trend of increasing reactivity with molecular weight carries through into very high molecular—weight components. Figure 4—5 shows data from a process designed to convert asphaltenes in petroleum residues and bitumens. The trend of decreasing reactivity with decreasing molecular weight is very striking in this case.

4.2 Catalytic Reactions

By definition, a catalyst lowers the energy barrier for a particular reaction pathway. In practice, a catalyst increases the rate of conversion of a reaction by direct participation in the reaction mechanism. Any catalytic reaction occurs in the following steps:

1. Diffusion of reactants to catalyst site, where the site may be an ion, metal particles supported on a surface, or a chemical compound.
2. Coordination of the reactants with the catalyst site.
3. Reaction at the site.
4. Diffusion of the products away from the site.

If the active site is supported on a solid surface, e.g. γ—alumina, then diffusive transport of reactants and products will be through the bulk liquid and into the solid matrix. Most of the commercial catalysts for hydroconversion are solids (usually metal sulfides or sulfides supported on a porous ceramic support), so that coordination of the reactants with the catalyst involves adsorption on the surface.

4.2.1 Catalytic kinetics

In the case of reactions with a solid catalyst, the reactants must adsorb to active sites on the solid surface in order to react. The derivation of the Langmuir isotherm illustrates the kinetics of adsorption and reaction. Consider adsorption of hydrogen onto a catalyst surface:

$$H_2 + S \underset{k_{-A}}{\overset{k_A}{\rightleftharpoons}} H_2 \cdot S \tag{4.20}$$

Here S is a vacant adsorption site on the catalyst surface and $H_2 \cdot S$ is an occupied surface site. The rate of adsorption is proportional to the concentration of vacant sites (C_V) and the partial pressure of hydrogen:

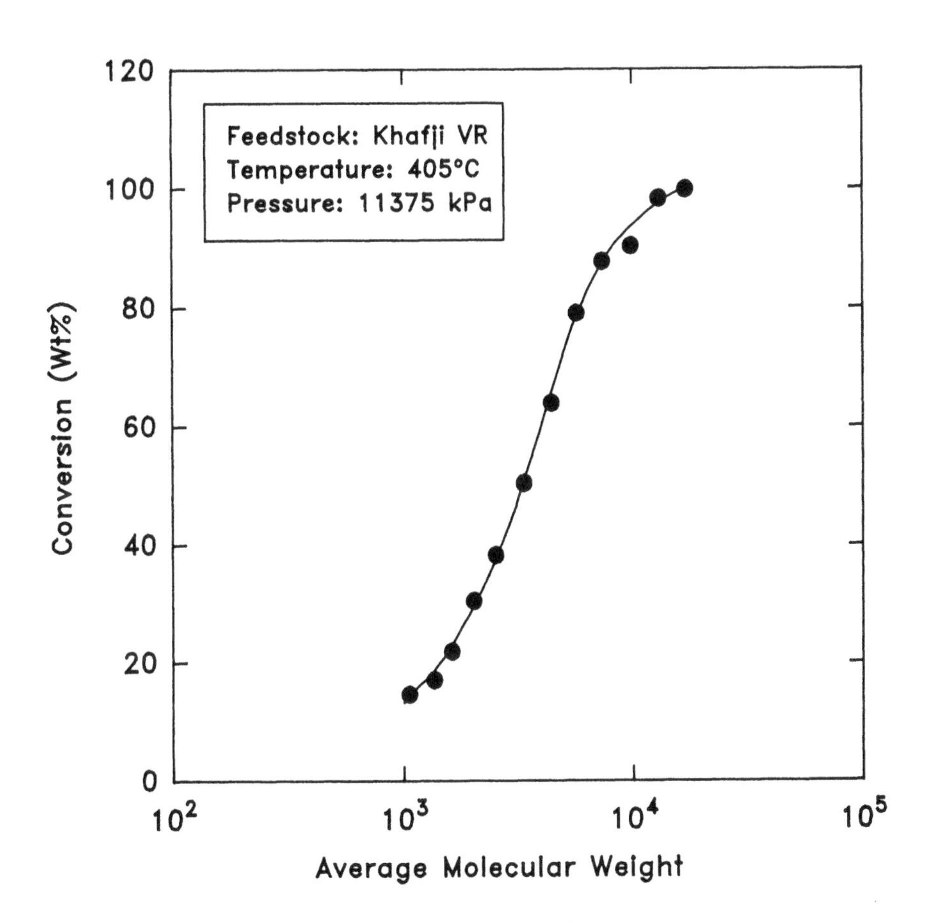

Figure 4—5
Conversion of asphaltene fractions (Data from Takeuchi *et al.*, 1983)

$$r_{ads} = \text{rate of adsorption} = k_A p_{H_2} C_v \tag{4.21}$$

$$r_{des} = \text{rate of desorption} = k_{-A} C_{H_2 \cdot S} \tag{4.22}$$

$$\text{Net Adsorption rate} = r_{ads} - r_{des} = k_A p_{H_2} C_v - k_{-A} C_{H_2 \cdot S} \tag{4.23}$$

The total concentration of surface sites is constant, so that a balance on surface sites gives:

$$C_t = C_v + C_{H_2 \cdot S} \tag{4.24}$$

At equilibrium, the net rate of adsorption will be zero, with $r_{ads} = r_{des}$. Substituting the balance on surface sites into equation 4.23 gives:

$$C^e_{H_2 \cdot S} = \frac{K_A p^e_{H_2} C_t}{1 + K_A p^e_{H_2}} \tag{4.25}$$

$$K_A = k_A / k_{-A} \tag{4.25a}$$

where the superscript *e* represents equilibrium concentrations. Equation (4.25) is the adsorption isotherm, which relates the occupancy of the surface to the partial pressure of the adsorbing species. Other forms of the adsorption isotherm are observed if the adsorption is accompanied by dissociation on the surface, or if the surface is nonuniform and the sites are not all equivalent. The equilibrium constant, K_A, is a function of temperature and normally decreases with increasing temperature.

Once species have adsorbed onto the catalyst surface, then several possible types of surface reactions are possible, depending on the reactants:

i) Adsorbed species isomerizes directly, as in rearrangement of cis—2—butene to trans—2—butene.

ii) The adsorbed compound reacts with another active site or with another adsorbed species.

$$A \cdot S + B \cdot S \rightarrow C \cdot S \tag{4.26}$$

This type of surface reaction is of the Langmuir—Hinshelwood type.

iii) Surface species react directly with gas—phase compounds (Eley—Rideal kinetics):

$$A \cdot S + H_2 \rightarrow AH_2 \cdot S \tag{4.26a}$$

Following surface reaction, the products must desorb from the surface and diffuse into the bulk fluid. In steady state, the rates of the different steps must be equal:

$$-r_{overall} = r_{ads} = -r_{surface} = r_{desorption} \tag{4.27}$$

Usually only one step is rate—limiting, and a knowledge of the limiting step is very important in designing the best reaction conditions.

Surface reactions are complex, and in many cases the available data can be described by more than one rate law. A systematic approach to deriving rate expressions, such as the Langmuir-Hinshelwood or Hougen-Watson approach is very helpful:

(i) Assume that one step is rate limiting

(ii) All the other steps are in equilibrium

(iii) Derive a rate expression

(iv) Check the fit to the data; if good then accept rate expression as acceptable, if not good then go to (i).

The hydrogenation of dibenzothiophene provides a good model of how the overall kinetics of conversion depend on the underlying fundamental reactions.

Example: Desulfurization of Dibenzothiophene

Hydrogenation of dibenzothiophene is commonly used to study the kinetics of desulfurization on different catalysts. Dibenzothiophene competes for surface sites with sulfur compounds, including hydrogen sulfide, as well as nitrogenous compounds. Catalysts such as Co/Mo on γ—alumina give direct removal of sulfur, forming biphenyl, as well as hydrogenation of the side rings as shown below [Sundaram *et al.*, 1988]

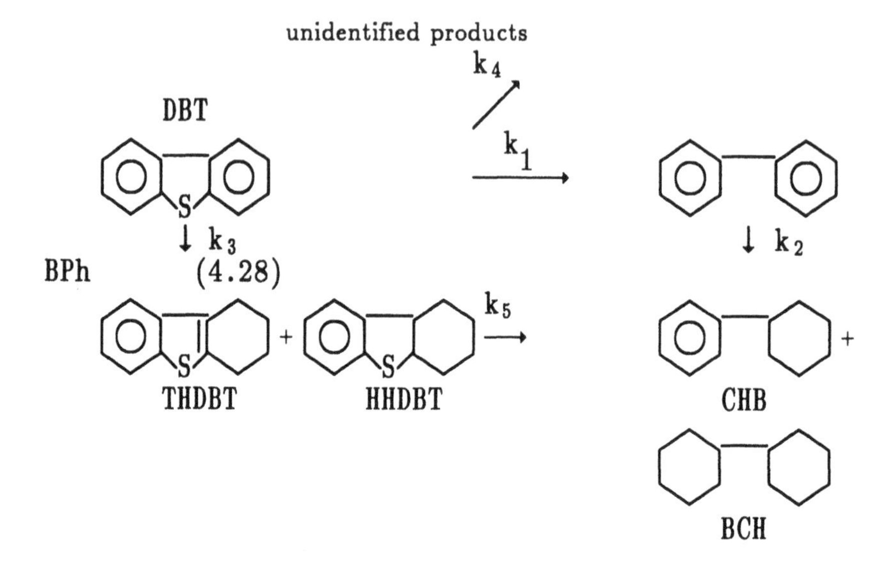

DBT = dibenzothiophene
BPh = biphenyl
THDBT = tetrahydrodibenzothiophene
HHDBT = hexahydrodibenzothiophene
CHB = cyclohexyl benzene
BCH = bicyclohexane

Each of the reaction steps involves the consumption of hydrogen, so that rate constants k_1 through k_5 include the contribution of hydrogen, which is held constant in model experiments. A kinetic expression can be developed based on two assumptions:

1. The surface reaction is rate limiting. This assumption implies that both adsorption of reactants and desorption of products is in equilibrium. If the surface reaction were not rate limiting, then it would be in equilibrium and hence reversible, which is not observed for HDS reactions.

2. The catalyst surface has two types of sites, one for sulfur compounds and one for hydrogen. DBT will compete, therefore, with hydrogen sulfide but not with hydrogen.

If the adsorption process is in equilibrium, then for sulfur sites

$$C_{t1} = C_{v1} + [DBT\text{–}S_1] + [THDBT\text{–}S_1] + [HHDBT\text{–}S_1] + [H_2S\text{–}S_1] \quad (4.29)$$

$$[DBT{-}S_1] = K_{DBT}\,[DBT]\,C_{v1} \tag{4.30}$$

$$[THDBT{-}S_1] = K_{DBT}\,[THDBT]\,C_{v1} \tag{4.31}$$

$$[HHDBT{-}S_1] = K_{DBT}\,[HHDBT]\,C_{v1} \tag{4.32}$$

$$[H_2S{-}S_1] = K_{H_2S}\,[H_2S]\,C_{v1} \tag{4.33}$$

Collecting the terms

$$C_{v1} = \frac{C_{t1}}{1.0 + K_{DBT}([DBT]+[THDBT]+[HHDBT]) + K_{H_2S}[H_2S]} \tag{4.34}$$

Similarly for the hydrogen sites

$$C_{v2} = \frac{C_{t2}}{1.0 + K_{H_2}[H_2]} \tag{4.35}$$

The surface reaction can be written, assuming the first hydrogenation step to be rate limiting

$$DBT{-}S_1 + H_2{-}S_2 \xrightarrow{\Sigma k_i'} \text{products} \tag{4.36}$$

$$i = 1,3,4$$

$$-r = \Sigma k_i'[DBT{-}S_1][H_2{-}S_2] \tag{4.37}$$

$$-r = \frac{(\Sigma k_i K_{DBT} K_{H_2} C_{t1} C_{t2}) \quad [DBT] \quad [H_2]}{\{1+K_{DBT}([DBT]+[THDBT]+[HHDBT])+K_{H_2S}[H_2S]\}}$$

$$X\,\frac{1}{\{1+K_{H_2}[H_2]\}} \tag{4.37a}$$

Data from batch and flow reactor experiments on DBT and its products at constant hydrogen pressure of 500 psig and 350° C were fitted better by a form of this equation than by other similar rate expressions.

$$k_1 = k_1' K_{DBT} K_{H_2} C_{t1} C_{t2} [H_2] = 3.24 \text{ g/g cat—min} \tag{4.38}$$

$$k_3 = k_3' K_{DBT} K_{H_2} C_{t1} C_{t2} [H_2] = 1.26 \text{ g/g cat—min} \tag{4.39}$$

$$k_4 = k_4' K_{DBT} K_{H_2} C_{t1} C_{t2} [H_2] = 0.69 \text{ g/g cat—min} \tag{4.40}$$

$$K_{DBT} = 0.13 \times 10^4 \text{ g/mol} \tag{4.41}$$

$$K_{H_2S} = 0.83 \times 10^4 \text{ g/mol} \tag{4.42}$$

The fit of the model to batch data is illustrated in Figure 4—6. At DBT concentrations below 4 x 10^{-5} mol/g, and hence comparable H_2S levels, the effect of adsorbed DBT was negligible. The rate equation for DBT disappearance at 500 psig hydrogen pressure and 350° C was

$$-r = \frac{k\ [DBT]}{1 + K_{H_2S} [H_2S]} \tag{4.43}$$

$$k = 5.19 \text{ g/g cat—min}$$

The order of reaction with respect to DBT was less than one at all concentrations. If we consider a power law form for the rate expression

$$-r = k\ [DBT]^{\alpha}\ [H_2S]^{\beta} \tag{4.44}$$

then β will be negative because H_2S appears in the denominator of the full rate expression, and acts as an inhibitor. Fitting the values of equation (4.43) to the power law expression over the range of H_2S concentrations gave a value of $\beta = -0.12$. Catalytic reactions commonly fit a power—law expression of the same general type as equation (4.44); such expressions are commonly used to fit experimental data directly when the details of the rate equation are unknown.

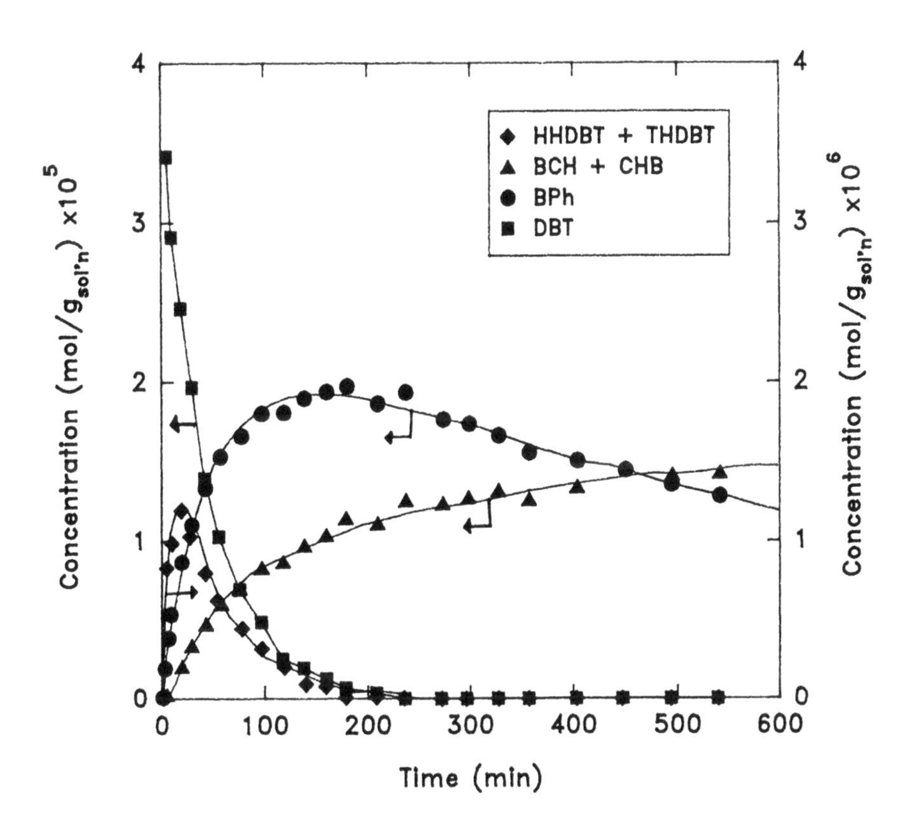

Figure 4–6

Fit of kinetic model for DBT to batch reactor data at 350° C and 3.45 MPa (Data from Sundaram *et al.*, 1988).

Rate equations such as (4.37) and (4.44) can be tested statistically against the available data for consistency. Because the surface reaction is often not well defined, a number of rate expressions can be derived, each with an equal likelyhood. Broderick [1980], for example, tested 14 different rate expressions, each based on a different surface reaction. He found that a rate expression based on a two–point adsorption of DBT also fitted the data well. The rate expression for this mechanism was

$$-r = \frac{k \; [DBT] \; [H_2]}{\{1 + K_{DBT} \; [DBT] + K_{H_2S} [H_2S] \}^2 \{ 1 + K_{H_2} [H_2] \}} \tag{4.45}$$

The tools for discriminating between such rate expressions are limited given normal kinetic data, so we must rely on statistical information and internal consistency checks. For example, adsorption and rate constants must be positive, rate constants must follow the Arrhenius temperature dependence, and adsorption constants must follow the van't Hoff equation and give an exothermic adsorption.

A more definite confirmation of a mechanism requires more sensitive probes of the surface reaction. Examples are studies of competitive adsorption with nitrogen compounds such as quinoline, which suggest that DBT (and quinoline) are each adsorbed on a single site [Sundaram *et al.*, 1988].

4.2.2 Steady–state kinetics of catalytic reactions

The reactions of model compounds are often studied in batch reactors, but the reactions of actual feeds are normally tested in continuous reactor systems operating at steady state to ensure representative catalyst activity. Catalytic continuous stirred tank reactors (catalytic CSTRs) are commonly used to measure the kinetics of hydroconversion without interferance from mass transfer or contacting effects (see section 4.3.3). A typical reactor of this type is illustrated in Figure 4–7. Trickle–bed reactors (Figure 4–8) are used for catalyst testing, measurement of kinetics, and scale–up, despite common problems with incomplete catalyst contacting with the liquid phase [e.g. Mears, 1974]. Such a trickle–bed can be approximated as a plug–flow reactor (PFR).

4.2.2.1 Kinetics of Catalytic CSTR

Rangwala *et al.* [1986] measured the kinetics of hydrotreating a coker gas oil in a well–stirred catalytic reactor. The flow reactor

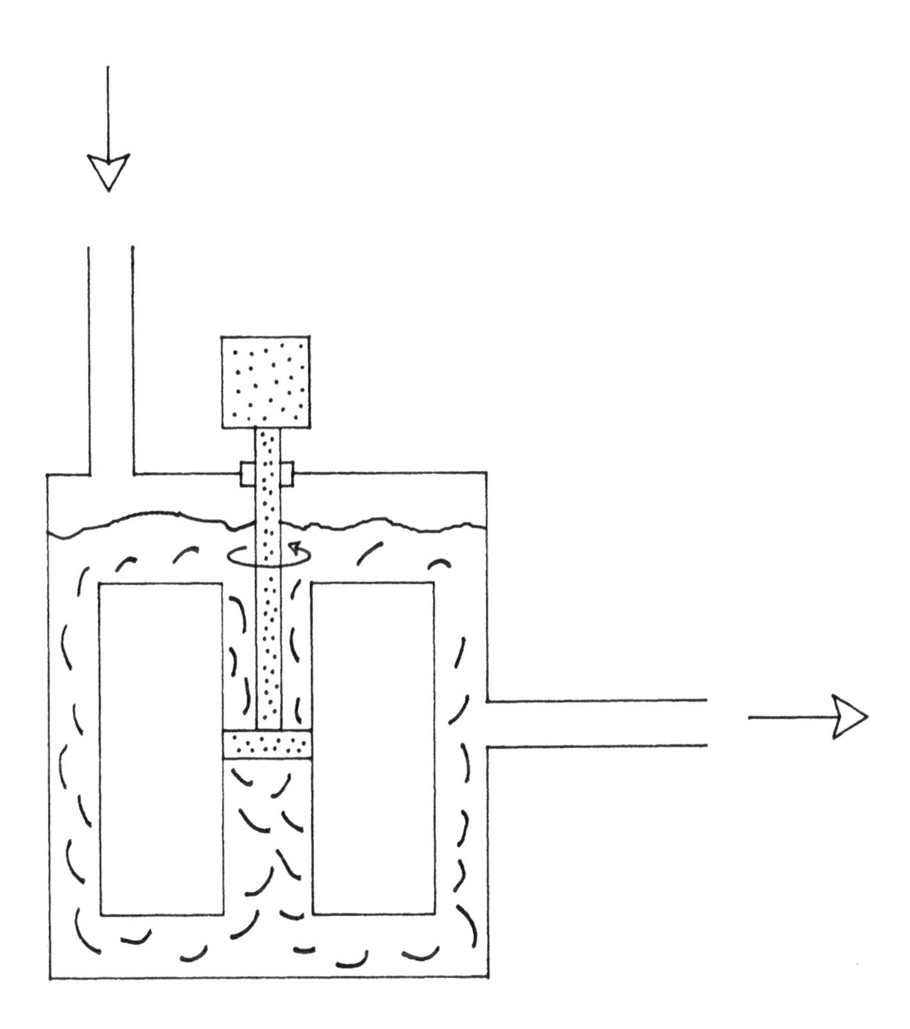

Figure 4–7
Spinning basket reactor for measuring kinetics of catalytic reactions

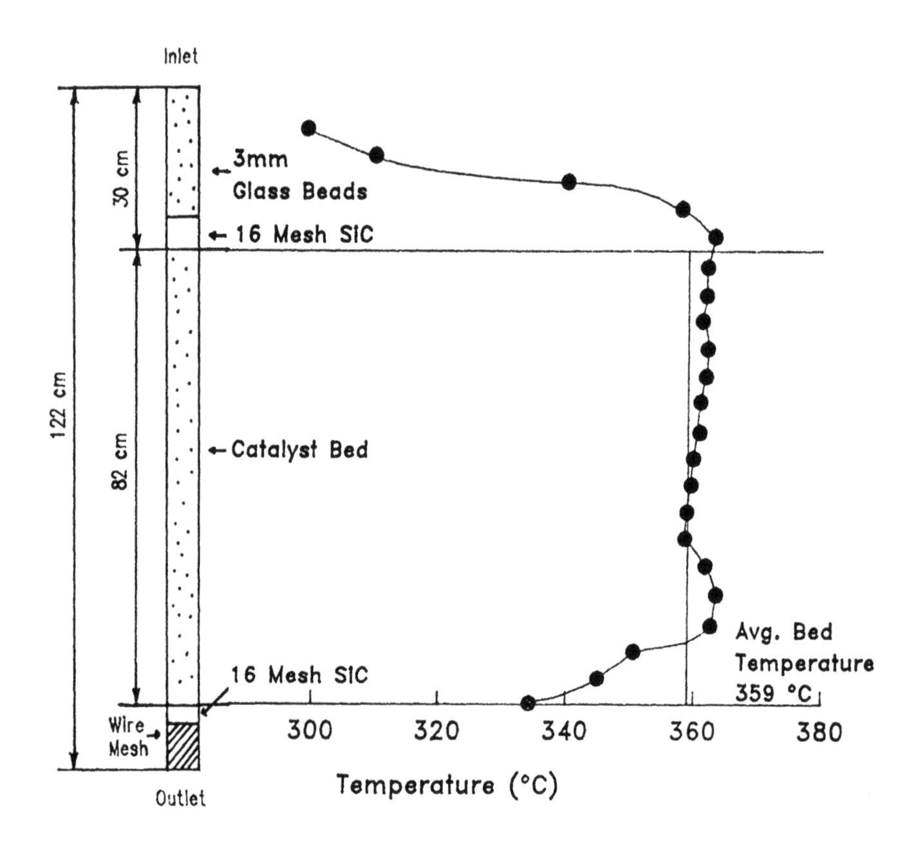

Figure 4–8
Trickle bed reactor for testing catalyst activity (From Yui, 1989)

consisted of baskets, holding 8 g of catalyst pellets, in a 150 mL volume. An impeller below the baskets was operated at 1100 RPM to eliminate concentration gradients outside the catalyst pellets. Previous tests showed that conversion was constant for agitation speeds above 800 RPM, so that external mass transfer effects were negligible.

For an irreversible first–order reaction, we can write a mass balance for the reactor for species i:

$$F_{io} - F_i = -r_i' W_t \tag{4.46}$$

$$Q_o C_{io} - Q_e C_i = k_i C_i W_t \tag{4.47}$$

Any dependence of the rate on hydrogen pressure (as in equation (4.44)) is implicit in the rate constant k_i. If we use X_i to indicate fractional conversion and $\tau = W_t/Q_o$ = 1/space velocity, then

$$k_i \tau = \frac{Q_e X_i}{Q_o (1 - X_i)} \tag{4.48}$$

If a reaction follows pseudo first–order kinetics in this type of reactor, it will give a straight line through the origin if τ (measurements at different liquid flow rates Q_0) is plotted against the right–hand side of Equation (4.48), which is a function of the measured conversion and flow rate. This plot is illustrated in Figure 4–9 for HDS, HDN, cracking of atmospheric residue (343° C+) and formation of C_1–C_4 gases. Over this range of conversions all four reactions gave apparent first–order kinetics within experimental error. Other apparent reaction orders would require a different solution for equation (4.47).

4.2.2.2 Kinetics of Ideal Trickle–Bed Reactor

An ideal trickle–bed reactor would provide plug–flow, isothermal operation, and complete wetting of the catalyst, so that the standard design equation can be applied. The design equation for an ideal plug–flow reactor (PFR) is as follows [Fogler, 1986]:

$$\int_0^{W_t} dW = F_{Ao} \int_0^{X_e} \frac{dX}{-r_A} \tag{4.49}$$

where F_{Ao} is the molar flow rate of reactant *A* into reactor, and $-r_A$ is

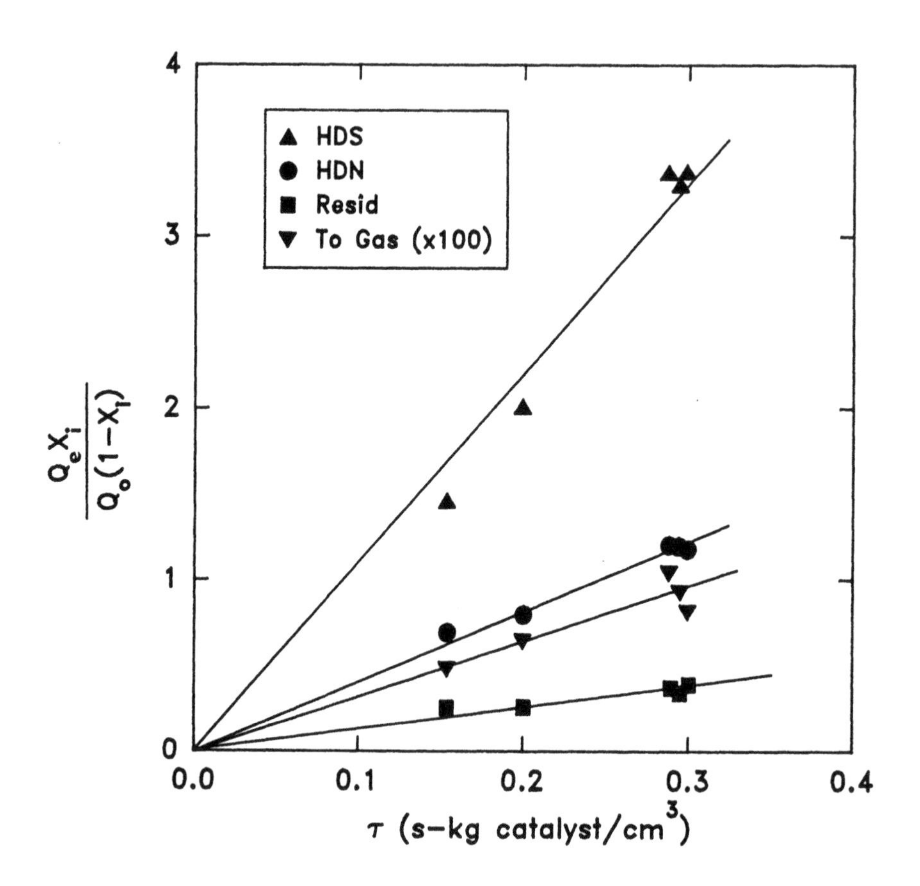

Figure 4–9
Kinetics of gas oil reactions in a 150 mL CSTR at 400° C, 13.9 MPa over a Ni–Mo/γ–alumina catalyst. LHSV ranged from 12.0 to 23.5 mL/h–g catalyst (Data from Rangwala *et al.*, 1986)

the reaction rate as a function of conversion X in units mol/[time—mass catalyst]. Substituting an n—th order rate expression for HDS gives:

$$k'[S]_o^{n-1} \int_0^{W_t} dW = Q_o \int_0^{X_e} \frac{dX}{(1 - X)^n} \tag{4.50}$$

where $[S]_0$ is the inlet sulfur concentration, W_t is the mass of catalyst in the reactor, and X_e is the outlet conversion of sulfur. Integration of the equation, for $n > 1$, gives:

$$\frac{1}{[S]_o^{n-1}(1 - X_e)^{n-1}} - 1/[S]_o^{n-1} = (n-1)\, W_t k'/Q_o \tag{4.51}$$

In this case, a plot of the left—hand side of equation (4.51) versus W_t/Q_o will give a straight line if the reaction order has been selected to best fit the data. This type of plot is illustrated in Figure 4—10 for a pilot test reactor for HDS of a gas oil stream; in this case the best fit of the data was to 2nd—order kinetics. A similar analysis by Yui [1989] used power—law kinetics to account for the effects of sulfur concentration and the partial pressure of hydrogen. His pilot and commercial data gave the following rate equation:

$$r_{HDS} = A\, e^{-E/RT}\, [S]^{1.5}\, p_{H_2}^{0.8} \tag{4.52}$$

$$A = 5.22 \times 10^{11}\ [cm^{4.5}/(h\text{—}g\ catalyst\text{—}mol^{0.5}\text{—}MPa^{0.8}]$$
$$E/R = 15{,}100\ K$$

If the reaction occurs at 400°C and 13.9 MPa hydrogen partial pressure, then the rate expression is:

$$r_{HDS} = k'[S]^{1.5} \tag{4.53}$$

$$k' = 772.5\ [cm^{4.5}/(h\text{—}g\ catalyst\text{—}mol^{0.5}]$$

4.2.3 Kinetics of reversible reactions

The previous calculations for desulfurization assumed that the reaction was irreversible, i.e. that if the reaction were complete all the

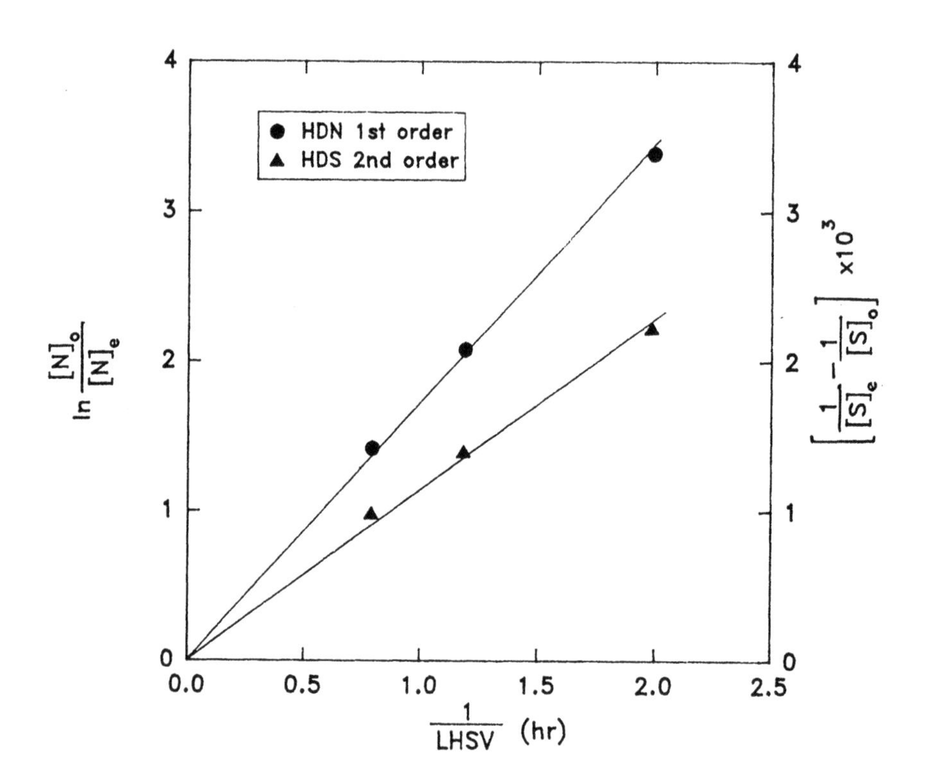

Figure 4—10
Plot to verify reaction order with a plug flow reactor
(Data from Oballa *et al.*, 1991)

sulfur would be removed. Hydrogenation reactions are, however, reversible so long as the ring is intact. Such reactions will exhibit more complex kinetics because the net conversion may include significant forward and reverse reactions. This reversibility can affect HDN reactions [Ho, 1988] and hydrogenation of aromatic ring compounds. An excellent example is the conversion of aromatics in distillates to enhance cetane number.

Conversion of aromatics in middle distillates from syncrudes to cycloalkanes is necessary to enhance the cetane number. Consider the following example reaction:

$$\underset{(PB)}{\text{n–propylbenzene}} + 3H_2 \underset{k_2}{\overset{k_1}{\rightleftharpoons}} \underset{(PC)}{\text{n–propylcyclohexane}} \tag{4.54}$$

If the hydrogen is present in large excess then both the forward and reverse reactions will be pseudo first–order:

$$\text{Net rate of reaction} = -k_1[PB] + k_2[PC] \tag{4.55}$$

When the reaction goes to completion then the equilibirum constant will apply to the reactants and product:

$$K = \frac{f_{PC}}{f_{PB}\ f^3_{H_2}} \simeq \frac{[PC]}{[PB]\ f^3_{H_2}} = \frac{k_1}{k_2\ f^3_{H_2}} \tag{4.56}$$

Equation (4.55) can then be rewritten:

$$\text{Net rate of reaction} = -k_1[PB] + \frac{k_1}{K\ f^3_{H_2}}[PC] \tag{4.57}$$

This kinetic relationship can be used to analyze the kinetics of aromatics conversion in a tubular reactor over a Ni–W on γ–alumina catalyst at 340–440° C and 50–170 atm hydrogen pressure. Wilson *et al.* [1985]

reported rate constants for conversion of alkylbenzenes in a syncrude middle distillate:

$$k_1 = A \exp(-E/RT) \tag{4.58}$$

$$A = e^{10.1}$$
$$E = 57 \text{ kJ/mol}$$

$$[PC]_0 \simeq [PB]_0 = 17 \text{ mass\%}$$

In this analysis the alkylbenzenes are represented as the n—propyl species as a model. Miki [1975] measured the equilibrium constant for reaction (4.54), which is representative of the conversion of alkylbenzenes:

$$\log K = 10975/T - 20.58 \tag{4.59}$$
$$(518 \text{ K} < T < 717 \text{ K})$$

Putting the rate expression into an expression for a tubular reactor with space time τ and conversion X we obtain:

$$-\frac{dX}{d\tau} = -k_1(1-X) + \frac{k_1}{K f_{H_2}^3}\left([PC]_0/[PB]_0 + X\right) \tag{4.60}$$

which is integrated to obtain

$$k_1\tau = -1/\beta \ln\left[(\alpha+\beta X)/\alpha\right] \tag{4.61}$$

$$\alpha = \frac{1}{f_{H_2}^3 \ K} - 1 \tag{4.61a}$$

$$\beta = \frac{1}{f_{H_2}^3 \ K} + 1 \tag{4.61b}$$

At the pressure of 170 atm used by Wilson *et al.* [1985] the equilibrium conversion of alkylbenzenes should be almost 100% over the entire temperature range according to the data of Miki [1975]. At a nominal pressure of 100 atm the control of the reaction passes from kinetics to equilibrium for LHSV from 0.75 to 2 h^{-1} ($\tau = 2.5$ to 7.5 min at reactor conditions, assuming 100% vapor in reactor). The concentration of alkylbenzenes in the product at 100 atm is illustrated in Figure 4—11.

4.3 Mass Transfer and Catalytic Reactions

The sequence of steps in a catalytic reaction involves diffusion of reactants from the bulk fluid to the active catalytic site. If the catalytic site is supported on a ceramic support, such as γ—alumina, then the diffusive process occurs in two steps:

1. External Mass Transfer
   Diffusion through the fluid boundary layer surrounding the catalyst pellet to reach the outer surface of the ceramic support.

2. Intrapellet Mass Transfer
   Diffusion within the porous solid to reach the catalytic sites.

Of the two resistances, the diffusion within the catalyst is greatest concern because the reactant molecules in a residue can be large in relation to the pores in the catalyst, giving rise to hindered diffusion and a reduction in catalyst effectiveness.

4.3.1 Hindered diffusion in porous catalysts

The common equation for diffusion in a catalyst relates the effective diffusivity to the porosity (ϵ_p) and tortuosity factor (ζ):

$$D_e = \frac{D_b \epsilon_p}{\zeta} \tag{4.62}$$

The porosity and tortuosity factors are used in this equation to correct for the diffusion within a solid, wherein only a fraction of the cross-sectional area is available for fluid diffusion, and the diffusion paths are tortuous and winding. For γ—alumina typical values are $\epsilon_p = 0.64$ to 0.78 and $\zeta = 1.3$, so that $D_e = 0.49\ D_b$ to $0.6\ D_b$. Bulk diffusivities are typically of order 10^{-5} cm^2/s in liquids and 10^{-1} cm^2/s in gases.

Equation (4.62), however, assumes that the diffusing molecules are much smaller than the radii of the pores in the solid, so that the microscopic rate of diffusion within a pore is the same as in a bulk fluid. The overall rate of diffusive flux within the solid is reduced by the tortuous path inside the solid. When the radius of the molecule is similar to the pore, then diffusion will be hindered, and the effective diffusivity can be much smaller than indicated by equation (4.62).

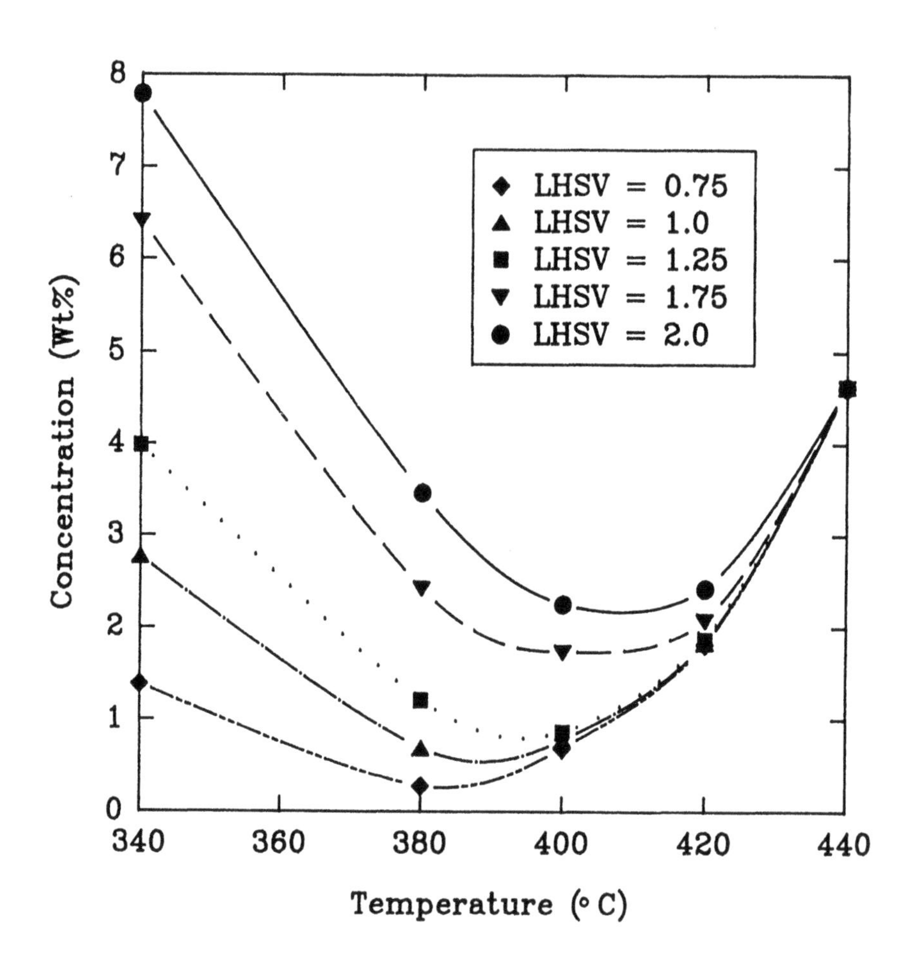

Figure 4–11

Predicted product concentration of alkylbenzenes from hydrogenation of synthetic crude fraction at 100 atm hydrogen pressure. (Kinetic data are from Wilson *et al.*, 1985)

Baltus and Anderson [1983] studied the diffusion of asphaltenes through pores in membranes. These pores were cylindrical and had a controlled diameter, r_p. A common equation for hindered diffusion is of the form:

$$D_e/D_b = \frac{\epsilon_p}{\zeta} e^{-B\lambda} \qquad (4.63)$$
$$\lambda = a/r_p$$

where B is a constant and a is the hydrodynamic radius of the diffusing molecule. As $\lambda \to 0$ then equation (4.63) becomes equivalent to equation (4.62). As $\lambda \to 1$ then D_e becomes small. For the porous membranes of Baltus and Anderson, $\epsilon_p = \zeta = 1.0$. Asphaltenes fractions with molecular weights ranging from 3000 to 48000 were diffused through pores of 70—500 Å radius. The bulk fluid diffusivity, D_b, was estimated by extrapolating the diffusion through the membrane to very large pore diameter ($\lambda \to 0$).

The hydrodynamic radius of the molecule was estimated from the Stokes—Einstein equation

$$D_b = k_b T/6\pi\mu a \qquad (4.64)$$

where k_b is the Boltzmann constant, T is the temperature, and μ is the viscosity of the solvent (THF in these experiments). The results for the asphaltene fractions are listed in Table 4.1 below.

Table 4.1
Diffusivity of Asphaltenes

Fraction	MW	$D_b \times 10^7$ cm^2/s	Molecular radius, Å
A	3000	16.1	26
B	6000	13.1	32
C	12000	8.85	47
D	24000	5.24	79
E	48000	2.71	153

(MW based on GPC with polystyrene standards)

The results for diffusion in pores of different radii are illustrated in

Figure 4—12, for the asphaltene fractions A—E. These data were regressed to Equation (4.63) to give Figure 4—13, with a constant of 3.89. Other studies in alumina and membranes have observed B = 4.5—4.6, for λ = 0.05 to 0.5 [Ternan, 1986]. Allowing for errors in extrapolating to obtain D_b, Figure 4—13 shows that hindrance of diffusion will be significant for $\lambda > 0.1$.

Lee *et al.* [1991b] measured diffusivities of nitrogen compunds in decalin solvent with reaction on Ni/Mo on γ—alumina catalysts at 350° C and 5.27 MPa hydrogen pressure. The catalysts were bidisperse, with macropore diameters of ca. 1000 nm and micropores in the range 6—10 nm. The results are listed below in Table 4.2.

Table 4.2
Effective Diffusivity of Nitrogen Compounds
(Diffusivity at 350° C, 5.27 MPa in decalin, data from Lee *et al.*, 1991b)

Reactant	Pore Diam, nm Macro	Micro	D_e/D_b	D_b cm^2/s x10^5
Indole	1217	17	0.25	25.6
	632	8.3	0.18	
	911	6.2	0.15	
9—PA[1]	1217	17	0.18	16.7
	632	8.3	0.12	
	911	6.2	0.10	

1. 9—phenyl acridine

The effective diffusivity was fitted to the following equation:

$$D_e = \frac{D_b \epsilon}{\zeta}(1-\lambda)^B \tag{4.65}$$

where the value of B was 4.9. Indole had a critical diameter of 0.69 nm, and 9—PA a critical diameter of 1.13 nm, so that $0.04 < \lambda < 0.18$ in Equation (4.65). Note that this equation satisfies the key limiting condition that as $\lambda \rightarrow 1$, then $D_e \rightarrow 0$, unlike equation (4.63) [Ternan, 1987]. Both equations (4.63) and equation (4.65) take the same approximate form when λ is small:

$$D_e \approx \frac{D_b \epsilon}{\zeta}(1 - B\lambda) \tag{4.66}$$

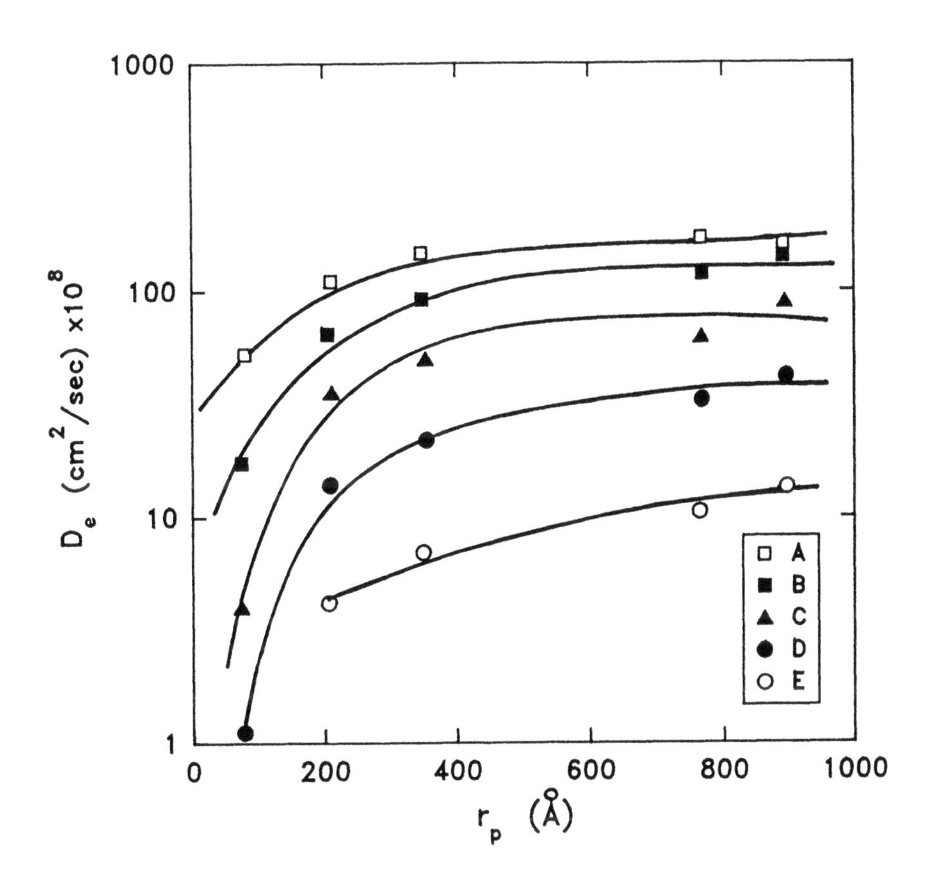

Figure 4—12
Effective diffusivity of asphaltenes in THF verus pore radius. Asphaltene fractions A through E are defined in Table 4—1 (Data from Baltus and Anderson, 1983)

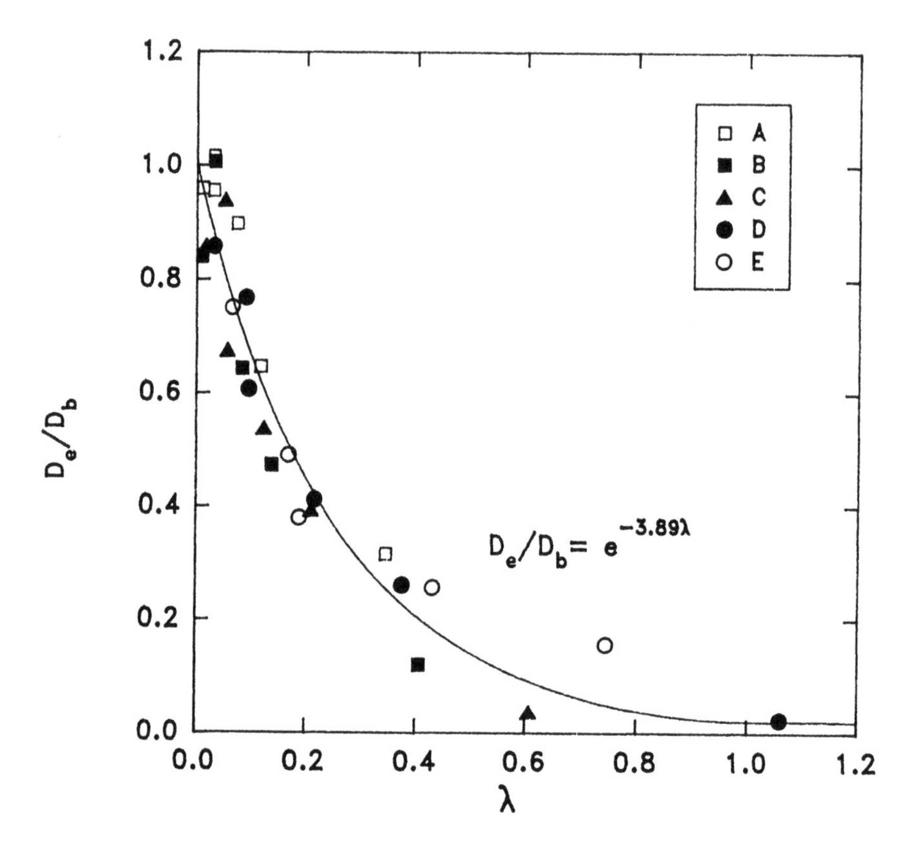

Figure 4—13
Fit of effective diffusivity data to equation 4—63. Asphaltene fractions A through E are defined in Table 4—1 (Data from Baltus and Anderson, 1983)

The similarity between the coefficient measured for equations (4.63) and (4.65) using the data of Baltus and Anderson [1983] and Lee *et al.* [1991a] suggest that a value of B = 4–5 is a reasonable first approximation for hindered diffusion in hydroconversion catalysts based on γ–alumina. Other empirical equations, such as Ternan [1987], have a similar form to equation (4.65) but are less useful for making estimates due to the wider range of values for the adjustable parameter.

In the conversion of nitrogen compounds, Lee *et al.* [1991b] found that the effective diffusivity was dominated by the micropore characteristics of the catalyst; the presence of macropores gave an effective diffusivity only slightly larger than for micropores alone. A related study by Lee and coworkers [1991a] showed that the effects of temperature and solvent phase on hindered diffusion were small at temperatures > 300 °C. These results, however, were for small values of λ. In processing of residues, a range of molecular sizes would be present such that some components would have $\lambda \geq 1$. With such severe hindrance of diffusion, we would expect the presence of macropores to have a large effect on the apparent diffusivity by reducing the length of the diffusion path [Ternan and Menashi, 1993]. Bimodal catalysts with significant pore volume in large pores, therefore, are desirable in processing residues (Section 5.4).

4.3.2 Catalyst effectiveness due to internal diffusion

The effectiveness factor, η, relates the observed reaction rate to the rate in the absence of a diffusion limitation on the reaction. For a first–order reaction, the equation is written

$$r_{obs} = \text{Observed Rate} = k\eta C_s \tag{4.67}$$

$$\text{where } \eta = \frac{\text{actual overall rate of reaction}}{\text{rate if } C_s \text{ were present throughout solid}}$$

The effect of diffusion on a reaction will depend on the diffusion coefficient and the path length for diffusion within the solid. A convenient definition for the length of the path of diffusion is

$$L_p = V_p/S_x = \text{volume/external area} \tag{4.68}$$

The relationship between diffusion and the rate of reaction is governed by the Thiele modulus, which is the ratio of the reaction rate to the rate of diffusion. Fogler [1986] provides the derivation of the Thiele modulus from the equations for reaction within a porous solid catalyst. The Thiele modulus for a first—order reaction with generalized geometry is given by

$$\phi = L_p \, (k_1/D_e)^{1/2} \tag{4.69}$$

where k_1 is the first—order rate constant on a volumetric basis. Using the Thiele modulus from equation (4.69), the effectiveness factor for a first order reaction will be

$$\eta = (3\phi \coth(3\phi) - 1)/3\phi^2 \tag{4.70}$$

As the length of the diffusion path increases, or as the diffusivity decreases (larger ϕ), then the effectiveness factor (η) for a given reaction will decrease. When the rate of diffusion significantly limits the reaction ($\eta < 0.4$ and $\phi > 2$), then the value of η is log—linear with the Theile modulus ϕ.

A convenient method for evaluating diffusion effects on catalytic reactions is to conduct a reaction using identical catalysts of two different sizes (Size 1 and Size 2). If the mass of catalyst and the reaction conditions are identical, then the ratio of the catalyst sizes and the observed reaction rates can be used to estimate the effectiveness factors. The ratio of the lengths of the catalyst particles is proportional to the ratio of the Thiele moduli

$$L_{p1}/L_{p2} = \phi_1/\phi_2 \tag{4.71}$$

The ratio observed rates is related to the effectiveness factors for the two sizes of catalyst as follows:

$$\text{rate 1/rate 2} = \eta_1/\eta_2 \tag{4.72}$$

Two experiments give two points on the η—ϕ relationship, so that the values can be solved as follows:

1. Guess ϕ_2, then calculate η_2 from equation (4.70)

2. Calculate ϕ_1 from equation (4.71), then calculate η_1 from equation (4.70)

3. Check the ratio of η_1/η_2 against the observed rates, eqn. (4.72)

4. Adjust the value of ϕ_2 until the ratio η_1/η_2 converges on the correct (i.e. measured) value

For example, the role of internal diffusion resistance was studied by hydrotreating coker gas oil in a continuous–flow stirred catalytic reactor, using whole catalyst pellets and ground and sieved material [Trytten, 1989]. The reaction conditions were LHSV 12.5 mL/h– g catalyst, 400°C, and 13.9 MPa of hydrogen pressure. The agitation speed was 1100 RPM to eliminate external mass transfer resistance. The catalyst was Ni/Mo on γ–alumina. The data from the two runs were as follows:

Table 4.3
Catalytic Conversion of Gas Oil
(Data from Trytten, 1989)

Catalyst	Whole Pellets (1)	Ground (2)
Diameter, mm	0.913	0.5
HDS, %	76.4	81.1
HDN, %	53.4	61.4
k_{HDS}, mL/s–kg cat	11.77	15.66
k_{HDN}, mL/s–kg cat	4.17	5.82

The above iterative calculation procedure (Steps 1–4) gave the following results:

Table 4.4
Calculated Effectiveness Factors for Gas Oil

	Whole Pellets	Ground
η_{HDS}	0.701	0.936
ϕ_{HDS}	0.92	0.341
η_{HDN}	0.848	0.974
ϕ_{HDN}	0.567	0.21

Note that the values of the Thiele modulus are different for each reaction in the same catalyst becuase each reaction has its own intrinsic rate. The intrinsic rates were 16.7 mL/s–kg catalyst for HDS and 5.98 mL/s–kg catalyst for HDN. The catalyst would have to be ground smaller than 0.5 mm to eliminate diffusion resistance in the HDS reaction (i.e. $\eta > 0.95$). The HDN reaction was not subject to any significant diffusion limitation in the ground catalyst.

Ammus and Androutsopoulos [1987] studied the desulfurization of a Greek crude oil residue in a spinning basket reactor, using Co/Mo on γ–alumina catalysts. The order of the desulfurization reaction ranged from 1.9 to 2.3 with respect to total sulfur. They compared the activity of whole and ground pellets to determine internal effectiveness factors.

Table 4.5
Effectiveness Factors for HDS of Thasos Atmospheric Residue
(Data from Ammus and Androutsopoulos, 1987)

Catalyst	Whole	Ground	Powdered
HT–400			
L_p, cm	0.0589	0.0326	0.00567
η	0.21	0.46	0.97
ϕ	4.7	2.0	0.42
ICI–41–6			
L_p, cm	0.0294		0.00567
η	0.26		0.90
ϕ	3.4		0.65

Note that this analysis assumes that the true reaction kinetics of HDS were 1.9 and 2.3 order on the two catalysts respectively.

The main limitation of much of these analyses is the assumption of simple first–order of power law kinetics in evaluating the Thiele modulus and effectiveness factor. Froment and Bischoff [1979] provide a derivation of a rigorous analysis for any kinetic rate law.

The main problem with the Thiele modulus approach is that it requires iterative solution, since the intrinsic reaction rate k_1 is not normally known at the outset. An alternative approach is the Weisz–Prater equation, which rearranges the equation to separate the measurable quantities:

$$\Phi = \eta\phi^2 = \frac{r_{obs} L_p^2}{D_e C_s} \tag{4.73}$$

The criterion for insignificant diffusion resistance then becomes $\phi << 1$, $\eta = 1$, so that $\Phi << 1$. The Weisz—Prater equation shares the same weakness as the Theile modulus approach; if the reaction rate expression is unknown, then an analysis that assumes first—order kinetics can give misleading results.

Rangwala *et al.* [1991] used the Weisz—Prater approach to analyze diffusion resistances in the catalytic HDS and HDN of distillate liquids from coprocessing of bituminuous coal with vacuum residue. As pellet diameter increased from 0.24 mm to 1.59 mm, the value of Φ for HDS increased from 0.06 to 1.13 and the value of Φ for HDN increased from 0.02 to 0.61. Calculated values of η, from the Theile modulus approach, ranged from 1.0 to 0.43 for HDS. This study illustrated that for pseudo first—order reactions, either approach will give a consistent estimate of the significance of diffusion resistances. The main limitation of confidently applying either the Theile modulus or Weisz—Prater approach to hydrotreating was the nature of the feed, which ranged in boiling point from naphtha up to heavy gas oil (525° C cut point). Such a feed would have sulfur compounds ranging in molecular weight from 84 to over 500, with a corresponding range of reaction rate and diffusion coefficient. Treating the feed as a single component, therefore, involves several approximations which will be discussed in more detail in section 4.4.

A very important side effect of diffusion resistances is that the reaction kinetics can be falsified when diffusion controls the reaction rate. Consider the case of an n—th order reaction:

$$-r_A = \eta\, k_n\, C_{As}^{n} = k'_n\, C_{As}^{n'} \tag{4.74}$$

where n is the actual order of the reaction, allowing for $\eta \leq 1$, and n' is the apparent order of the reaction. When diffusion in the pellet is controlling the reaction ($\eta <$ ca. 0.5), the value of η is log—linear with ϕ, from the following equation:

$$\eta = \left(\frac{2}{n+1}\right)^{1/2} \cdot 3/\phi_n \tag{4.75}$$

Using the equation for Theile modulus for an n—th order reaction [Fogler, 1986] gives:

$$n' = \frac{n+1}{2} \tag{4.76}$$

$$E_{app} = E_{true}/2 \tag{4.77}$$

The effect of diffusion control, therefore, can be to reduce the apparent order (if n>1) and to reduce the apparent activation energy. As the data on η from Ammus and Androutsopoulos [1987] show, some reactions of heavy residue may be limited by diffusion so that kinetics will be falsified. A much more significant effect in estimating reaction order, however, is the effect of grouping mixtures of components as a single pseudocomponent. This procedure is almost always part of the kinetic analysis of reactions of hydrocarbon mixtures, and the implications of dealing with mixtures are dealt with in Section 4.4.

4.3.3 External mass transfer resistances

A fluid flowing around a catalyst pellet will have a stagnant layer immediately adjacent to the surface. This layer is a potential resistance to mass transfer, depending on its thickness, which in turn depends on the fluid conditions. The flux of reactant through the stagnant layer is given by:

$$\omega_r = k_c(C_b - C_s) \tag{4.78}$$

where k_c is a mass transfer coefficient, C_b is the bulk liquid concentration, and C_s is the concentration at the surface of the catalyst pellet. A variety of correlations are available to relate the mass transfer coefficient, k_c, to the fluid properties for single pellets and for packed beds [see for example, Fogler, 1986].

As an example for calculations of external mass transfer, consider the desulfurization of Athabasca bitumen during catalytic hydrocracking. In order to evaluate the possibility of an external limitation to mass transfer, we need a measurement of the rate of reaction in a well defined system, i.e. a stirred catalytic reactor.

Table 4.6
Example Conditions in a CSTR for Hydrocracking

Reactor Conditions:	
Temperature	430° C
Pressure of Hydrogen	13.9 MPa
Impeller rotation	1100 RPM
Total Reactor Volume	150 mL
Catalyst	8 g Ni/Mo on γ–alumina
Liquid Hold–Up	104 mL

Table 4.6 — Continued

Liquid feed rate	1.50 mL/min
Liquid product rate	1.54 mL/min
Hydrogen feed rate	1.035 L(STP)/min
Feed	Athabasca bitumen (424° C+ fraction)
Reactor Performance:	
Overall reactor mass balance	99.8%
Conversion of 525° C+ fraction	55%
Conversion of Sulfur	60%

In order to evaluate possible limitations on reaction, the rate of reaction is calculated:

$$-r'_{HDS} = \frac{F_{s,in} - F_{s,out}}{W} \tag{4.79}$$

$$= \frac{1.50 \ \text{mL/min} \cdot 1.034 \ \text{g/mL} \cdot 0.0514 \ \text{S} - 1.54 \cdot 0.95 \cdot 0.0213}{8 \ \text{g catalyst}}$$

$$-r'_{HDS} = 0.006 \ \text{kg S/min—kg catalyst}$$

$$= 1.01 \times 10^{-4} \ \text{kg S/s—kg cat.}$$

This rate is the maximum achievable in the absence of any external limitations. The next step is to estimate the mass transfer coefficient under different reactor flow conditions. In order to use mass transfer correlations, we must estimate the properties if the fluid at reactor conditions (430° C, 13.9 MPa).

Using the Peng—Robinson equation of state, the phase behavior of the liquid can be predicted. Since the flow of hydrogen is usually well above the consumption rate, we can assume to a first approximation that the liquid phase is saturated with hydrogen, giving a product liquid that is about 30 mol% hydrogen, balance bitumen.

Fluid properties:

Viscosity (μ)	0.424 cp [from Twu, 1985]
Density (ρ_L)	847 kg/m^3 (corrected PR)
Kinematic viscosity (ν)	5.01 x 10^{-7} m^2/s
Diffusivity (D_b) (Wilke—Chang)	1.04 x 10^{-8} m^2/s

Schmidt Number, $Sc = \nu/D_i = \mu/\rho_f D_i = 47.9$

Catalyst Properties:

Cylindrical pellets, 1/32" diameter	
Particle diameter	0.794 mm
Length	5 mm
Characteristic length, d_p (equivalent diameter of sphere of same volume)	1.68×10^{-3} m
Shape factor (γ) (external area/πd_p^2)	0.41
Pellet density (ρ_p)	1200 kg/m^3
Porosity of bulk pellets (ϵ)	0.3

<u>Case 1: Single Catalyst Pellet</u>

The limiting case for mass transfer will be a single pellet in flowing fluid. Using the Frossling correlation to estimate the mass transfer coefficient, k_c:

$$\text{Sherwood Number (Sh)} = \frac{k_c d_p}{D_b} = 2 + 0.6\text{Re}^{1/2}\text{Sc}^{1/3} \tag{4.80}$$

The Reynolds number depends on the velocity of the fluid relative to the catalyst pellet, so we can calculate mass transfer as a function of fluid velocity. The maximum flux of sulfur compounds into the catalyst is given by the transport equation (4.78). The maximum flux would occur when the surface concentration is zero. The product contains 2.13% S, giving a concentration of 18 kg/m^3. The results from equation (4.80) as a function of $Re=U\rho_L d_p/\mu$ are given in Table 4.1.

TABLE 4.7
Mass Transfer to a Single Catalyst Pellet

Velocity m/s	Re[1]	Sh	k_c m/s	Max Flux kg S/s—kg cat
0.00001	.016	2	0.000026	0.0030
0.001	1.6	4.7	0.000062	0.0071
0.01	15.8	10.7	0.00014	0.016
0.05	79.2	21.4	0.00028	0.032
0.1	158	29.4	0.0004	0.044

Recalling that the rate of reaction was only 1 x 10^{-4} kg S/s—kg catalyst, we can see that the maximum flux for mass transfer is always much larger, even under completely laminar conditions. For a single catalyst pellet, therefore, we do not expect the reaction to be limited at all by external mass transfer.

The actual velocity of the catalyst pellet with respect to the fluid depends on the reactor type. In an ebullated—bed reactor, where the pellets are fluidized by flowing liquid, we can calculate the terminal settling velocity of a cylindrical pellet as a first approximation.

$$U_{max} = \text{terminal velocity} = \left[\frac{2gm_p(\rho_p - \rho_L)}{\rho_L \rho_p \chi_p C_d}\right] \tag{4.81}$$

Here m_p is the mass of a pellet, χ_p is the cross—sectional area to flow, C_d is the drag coefficient and g is the gravitational constant (9.81 m/s^2). For purely laminar flow, C_d= 24/Re, and in the transition to turbulent flow [Perry *et al.*, 1984]

$$C_d = (24/Re)(1 + 0.14\, Re^{0.7}) \tag{4.82}$$

Solving for the pellet of γ—alumina, we find the following:

C_d= 1.5 Re = 95 U_{max} = 0.06 m/s

Referring to Table 4.1, we see that this velocity gives a maximum mass transfer flux which is two orders of magnitude larger than the reaction rate.

<u>**Case 2: Packed Bed**</u>

A given reactor type will have its own hydrodynamic conditions, and hence its own relationship between mass transfer and fluid properties. For packed beds, the Thoenes—Kramers correlation is commonly used [Fogler, 1986].

$$Sh' = 1.0 \cdot (Re')^{1/2}(Sc)^{1/3} \tag{4.83}$$

The functional form of the correlation is very similar to the Frossling correlation, except that the groups are defined for a packed bed:

$$Sh' = \frac{\epsilon}{(1-\epsilon)\gamma} Sh \tag{4.84}$$

$$Re' = \frac{Re}{(1-\epsilon)\gamma} \tag{4.85}$$

where ϵ is the porosity of the bed (between the pellets) and γ is the shape factor. The velocity used to calculate the Reynolds number is the superficial velocity, i.e. the volumetric flow throught the reactor divided by the cross–sectional area of the bed. The value of Re' indicates the fluid characteristics in the interstices between the pellets. For the same fluid and pellets as before, we can calculate the mass transfer as a function of superficial velocity.

Table 4.8
Mass Transfer in a Packed Bed

Superficial Velocity m/s	Re'	Sh'	k_C m/s	Max Flux kg S/s–kg cat
0.00001	0.03	0.67	0.00025	0.0010
0.0001	0.34	2.11	0.000043	0.0033
0.001	3.4	6.7	0.00014	0.0104

At the lowest velocity of 10^{-5} m/s the maximum mass transfer rate is about ten times the reaction rate. If we apply the Mear's criterion, that mass transfer becomes a factor when the maximum transfer rate is ten times the reaction rate, then external mass transfer would only be a factor at velocities below 10^{-5} m/s.

Case 3: Trickle–Bed Reactor

Trickle beds, with simultaneous flow of gas and liquid through a packed bed, give much more complex hydrodynamics. For mass transfer from the bulk fluid to the catalyst pellets, Fogler [1986] suggests correlations of the form:

For Re < 60

$$Sh' = 0.815 \cdot Re^{0.822} Sc^{1/3} \tag{4.86}$$

$$Sh' = \frac{k_c d_p \psi}{D_i}$$

where ψ is the fraction of surface wetted. This correlation suggests a stronger dependence of mass transfer on the fluid velocity, as well as a limitation due to the wetting of the catalyst surface by the liquid phase. Assuming $\psi = 1.0$, this correlation indicates that a trickle bed would give mass transfer limitations at liquid flows of 10^{-4} m/s, using Mear's criterion. This result indicates that trickle beds may be affected by liquid—phase mass transfer at velocities 10 times higher than in a packed bed. The analysis of a trickle bed must also consider gas—liquid mass transfer resistances.

4.4 Lumping of Complex Reaction Mixtures

The observable reaction kinetics for a single component can be affected by a variety of factors, as outlined in the previous sections of this chapter, including complex catalytic or chain reactions, diffusion and mass transfer. Other effects can arise in real reactors, due to catalyst deactivation, nonisothermal conditions, and nonideal flow distribution. These reactor—specific topics will be dealt with in subsequent chapters. In almost every application of kinetics to upgrading, however, we are not dealing with single compounds. This section analyzes the implications of this fact for kinetic analysis, indicates how the apparent reaction kinetics are related to the reactions of the components in a complex mixture, then suggests how to minimize the error by grouping or lumping components properly for a given reaction analysis.

Mixtures of reactants are ubiquitous in any processing of petroleum fractions, which are made up of many different homologous series of components. Consider benzene, with a boiling point of ca. 80° C. Successive substitution gives a family of alkylbenzenes (Table 4.9), where each step corresponds to addition of a CH_2 group as well as higher boiling point (by about 14C° for C_8 and higher members of the series). Table 4.9 does not show all possible isomers, i.e. o—xylene is shown but not m—xylene or p—xylene, but it does illustrate that increasing the carbon number from a base species increases the number of possible combinations. The alkylbenzene series is normally considered to be the lower diagonal of this table, from benzene through to n—butyl benzene.

Just as the possible number of arrangements increases through the alkylbenzenes, so does the number of possible homologous series increase at higher molecular weights or boiling points. The result of this increasing complexity is that distillates with an IBP over about 170° C

Table 4.9
Alkylbenzene Isomers with Six to Ten Carbons

Carbon Number				
6	7	8	9	10

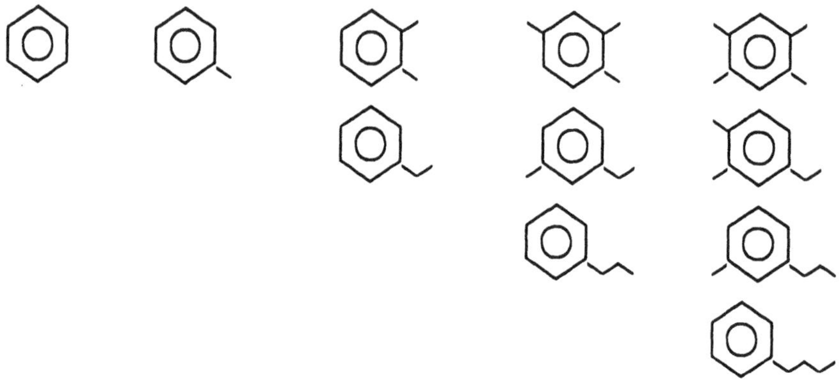

cannot be completely analyzed by instrumental methods, and we are forced to simplify our approach.

In thermodynamics this problem is dealt with by defining pseudocomponents, which represent the components within a given boiling range. For example, the components boiling between 300 and 350° C are treated as a single component, say NBP325. The vapor—liquid equilibrium and enthalpy can be estimated approximately if the specific gravity is known, and estimated even more accurately if other data are available such as paraffin—naphthene—aromatic (PNA) distribution (Chapter 2). These minimal data are sufficient for distillation, or heat exchange calculations.

The situation with regards to reactivity is very different. Trytten *et al.* [1990] reacted narrow boiling fractions of coker gas oil in a CSTR reactor over a Ni/Mo catalyst. Although the average molecular weights of the fractions ranged from 183 to 407, the rate constants for HDS at 400° C ranged from 93.7 to 1.8 mL/s—kg catalyst (Figure 4—14). The rate constants for HDN ranged from in excess of 17 mL/s—kg catalyst to 0.7 mL/s—kg catalyst. As indicated by the figure, the rate constants declined logarithmically with increasing molecular weight.

Whenever molecular weight changes, diffusional effects may influence the apparent kinetics. For this series of fractions, estimates of the effectiveness factor gave an unexpected result: the lighter fractions were more limited by diffusion than the heavier fractions.

Table 4.10
Effectiveness Factors for Coker Gas Oil Fractions at 400° C

Fraction T_b, ° C	AMW	ϕ_{HDS}	η_{HDS}	ϕ_{HDN}	η_{HDN}
278	183	7.5	0.35	—	—
314	215	5.6	0.44	2.6	0.73
360	250	4.3	0.54	2.7	0.71
421	300	4.5	0.52	1.6	0.86
481	363	2.9	0.70	1.5	0.88

The effectiveness factors for HDN were larger than HDS in all cases because the smaller reaction rate gave a smaller value of the Thiele modulus (ϕ). The effective diffusivities of the reactant species did not change significantly in this series. The molecular weights varied by a factor of 2, so that the hydrodynamic radius would vary by a factor of between 1.2 and 1.4. The main cause for the decrease in ϕ with boiling

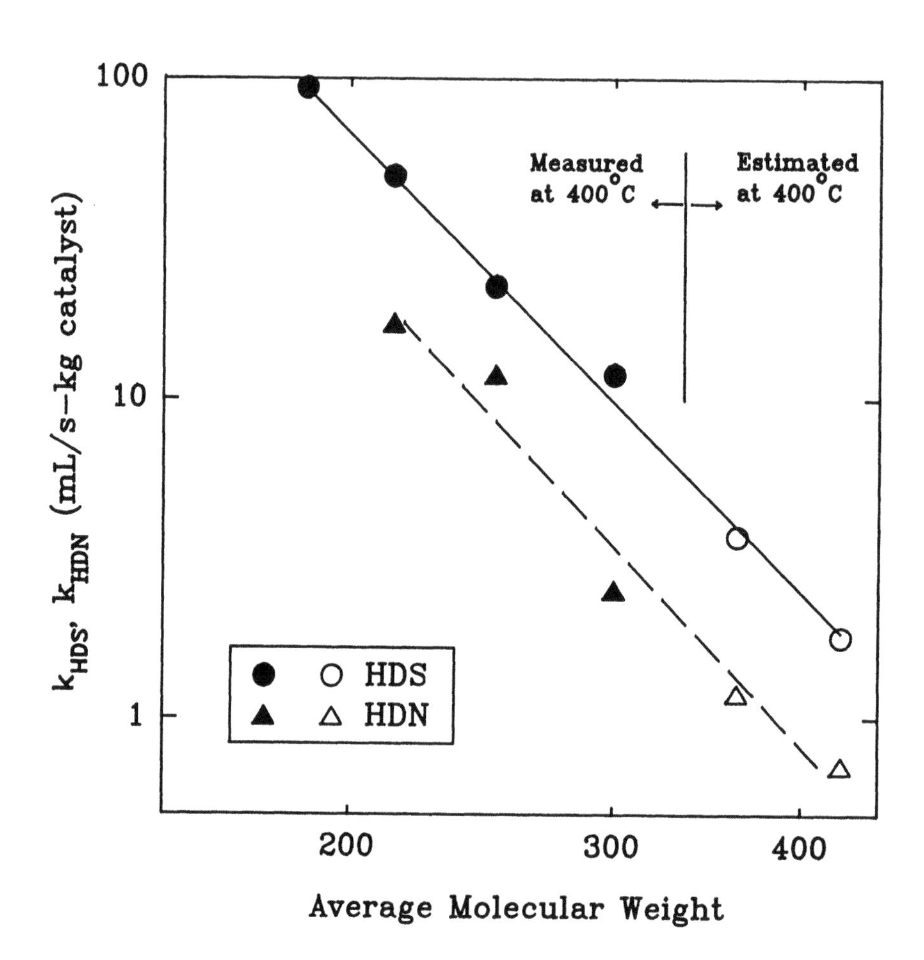

Figure 4—14
Rate constants for HDS and HDN of coker gas oil fractions. 150 mL CSTR reactor, 13.9 MPa, 400° C, 8g Ni/Mo on γ—alumina catalyst (Data from Trytten *et al.*, 1990)

point was the rate constant.

The wide range of reactivity observed in these distillates has very significant implications for processing of synthetic crudes, and leads into the problem of how to define rates of reaction for complex mixtures. As a first step, we will consider the conversion of sulfur in a gas oil which follows the distribution of rate constants from Figure 4—14.

4.4.1 Desulfurization of gas oil in a PFR

Consider the conversion of sulfur in a gas oil which has sulfur species exhibiting a distribution of reactivity following Figure 4—14. The gas oil was split into 6 fractions, designated by the subscript i. For each fraction we can write the design equation for the PFR:

$$W_t/Q_o = \int_0^{X_{e,i}} dX_i/(-r_i) \tag{4.87}$$

where $X_{e,i}$ is the outlet converison of component i. If each component follows first—order kinetics (like the pure compounds discussed previously) then

$$W_t/Q_o = \int_0^{X_{e,i}} dX_i/[k_i C_{o,i}(1-X_i)] \tag{4.88}$$

where $C_{o,i}$ is the inlet concentration of sulfur in boiling fraction i. Solving for the conversion and the outlet concentration

$$X_{e,i} = 1 - \exp(-k_i W_t/Q_o) \tag{4.89}$$

$$C_{e,i} = C_{o,i} \exp(-k_i W_t/Q_o) \tag{4.90}$$

The following equations give the overall sulfur contents:

$$C_o = \sum_{i=1}^{6} C_{o,i} \tag{4.91}$$

$$C_e = \sum_{i=1}^{6} C_{o,i} \exp(-k_i W_t/Q_o) \tag{4.92}$$

The integral design equation for a PFR (equation (4.87)) precludes derivation of a rigorous relationship between the apparent reaction order and the intrinsic rate constants (k_i).

The data from Trytten *et al.* [1990] can be used to give a numerical example. If desulfurization is the only reaction to occur, then equation (4.90) can be used to calculate the remaining concentration of sulfur in each boiling fraction. Figure 4—15 illustrates the initial distribution of sulfur concentration in each fraction (as % sulfur within each boiling range). When the conversion is 95% or 99%, all the remaining sulfur is in the two least reactive fractions. The high conversion performance of the reactor actually concerns only a portion of the total feed material. Note that in an actual hydroprocessing reactor other reactions occur to shift the distribution of sulfur compounds. Hydrogenation of aromatics and cracking will reduce molecular weight (and boiling point) and so the residual sulfur in Figure 4.15 would actually be more broadly distributed, though still concentrated in the high—boiling cuts.

The design equation for the PFR with n—th order kinetics is

$$W_t k_n C_o^{n-1}/Q_o = \int_0^{X_e} dX/(1-X)^n \tag{4.93}$$

Solving and rearranging

$$1/C_e^{n-1} - 1/C_o^{n-1} = (n-1)W_t k/Q_o \tag{4.94}$$

A plot of the LHS versus LHSV, therefore, will be linear with the appropriate choice of n. In practice, the choice of n is based on the best regression of the available data. For the example of the Coker Gas Oil from Trytten *et al.* [1990] the best fit was for n=1.5 when the conversion was in the range 0—99% (Figure 4—16).

This result is consistent with the observation of 1.5—order kinetics of desulfurization of the same coker gas oil feed in commercial trickle reactors [Yui, 1989]. The derivation does not prove the origin of fractional—order kinetics in these reactors, but it provides persuasive evidence that the distribution of reactivity in the feed is responsible.

4.4.2 Desulfurization of gas oil in a CSTR

Section 4.4.1 considered data for hydroprocessing of a gas oil in a CSTR over a similar Ni/Mo catalyst to Trytten *et al.* [1990]. The design equation for a CSTR is

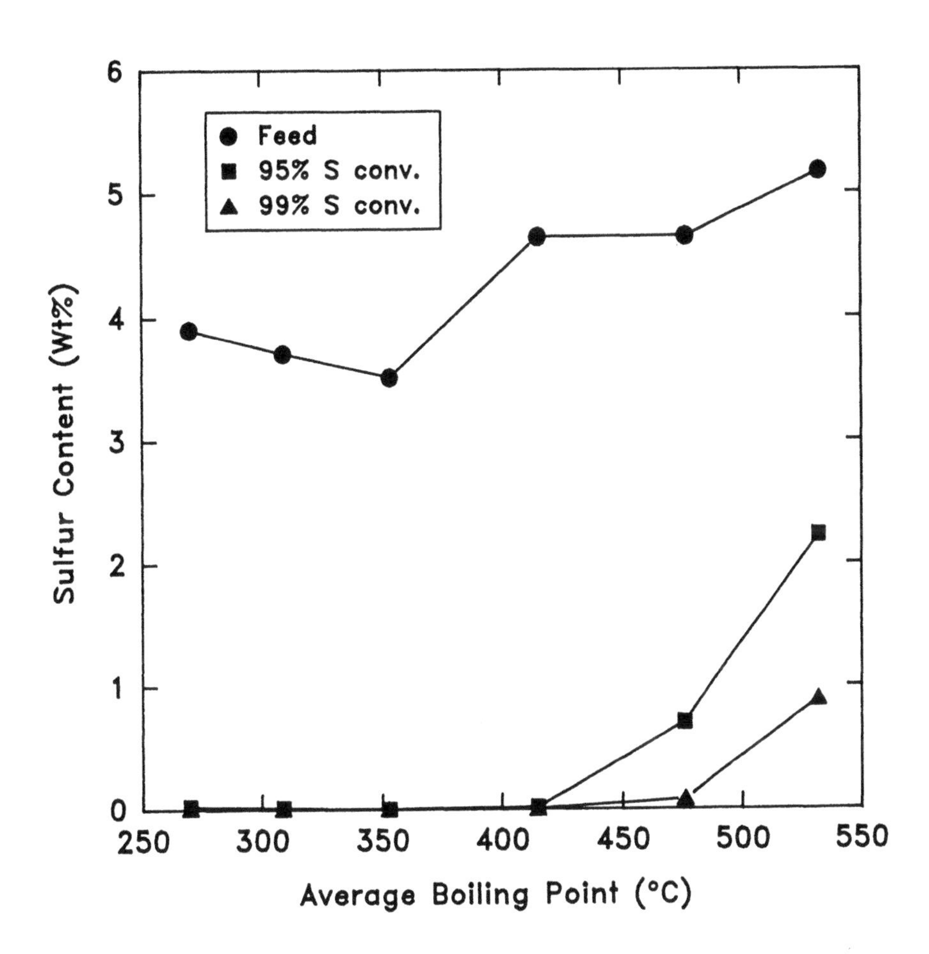

Figure 4—15
Predicted distribution of sulfur with molecular weight in a gas oil product from a plug flow reactor (Rate data from Figure 4—14)

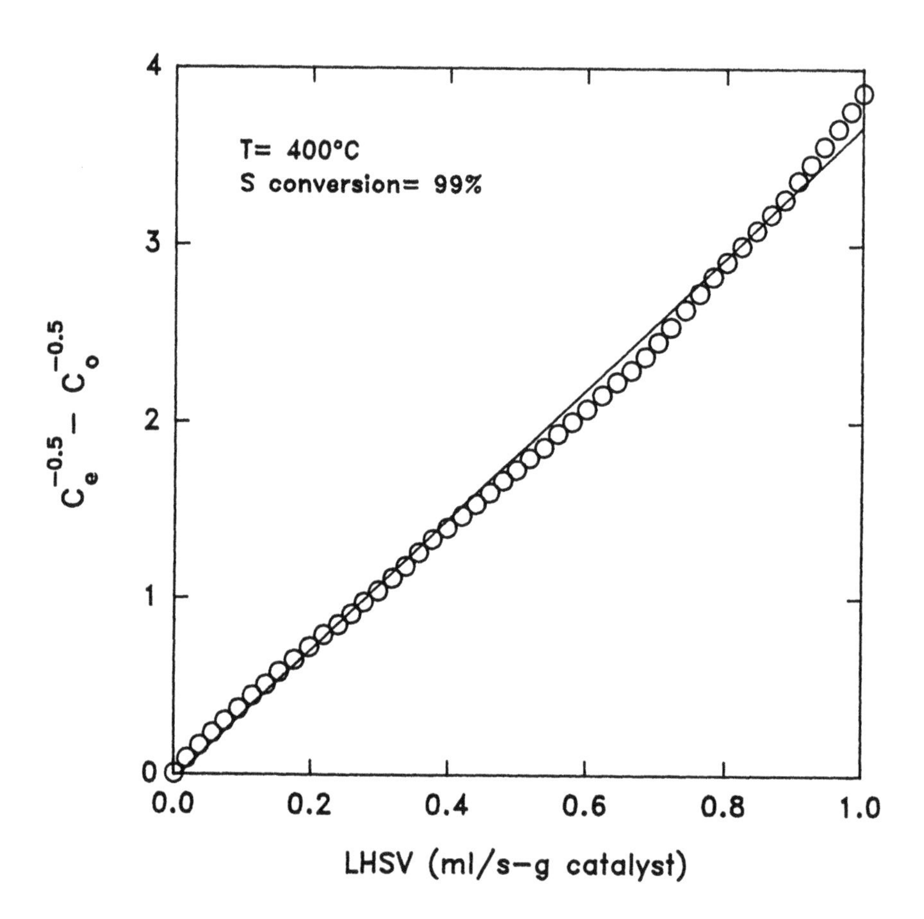

Figure 4–16
Fit of 1.5 order kinetics to predicted PFR conversion (Rate data from Figure 4–14)

$$(-r_i)W_t = Q_o C_{o,i} X_{e,i} \tag{4.95}$$

If the sulfur in each boiling cut follows first–order kinetics, as in the previous example, then

$$C_{e,i} = C_{o,i}/(1 + W_t k_i/Q_o) \tag{4.96}$$

For the overall outlet concentration

$$C_e = \sum_{i=1}^{6} C_{o,i}/(1+W_t k_i/Q_o) \tag{4.97}$$

The apparent overall rate is given by the design equation

$$(-r) = Q_o(C_o - C_e)/W_t = \sum_{i=1}^{6} C_{o,i}k_i/(1+W_t k_i/Q_o) \tag{4.98}$$

If the reaction is assumed to follow n–order kinetics, then

$$k_n C_f^n = (-r) \tag{4.99}$$

$$\ln k_n + n \ln C_e = \ln(-r) \tag{4.100}$$

If W_t or Q_o is varied, then the rate of reaction can be measured as a function of outlet concentration (C_e from equation (4.97)). If the overall reaction is first order, then a plot of rate versus C_e should be linear, after equation (4.99). Figure 4–17 shows the rate of reaction as a function of C_e from a CSTR reactor, as calculated from equation (4.98) using the data of Trytten *et al.* [1990]. The curvature indicates that the apparent reaction order is not unity througout. Using equation (4.100), the results in Table 4.11 were obtained.

In the case of a CSTR the apparent reaction order decreases with conversion. The data of Figure 4–18 illustrates that as conversion proceeds, more and more of the residual sulfur is in the least reactive fractions. The surviving sulfur is more evenly distributed with molecular weight than in the PFR, consistent with the characteristic that a CSTR gives lower conversion at a given retention time or LHSV. The kinetics of the CSTR at high conversion are dominated by a single fraction, so

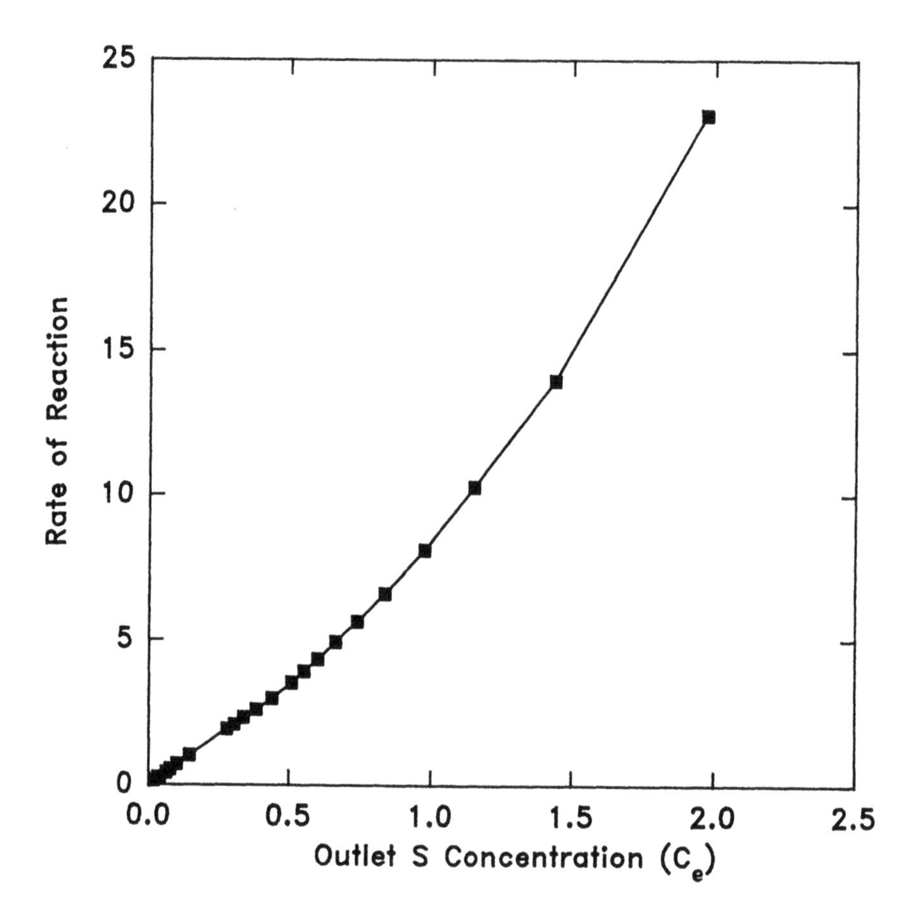

Figure 4—17
Rate of desulfurization of gas oil in a CSTR versus outlet concentration
(Rate data from Figure 4—14)

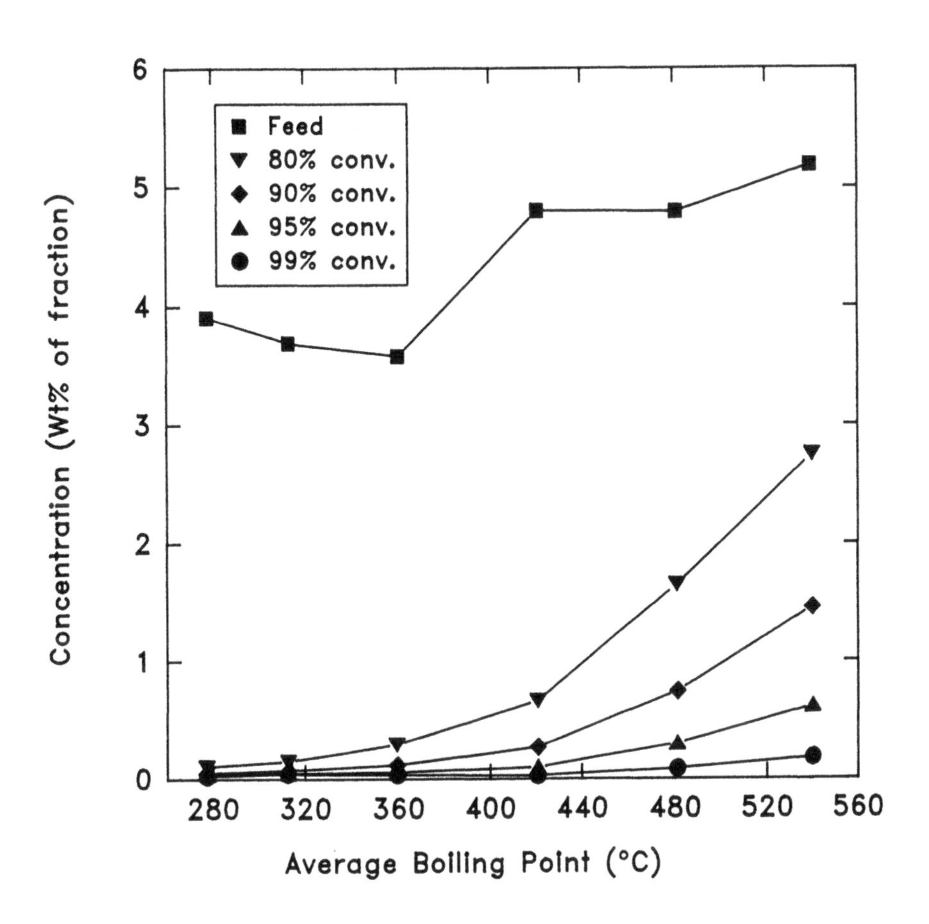

Figure 4—18
Distribution of sulfur with boiling point in hydrotreated gas oil from a CSTR (Rate data from Figure 4—14)

the order approaches unity.

$$(-r) \simeq C_{e,6}k_6 = C_{o,6}k_6/(1+W_t k_6/Q_o) \tag{4.101}$$

In contrast the PFR performance integrates over the entire range of reaction rates from inlet (high) to outlet (low).

Table 4.11
Apparent Reaction Order for Conversion in a CSTR

Conversion range	Order (n)	k_n
50—70	1.5	8.3
80—85	1.2	8.4
90—95	1.1	7.5

In section 4.2.2 the analysis of the CSTR data for the of fractional conversion 0.6 — 0.8 indicated that the reaction was best correlated as first order. How can we reconcile this result with the data of Trytten *et al.* [1990] and Table 4.4? One explanation is that this analysis has ignored the effect of cracking reactions. In the CSTR, with a content of 8 g catalyst/100 ml liquid holdup, cracking reactions would tend to redistribute the sulfur compounds between the boiling ranges, and make the high—boiling compounds more reactive to HDS.

4.4.3 Limits to fractional order for mixture reactions

Ho and Aris [1987] showed that reaction orders up to second—order kinetics would arise if a fraction of the feed was essentially unreactive, or if the concentration—reaction rate distribution followed an exponential distribution. If the feed followed the Γ—distribution, then the apparent reaction order could range from 1 to 2. In the case of coker gas oil, for example, the observed reaction order for HDS in trickle—bed reactors was 1.5 [Yui, 1989]. From Ho and Aris this apparent order is related to the parameter α in the Γ—distribution:

$$C(k,0)dk = [\alpha^{\alpha}/\Gamma(\alpha)][k/k_{av}]^{\alpha-1} \exp(-\alpha k/k_{av})/k_{av}\, dk \tag{4.102}$$

where C(k,0) dk is the fraction of the feed (time = 0) with reaction rate constant between k and k+dk. The value of α determines the shape of the distribution, while k_{av} is the mean rate constant for the mixture. If the apparent reaction order is β then Ho and Aris showed

$$\beta = (\alpha + 1)/\alpha \tag{4.103}$$

Hence if $\beta = 1.5$ for coker gas oil, then $\alpha = 2.0$ and Equation (4.102) can be used to calculate the initial distribution of material. Figure 4—19 shows the distribution of the feed reactivity (C(k,0) versus k) for $\alpha = 2.0$ by Equation (4.102). The value of k_{av} was selected to give a similar range of rate constants to the data of Trytten *et al.* [1990].

As the reaction proceeds, each component in the continuum reacts away, leaving the more refractory material in the outlet.

$$C(k,t) = C(k,0)\exp(-kt) \tag{4.104}$$

where t would be time in a batch system or residence time in a PFR. Figure 4—19 shows the distribution of sulfur compounds for conversion levels of 95 and 99%. As expected, the remaining material has the lowest reactivity and the distribution is shifted toward low values of k.

Given that Figure 4—18 exhibits a similar range of rate constants to the data of Trytten *et al.* [1990], is a Γ—distribution consistent with the experimental data for coker gas oil? In order to evaluate this question the rate constants for each boiling cut must be evaluated by integration.

$$\bar{k}_{1-2} = \left[\int_{k_1}^{k_2} kC(k,0)\, dk\right] / \left[\int_{k_1}^{k_2} C(k,0)\, dk\right] \tag{4.105}$$

As illustrated in Figure 4—20 the agreement between the experimental data for boiling cuts and the Γ—distribution with $k_{av} = 20$ and $\alpha = 2$ is quite reasonable at the extrema (low and high reactivity). The lack of agreement at intermediate k's either indicates that the actual distribution function was not the Γ—distribution, or that the value of α was too large. Note that Ho and Aris selected the Γ—distribution based on its common occurence, and not from experimental data for HDS reactions.

4.4.4 Error in lumping a distribution of reaction rates

Grouping of components, or lumping, is essential in the kinetic analysis of any upgrading process because of the following factors:

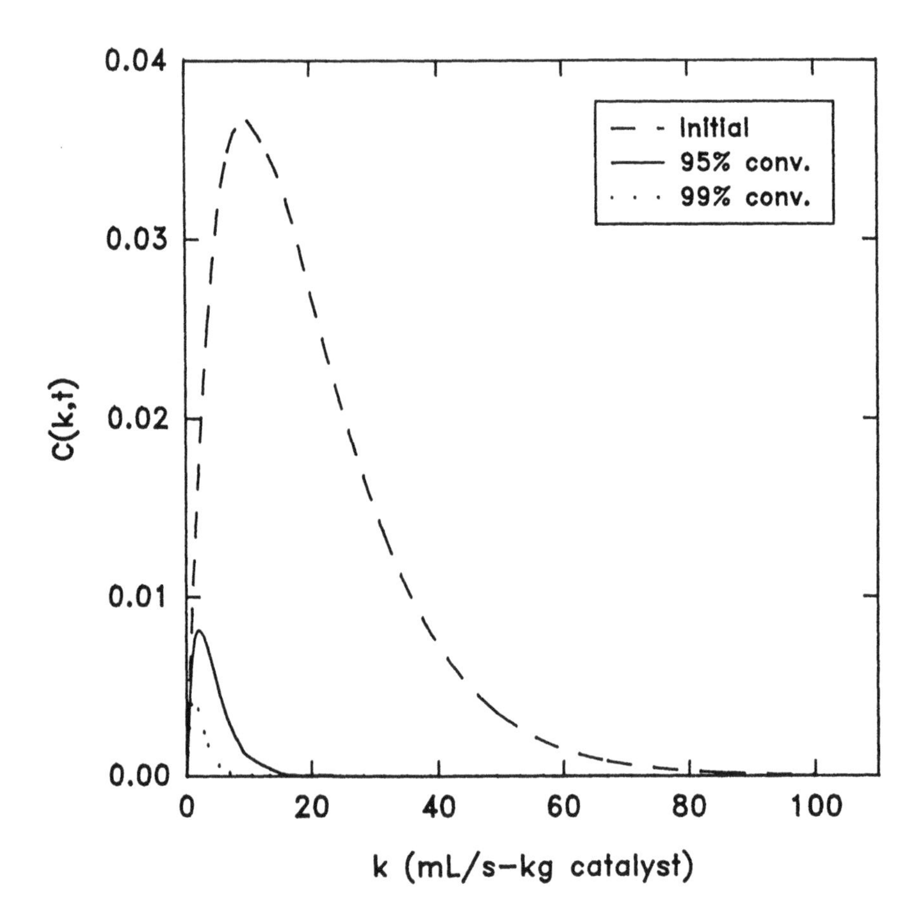

Figure 4—19
Distribution of reaction rates following a Γ—distribution

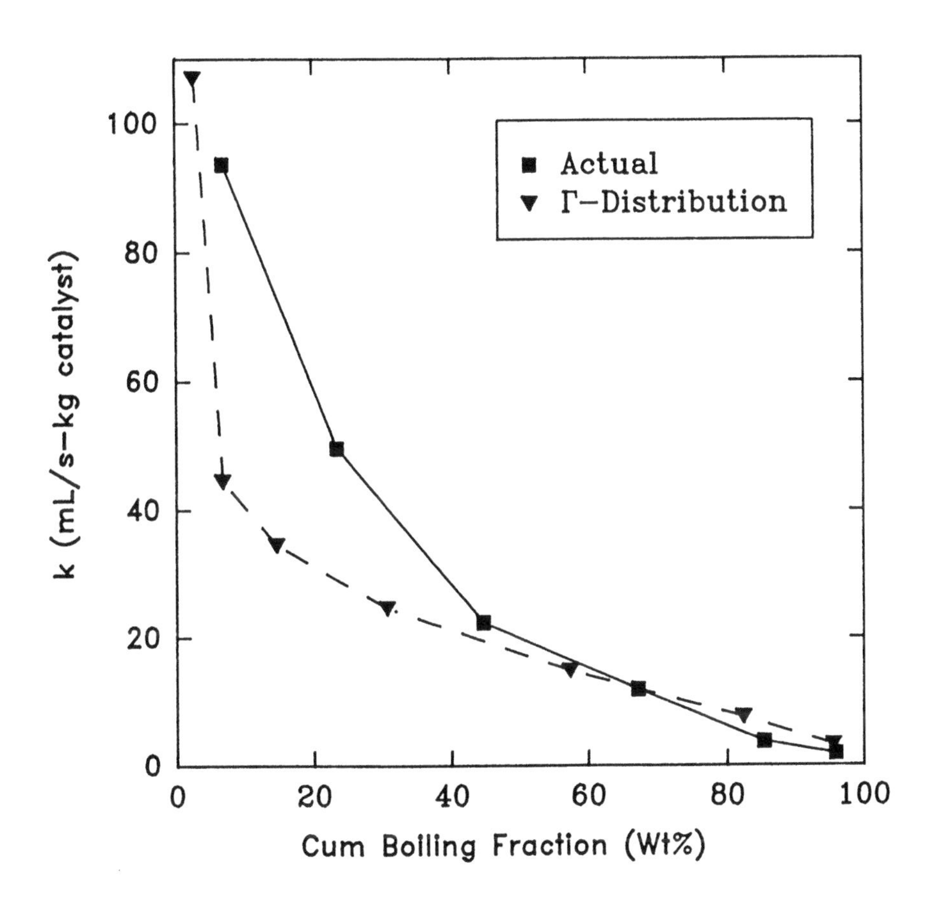

Figure 4–20
Distribution of rate constants vs. boiling fraction. Points indicate the mid–point of each boiling range; Rate constants are from Figure 4–14.

i) Concentration of individual components in the inlet stream is unknown.
ii) Concentration of components in product unknown.
iii) Stoichiometry and rate expressions are unknown.
iv) True rate parameters are unknown.

These uncertainties indicate that the kinetic analysis from the preceding sections should be considered as an analysis of the apparent behavior of mixtures, as opposed to the actual kinetic behavior of individual components. Two general approaches have been suggested: treat the mixture of components as a continuum with a distribution function to characterize reaction rates [Aris and Gavalas, 1966], or lump the mixture into a discrete set of pseudocomponents (see Wei and Prater [1962] for a rigorous analysis of this approach). The theory of discrete lumping has been more developed than continuous mixtures, but the rigorous application of either method has been limited by the difficulty of analyzing and reacting complex mixtures.

Weekman [1979] presented fluid catalytic cracking as a case study in the development of lumped kinetic models, and developed several guidelines:

i) The reactions of components within a lump must follow rate expressions that are similar in form (i.e. order), magnitude of rate constants, and temperature and concentration dependence.
ii) The lumps must be observable; if you cannot measure it, you cannot lump it.
iii) The kinetic parameters must be invariant with respect to conversion level and initial concentration, otherwise the model is not predictive.

In the example of hydrodesulfurization of a gas oil (section 4.4.1), the wide range of rate constants violated (i), and the observed behavior depended on reactor type and conversion level. Models based on solubility fractions (i.e., asphaltenes) or boiling fractions tend to contravene (iii) in that they must be reworked for each new feed mixture. This failure of thermodynamic lumping, which is based on solubility or boiling point, is due to conversion of one compenent in the same lump into another, and due to disparity between the chemical behavior of components within a lump. A successful lumping scheme is based on chemical classifications rather than just solubility, boiling point, or molecular weight.

Hutchinson and Luss [1971] developed methods to bound the conversion of lumped species, and thereby analyze the errors introduced by the choice of lumping. As a specific example, consider the concentration of sulfur in gas oil during catalytic desulfurization. To illustrate this approach on a realistic data set, consider the results of Trytten *et al.* [1990] to represent true HDS kinetics (Table 4.5). These data likely include considerable lumping of different sulfur types within a given boiling range, but serve to illustrate the point.

Table 4.12
Rate Constants for HDS of Coker Gas Oil Fractions at 400° C

Fraction ABP ABP, °C	AMW	k_{HDS} mL/s–kg cat	Sulfur wt%
278	183	93.7	3.9
314	215	49.6	3.7
360	250	22.4	3.6
421	300	11.9	4.8
481	363	3.71	4.8
539	407	1.79	5.2

For a set of parallel first–order reactions, Hutchinson and Luss showed that the concentration of a lump can be bounded as a function of time or residence time as follows:

$$\exp[-IR\cdot t] \leq \hat{C}_j(t)/\hat{C}_j(0) \leq \sigma^2/D + IR^2/D\cdot\exp[-D\cdot t/IR] \tag{4.106}$$

where $\hat{C}_j(t)$ is the concentration of a lumped species at time t, IR is the initial rate of reaction of the lump, and D is the second derivative. The strength of this analysis is that IR and D can, in principle, be determined from experimental data. For first–order reactions:

$$IR = -[d\,\hat{C}_j/dt\Big|_{t=0}]/\,\hat{C}_j(0) = \frac{\sum_{i=1}^{m} k_i C_i(0)}{\sum_{i=1}^{m} C_i(0)} \tag{4.107}$$

$$D = -[d^2 \hat{C}_j/dt^2\Big|_{t=0}]/\hat{C}_j(0) = \frac{\sum_{i=1}^{m} k_i^2 C_i(0)}{\sum_{i=1}^{m} C_i(0)} \tag{4.108}$$

The index i indicates the summation from i=1 to m of the m species that make up the lump. The variance for the lump is defined as follows:

$$\sigma^2 = D - IR^2 \tag{4.109}$$

In the case of lumping all sulfur into a single component, m=6 and we obtain the following:

IR = 24.5 mL/s—kg cat. (range 1.79 to 93.7 mL/s—kg cat.)
D = 1340 (mL/s—kg catalyst)^2
σ^2 = 738 (mL/s—kg cat)^2

Substituting these results into equation (4.106) provides:

$$\exp[-24.5\,\tau] \leq [S_t]/[S_o] \leq 0.55 + 0.45\exp[-54.6\,\tau] \tag{4.110}$$

where $[S_t]$ is the total sulfur concentration as a function of time or τ. For consistency of units, equation (4.110) has been written in terms of a residence time with units (s—kg cat/mL).

What does equation (4.110) say about the behavior of the lumped species? The lower bound shows that the conversion of the lump cannot fall below that of a pseudocomponent with the same initial concentration and initial reaction rate IR. The maximum value of the mean rate constant, k_{av}, is given by this lower bound, i.e. $k_{av} \leq 24.5$ mL/s—kg cat. Given the wide range of the rate constants, this bound is not very helpful. As conversion proceeds the instantaneous rate constant will tend toward the values for the less reactive components (1.8—3.7 mL/s—kg cat.). The bound from equation (4.106) only sets the maximum order of magnitude for the rate constant.

If we consider the upper bound, then as time proceeds

$$[S_t]/[S_o]\Big|_{t\to\infty} \lesssim 0.55 \qquad (4.111)$$

The upper bound, therefore, does not give a limiting estimate of the rate constant, but it does give an estimate of the minimum conversion to be expected at long times. By any criterion, 45% conversion is fairly poor and so the bound is not very helpful.

At this stage, keep in mind that grouping all of the sulfur types together in a single lump was an example of poor lumping, which gave rise to apparent 1.5–order kinetics. The poor estimates of the bounds from equation (4.105) result from the spread of rate constants in the lump, not from any fault in the equations.

Consider an alternative lumping scheme, where the two least reactive fractions form one lump, and the remaining four fractions a second.

Table 4.13
Two–Lump Model for HDS of Gas Oil

Parameter	Lump 1 Fractions 1–4	Lump 2 Fractions 5–6
IR	35.9	3.03
D	2041	10.1
σ^2	755	0.84
σ^2/D	0.37	0.08

Lump 2 is fairly tight in terms of its reactivity:

$$\exp[-3.03\ t] \lesssim [S_2(t)]/[S_2(0)] \lesssim 0.08 + 0.92\ \exp[-3.13\ t] \qquad (4.112)$$

Hence $k_{av} \lesssim 3.03$ mL/s–kg cat. and ultimate conversion $\gtrsim$ 92% at long times.

For a systematic approach to lumping, the best approach is to minimize the range between the upper and lower estimates. From equation (4.106) the range is given by

$$\begin{aligned}\text{Range} &= \text{upper bound} - \text{lower bound}\\ &= \sigma^2/D + IR^2/D\cdot\exp[-D\cdot t/IR] - \exp[-IR\cdot t]\end{aligned} \qquad (4.113)$$

As $t \to \infty$ then range $\to \sigma^2/D$. An objective for accurate lumping, therefore, is to minimize $\sigma^2/D = 1 - IR^2/D$. In order to achieve this each lump should contain species with similar rates of reaction.

Table 4.14
Range of Estimates for Lumped Model for HDS

Parameter	One Lump	Two—Lump	
σ^2/D	0.55	Lump 1	0.37
		Lump 2	0.08

Clearly the lumping of sulfur species in this example compromises the accuracy of any estimates, so that any available data for reactivity of fractions should be used as a first pass. Later optimization can be used to see if fewer lumps are adequate. An example of this type of analysis was given by Coxson and Bischoff [1987]. Given that analytical and kinetic data on sub—groups within the reacting system usually limit our ability to understand complex reactions, optimization of lumping is of secondary importance.

4.4.5 Activation energy of lumped reactions

Activation energies of apparent reactions such as HDS are frequently reported (see, for example Section 4.2), but how is the behavior of the lump related to the underlying activation energy of the individual components? Golikeri and Luss [1972] considered a simplified system of reactions, with m reactions occuring in parallel:

$$r_i = dC_i/dt = k_i C_i^n \tag{4.114}$$

Lumping the components implies that the apparent activation energy will depend on the sum of the reaction rates:

$$C = \sum_{i=1}^{m} C_i \tag{4.115}$$

$$-E''/R = \left[\frac{\partial \ln(-dC/dt)}{\partial(1/T)} \right]_{X_c} = \left[\frac{\partial \ln(\Sigma\ r_i)}{\partial(1/T)} \right]_{X_c} \tag{4.116}$$

where X_c is a constant level of conversion, and E" is the apparent activation energy. Golikeri and Luss showed that the apparent activation energy was given by:

$$E'' = E' + t \left[\frac{\partial E'}{\partial t} \right]_T \tag{4.117}$$

where E' is the mean activation energy weighted by reaction rate

$$E' = \Sigma\, r_i E_i / \Sigma r_i \tag{4.118}$$

The significance of equation (4.117) is that the apparent activation energy can change with time in a batch reactor (i.e. with conversion in a flow reactor), and that the value is not bounded by the activation energies of the individual components in the lump. Indeed, Golikeri and Luss gave examples where E" was negative at some levels of conversion! Clearly, lumping of components with dissimilar activation energies can give rise to apparent activation energies that are unrelated to the component behavior.

Notation

a	hydrodynamic radius, m
A	preexponential factor, s^{-1} or $m^3 kmol^{-1} s^{-1}$
AMW	average molecular weight, Da
B	empirical parameter
$\hat{c}$	concentration of a lumped species, $kmol/m^3$
C	concentration, $kmol/m^3$
C_d	drag coefficient
CSTR	continuous–flow stirred tank reactor
d_p	pellet diameter, m
D_e	effective diffusivity in catalyst pellets, m^2/s
D_b	diffusivity in bulk liquid at infinite dilution, m^2/s
D	empirical parameter in equation (4.105)
E	activation energy, kJ/kmol
f	fugacity
F	molar flow rate, kmol/s
g	gravitational acceleration, 9.81 m/s^2
H	hydrogen atom
IR	initial reaction rate, kmol s^{-1} m^{-3}
k	reaction rate constant, s^{-1} or $m^3 kmol^{-1} s^{-1}$
k_{av}	mean reaction rate in a lump

k_b	Boltzmann's constant
k_c	solid—liquid mass transfer coefficient, m/s
K	equilibrium constant
L_p	diffusion path length, m
LHSV	liquid hourly space velocity
m	number of reacting species or lumps
M	parent alkane
O	olefin
p	partial pressure
PFR	plug flow reactor
r	rate of reaction, kmol/s
r_p	pore radius, Å or m
R	alkyl group; gas constant
Re	Reynolds number
S	catalyst site
[S]	concentration of sulfur, kg/m^3
S_x	external surface area, m^2
Sc	Schmidt number
Sh	Sherwood number
t	time, s
T	temperature, K
T_b	boiling temperature, °F or °C
U	velocity, m/s
V	volume, m^3
W	mass of catalyst, kg
X	fractional conversion
α	empirical parameter in equations (4.61) and (4.101)
β	empirical parameter in equations (4.61) and (4.102)
γ	shape factor
ϵ	porosity
ζ	tortuosity
η	effectiveness factor
λ	dimensionless radius
μ	viscosity, cp or kg m s^{-1}
ν	kinematic viscosity, m^2/s
ρ	density, kg/m^3
σ	standard deviation
τ	mean residence time, s
ϕ	Thiele modulus
Φ	Weisz—Prater parameter
χ_p	cross sectional area of pellet, m^2
ψ	fraction of catalyst surface wetted

ω	flux, kmol m^{-1} s^{-1}

Superscripts

e	equilibrium condition
n	reaction order
α	reaction order in equation (4.44)
β	reaction order in equation (4.44)

Subscripts

ads, A	adsorption
app	apparent
b	bulk fluid phase
des	desorption
e	exit condition
HDN	hydrodenitrogenation
HDS	hydrodesulfurization
i, j	counter variables
max	maximum or limiting
o	initial condition
obs	observed results
p	pellet
s	surface
t	total
v	vacant

Further Reading

Fogler, H.S. *Elements of Chemical Reactor Design*, Prentice—Hall, Englewood Cliffs, NJ, 1986.

Weekman, V.W. "Lumps, Models, and Kinetics in Practice", AIChE Monograph Series 11(75), 29 pp., 1979.

CHAPTER

5

Residue Hydroconversion Processes

5.1 Introduction

The desirable products from processes for treating residues are distillates. The cost and suitability of a given process depends on a number of global factors:

1. Coke formation — Coke is a high-carbon solid which tends to form at high temperature, either as a deposit on catalysts or as a solid within the reactor. Any process that produces coke has a potential byproduct problem because the product value or heating value of the coke must be recovered. Combustion of coke may be limited by restrictions on sulfur emissions. The sulfur content of a petroleum coke tends to increase with the sulfur content of the residue or asphaltene fractions.

2. Production of high-boiling residuum or pitch — As in coke formation, unconverted high-boiling products are a liability if they contain significant amounts of sulfur. High-sulfur residual oils cannot be sold or burned without further treatment. Although some pitch may be used within a refinery as process fuel, excessive yields of high-sulfur pitch are difficult to sell. In general, pitches from

cracking processes are inferior for the manufacture of asphalts (known as bitumens in Europe) relative to the original residue. Pitches from thermal or catalytic hydrocracking have a high softening temperature and lack the ductility required for most asphalt applications.

3. Gas formation — The production of light ends (methane through butane or pentane) is undesirable because these products are less valuable than liquid products. Disposal is not a problem because the light ends are useful as clean process fuels, once they have been treated to remove hydrogen sulfide and ammonia. Gas formation depends on the severity of processing, which is a combination of time and temperature of processing.

4. Hydrogen consumption and pressure — Addition of hydrogen to a process represents an operating expense, as does any increase in hydrogen pressure. The hydrogen is not wasted unless it ends up in unconverted residue material or coke. The addition of hydrogen to the liquid products is highly desirable because it promotes the removal of heteroatoms and enhances product quality.

5. Catalyst consumption — Any catalysts or additives in upgrading processes represent an operating expense. In general, the higher the activity of the catalyst the greater the expense.

The selection of an upgrading process depends on the combination of cost, product slate, and byproduct considerations. Recently, process selection has tended to favor hydroconversion processes, which maximize distillate yield and minimize byproducts (especially coke). Thermal processes that produce coke may be attractive for processing unconverted residues from hydroconversion.

The presence of high–pressure hydrogen (> 5–7 MPa) during the conversion of a residue has a number of beneficial effects, as mentioned in Chapter 3. Hydrogen tends to suppress free–radical addition reactions and dehydrogenation. In the presence of a hydrogenation catalyst, the hydrogen converts aromatic and heteroaromatic species, and further converts heteroatoms to hydrogen sulfide, water, and ammonia. All hydroconversion processes involve a complex suite of series and parallel reactions; cracking, hydrogenation, sulfur removal, demetallation, etc. Almost all hydroconversion processes use a catalyst or additive to control the formation of coke, to serve as a surface for deposition of metals, and to enhance hydrogenation reactions. The challenge in analyzing these processes is to direct the reactions toward desirable product characteristics. The initial focus for residue hydroconversion was the removal of sulfur, but many recent processes have concentrated on the objective of achieving high conversion of residue fraction. The emphasis on sulfur removal, in these process designs, shifts from the primary hydroconversion step to the secondary hydroprocessing reactor.

5.2 Fixed–Bed Catalytic Processes

Fixed–bed processes have advantages in ease of scaleup and operation. A typical process flowsheet is given in Figure 5–1. The reactors operate in a downflow mode, with liquid feed trickling downward over the solid catalyst cocurrent with the hydrogen gas. The usual catalyst is Co/Mo or Ni/Mo on γ–alumina, containing 11–14% Mo and 2–3% of the promoter Ni or Co. The γ–alumina typically has a pore volume of ca. 0.5 mL/g. The catalyst is formed into pellets by extrusion, in shapes such as cylinders (ca. 2 mm diameter), lobed cylinders, or rings. Because these trickle–bed reactors normally operate in the downflow configuration, they have a number of operational problems, including maldistribution of liquid and pulsing operation at high liquid and gas loadings. Scaleup of these liquid–gas–solid reactors is much more difficult than a gas–solid or gas–liquid type. Nevertheless, the downflow regime is convenient when the bed is filled with small catalyst particles. Further discussion of trickle–bed operation is given in Chapter 8.

Because the catalyst particles are fairly small, these reactors are quite effective as filters of the incoming feed. Any solids, such as fine clays from production operations or iron sulfide scale, accumulate at the front end of the bed and can give high pressure drops.

The main limitation of this type of reactor is the gradual accumulation of metals when a typical residue is processed. The metals accumulate in the pores of the catalyst and gradually block access for hydrogenation, desulfurization etc. The length of operation is then dictated by the metal–holding capacity of the catalyst and the nickel and vanadium content of the feed. As the catalyst deactivates, the reactor feed temperature is gradually increased to maintain conversion. Toward the end of a run this mode of operation leads to accumulation of carbonaceous deposits on the catalyst, further reducing the activity.

At the end of a run cycle, the reactor must be shut down and the catalyst bed replaced. Because this operation is expensive, these units are sized for infrequent changes of catalyst. Many heavy oils and bitumens are not suited to fixed bed operation because of plugging and metals accumulation.

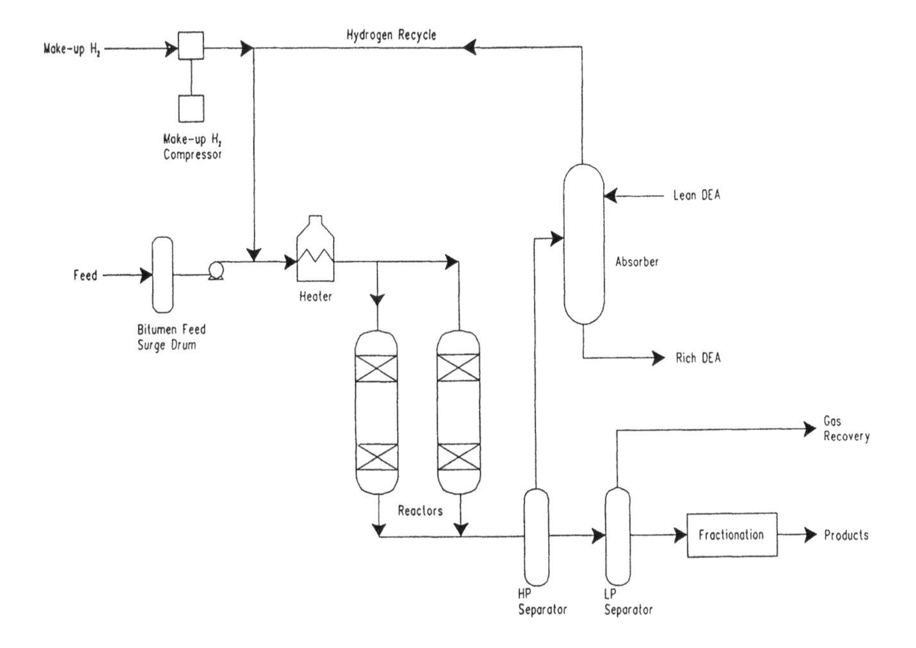

Figure 5–1
A typical fixed bed desulfurization process for residues

5.3 Catalytic Ebullated Bed Processes

The Co/Mo and Ni/Mo catalysts can be used to process feeds that contain solids if the catalyst is fluidized, to prevent plugging, and if fresh catalyst is added regularly to maintain constant activity. The resulting reactor uses an expanded or ebullated bed to treat the residue feed, and can handle solids and metals in the feed. The catalysts for expanded—bed operation are equivalent in composition to the catalysts for fixed—bed operation, but the pellet diameter is usually $\leq$ 1 mm to facilitate suspension by the liquid phase in the reactor.

5.3.1 Moderate conversion of residue

Two similar licensed processes of this type are available: H—Oil and LC—Fining. Differences between the two technologies are mainly in details of equipment design.

Figure 5—2 shows a schematic diagram of an ebullated—bed reactor. The hydrogen gas and liquid feeds are introduced at the bottom of the reactor, from whence they flow upward through the expanded catalyst bed. The hydrogen, hydrocarbon vapors, and liquid product exit from the top of the reactor. A recycle stream of liquid is drawn down a draft tube and pumped out of a distributor at the bottom of the reactor. This recycle flow of liquid is 5—10 times the feed rate, and it provides the energy to expand the bed. Typical operating conditions are 420—450° C and liquid hourly space velocities of 0.1—1.5 h^{-1}. Pressure is selected depending on cost and the desired product quality, but it must be above 7—10 MPa for the hydrogen to be effective. The hydrogen gas is recycled at approximately 3—4 times the consumption rate to ensure that excess hydrogen is always present in the liquid phase.

The use of an expanded bed also allows the addition and removal of catalyst while the reactor is in service. This operation is accomplished daily, and usually amounts to about 1% of the catalyst contained in the reactor. Rates of catalyst addition for processing a blend of Cold Lake and Lloydminster oil, for example, are 0.3—1.5 kg/m^3 of fresh feed [Chase, 1990]. Hence, the Husky Upgrader at Lloydminster will consume 1.5—7.6 tonnes/day of catalyst. The rate of addition of fresh catalyst is adjusted to maintain constant reactor conversion.

Figure 5—3 is the flowsheet of an ebullated—bed reactor system at Syncrude in Fort McMurray, Alberta, showing the ancillary process equipment. More than one reactor in series has been used in other plants, depending on the conversion level desired and the process capacity.

This type of reactor has been quite successful in processing heavy residues and bitumens for several reasons:

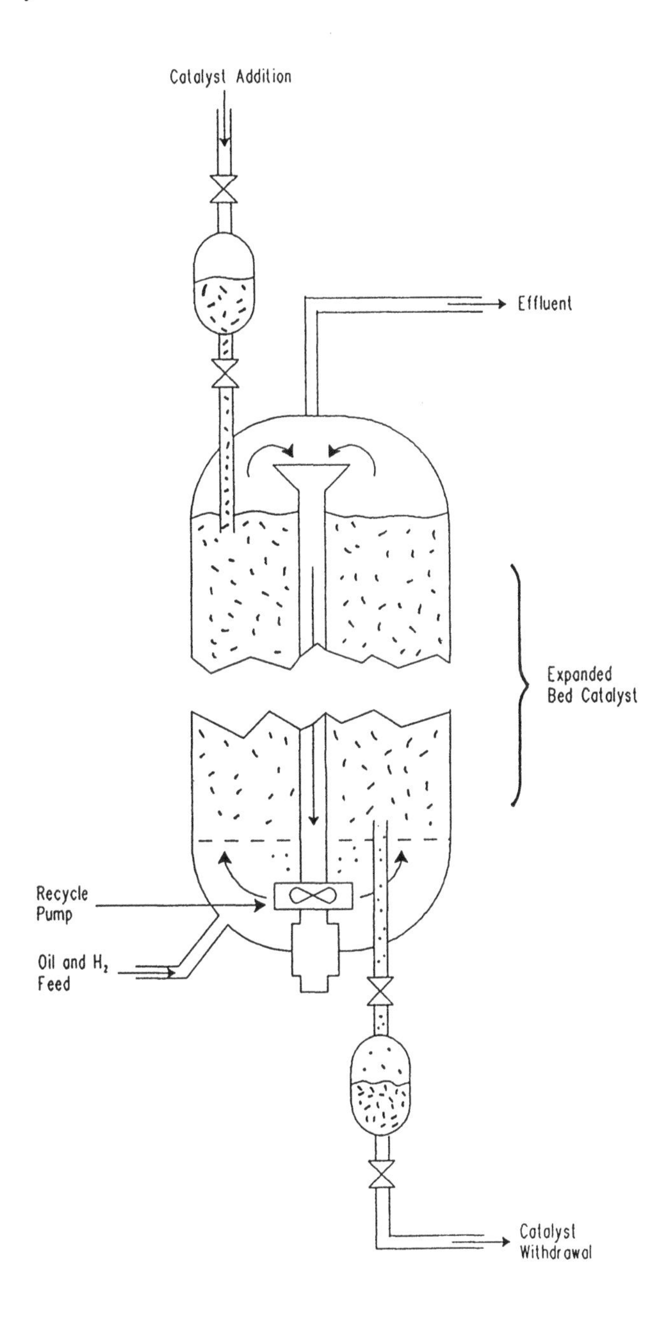

Figure 5–2
Schematic diagram of an ebullated bed reactor

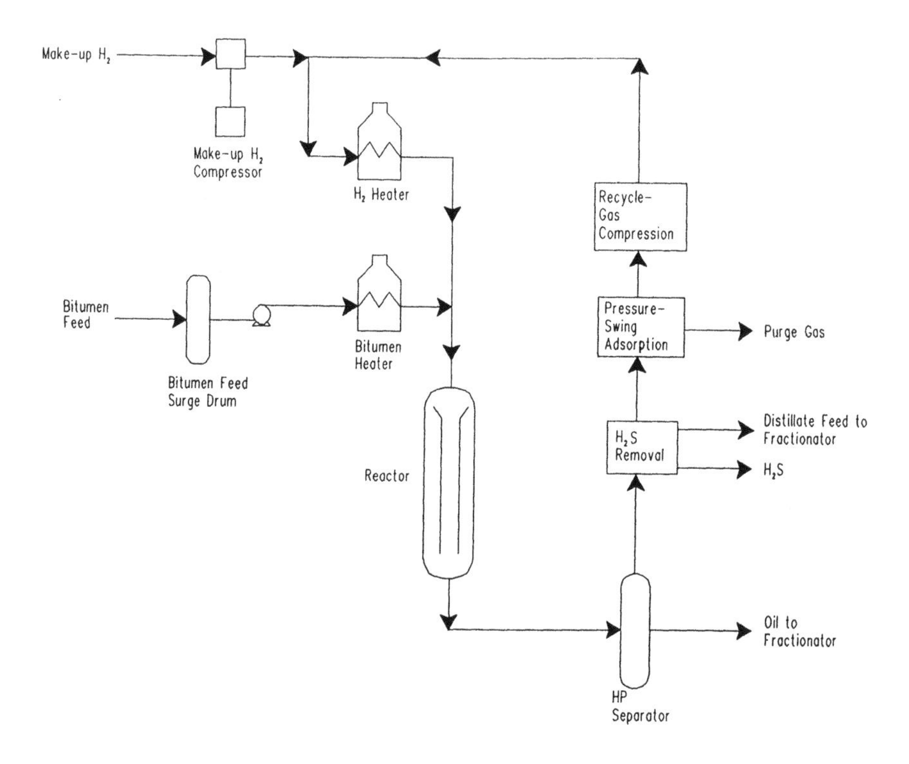

Figure 5–3
Process schematic of LC–Fining process at Syncrude (From Bishop, 1990)

1. The expanded bed is not plugged by solids in the feed, which is a common problem with Alberta bitumens.

2. The liquid recycle gives good mixing within the reactor, so that it approximates a continuous stirred tank reactor. The mixing ensures that temperature gradients are minimized. For example, a test on H—Oil reactors at Petroleos Mexicanos showed gradients on the order of 2° C within the reactor. The isothermal operation of these reactors is very important because the hydrogenation reactions are quite exothermic. If the catalyst bed "slumps", or settles then local hot spots can develop. Coke deposition then occurs rapidly because hydrogen becomes depleted in these zones. In one early unit, a slumped bed led to a serious explosion and destruction of the plant.

3. Catalysts can be added and withdrawn continuously, allowing long operating runs without shutting down the reactor.

Data are available for the performance of these systems for a variety of feeds. Table 5—1 lists the operating conditions and the actual process performance for the LC—Fining Plant at Syncrude during 1989 and 1990:

Table 5—1
Conversion of Athabasca bitumen by LC—Fining (From Bishop, 1990)

Temperature	425—450° C
Pressure	10—15 MPa
% Hydrogen	55—85% in gas
Capacity	40,000 bbl/d

Operating Performance

Conversion 524° C+ (vol%)	66.8
CCR Reduction, wt%	42
Desulfurization, wt%	65
Vanadium removal, %	71
Nickel removal, %	60
Hydrogen consumption, scf/bbl	938

Table 5—2
Syncrude LC—Fining Feed and Product Properties (From Bishop, 1990)

	Products					
	Feed	Naphtha		Gas oils		Residue
		Lt.	Hvy.	Lt.	Hvy.	
API Gravity	7.0	81.6	55	27.8	15.3	2.3
S, wt%	4.75	0.37	0.29	1.09	1.94	2.77
N, w%	0.42	—	0.04	0.16	0.37	0.72
CCR, wt%	13.8					21.4
V, wppm	202					155
Nickel, wppm	77					83
524° C+, vol%	58.4					54

Table 5—3
Syncrude LC—Fining Product Yields (From Bishop, 1990)

Product	Yield on feed bitumen Weight%	Volume%
$H_2S + NH_3$	3.39	
Fuel Gas	3.81	
C_4's	1.49	3.4
Light naphtha	2.92	4.0
Heavy naphtha	8.12	10.7
Light gas oil	35.38	41.2
Heavy gas oil	9.24	9.9
Residuum	37.20	34.8
Total	101.55	104.0

The data in Table 5—3 illustrate why the hydroconversion processes are attractive; the yields of liquid products actually exceed 100% on a volumetric basis due to the reduction in density caused by hydrogenation. Table 5—3 shows that the mass yield of liquid products is actually below 100% due to the formation of H_2S, NH_3, H_2O, and fuel gas (C_1—C_3 alkanes).

The data in Table 5—1 illustrate an important characteristic of hydroconversion processes. The cracking reactions give rise to methane, ethane, and propane which are not easily removed from the hydrogen recycle stream. The light ends are formed continuously, therefore, the hydrogen recycle loop must be purged and make–up hydrogen added to maintain the partial pressure of H_2. This purge results in a loss of hydrogen to fuel gas, unless secondary recovery of hydrogen is instituted by pressure–swing adsorption or membrane methods.

5.3.2 Processing for high conversion of residue

The Syncrude LC–Fining Plant is an example of medium–conversion operation, which is achieved by a single reactor. Addition of a second reactor in series (Figure 5–4) gives higher conversion, and values in the range of 60–80% conversion of the 524° C+ fraction are typical of ebullated–bed processes [Van Driesen *et al.*, 1987]. Higher conversions are limited by the formation of solids in the product lines, and particularly in the residue product of the process. In order to achieve conversions as high as 95%, two significant process modifications are required:

1. The unconverted residue must be recycled. This recycled material is less reactive than the virgin feed, and hence a larger reactor capacity is required than in single–pass operation. Recycling gives more reactor throughput with a lower average reactivity than does single pass operation.
2. The recycle stream must be treated to remove the fraction of the residue which leads to the formation of solids in the products.

One method of removing solid–formers from the recycle stream is to pass it through a bed of coke particles. The bed adsorbs the heaviest, most polar fractions of the residuum [Van Driesen *et al.*, 1987]. An alternate approach to controlling solids is to blend highly aromatic cuts with the feed to the reactor [Beaton and Bertolacini, 1991]. The factors that contribute to formation of solids are discussed in more detail in section 5.6.

5.3.3 Reactor hydrodynamics

The presence of gas, liquid, and solid contributes to the hydrodynamic complexity of ebullated–bed reactors. The mixing of liquid and solids in these reactors is quite good, so that the kinetics are approximated by the equations for a CSTR. The gas phase, however, is more complex. Reactor design requires a knowledge of the gas holdup in the reactor, which determines liquid retention time and the size of the gas bubbles which controls interfacial area and mass transfer. Unfortunately, very few studies have been conducted at sufficient pressure to determine the true effects of vapor density and interfacial tension on bubble behavior.

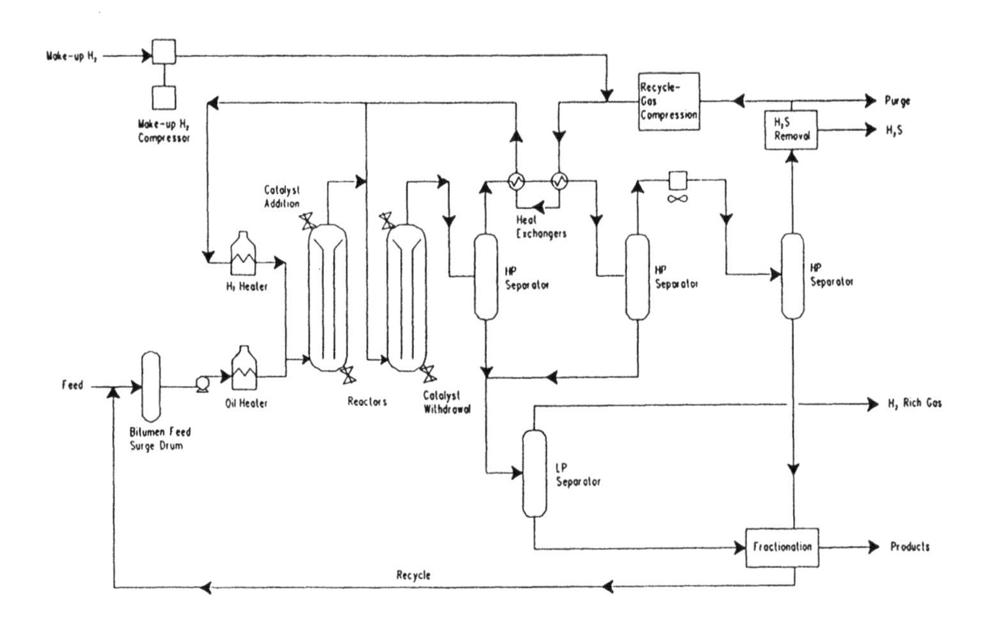

Figure 5–4
Process flow schematic for high–conversion hydroconversion of residue with ebullated–bed reactors (From van Driesen *et al.*, 1987)

Jiang *et al.* [1992] found that pressure had a significant effect on the operation of a three–phase bubble column. At 1 MPa, the bubble size was uniform at ca. 8 mm, and the gas holdup was uniform in the axial direction. With fine solids (0.46 mm) pressure only affected gas holdup at velocities over 2 cm/s. Above this velocity, more gas holdup was observed at higher pressure (up to h = 0.25; h=gas hold–up; volume gas/total volume) presumably due to a reduction in bubble coalescence. Coarse solids (6 mm) eliminated any effect of pressure on gas holdup, but a holdup of h = 0.3 was observed at high gas velocity.

5.4 Catalyst Deactivation

The catalysts used for processing residues inevitably deactivate with time due to accumulation of nickel and vanadium sulfides, and the accumulation of carbonaceous residues or "coke". The former deposits have been studied intensely, in part because the metal deposits tend to accumulate near the surface of a catalyst pellets, rendering the interior ineffective. Both metal sulfides and coke may contribute to loss of activity; therefore, a significant challenge is to determine their respective contributions.

5.4.1 Coke deposition on residue conversion catalysts

The Ni/Mi or Co/Mo on γ–alumina catalysts for hydroconversion processing give rapid accumulation of coke at short times, followed by constant concentrations (Figure 5–5). The amount of coke deposited on residue processing catalysts ranges from 15 wt% to 35 wt% on a carbon basis [Thakur and Thomas, 1985], and this level is achieved within 1 d in some cases [McKnight and Nowlan, 1993]. A significant loss in activity is associated with this rapid accumulation [Ternan and Kriz, 1980; Mosby *et al.*, 1986].

Coke deposits are thought to develop via surface adsorption of coke precursors, followed by a combination of oligomerization and aromatization reactions [Furimsky, 1979]. Since adsorption is the first step in the process, decreases in volatility (higher molecular weight) and increases in polarity would enhance the first step of the process. Residue fractions, therefore, give more coke deposition than an equivalent gas oil fraction due to the decrease in volatility of the reacting molecules. Although asphaltenes are commonly considered to be the heaviest, most polar components in residue, the asphaltene content of the feed does not correlate with coke deposition [Furimsky, 1979]. For example, deasphalting a residue only reduced the coke levels from 30 wt% of the catalyst to 20 wt% [Thakur and Thomas, 1985].

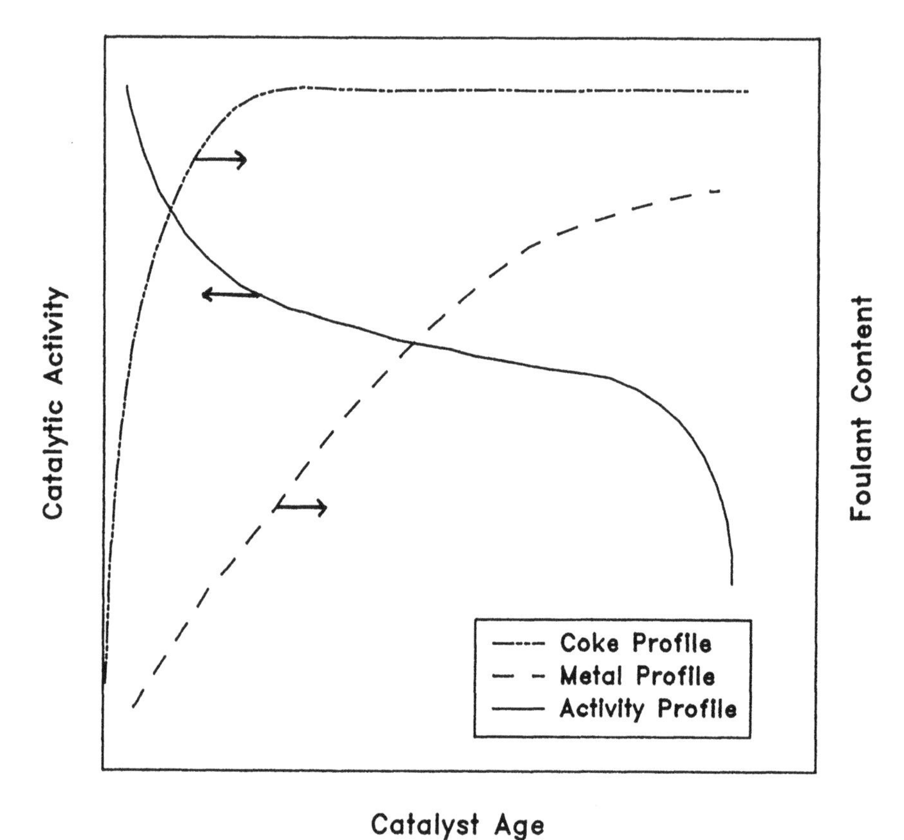

Figure 5–5

Schematic representation of time–dependent catalytic activity. Foulant content indicates level of metal sulfides or coke within the catalyst (from Thakur and Thomas, 1985)

Operating conditions are also very important. An increase in hydrogen pressure reduces coke accumulation by suppressing the oligomerization and by hydrogenating the adsorbed species. Higher temperatures give more coke deposition because the dehydrogenation reactions become more favored, even though adsorption would be reduced at higher temperature. Since increased temperature also speeds up the coking reactions, these catalysts are more likely to have a large coke concentration at the outer edge of the catalyst pellets. Similarly, larger pellet sizes favor an uneven distribution of coke within a pellet. In fixed reactors, larger extrudates are used to control pressure drop, and the reactors are operated in a rising–temperature mode to maintain constant catalyst activity. Both of these factors contribute to a higher concentration of coke at the outer edge of the catalyst pellets. In contrast, the smaller ebullated bed catalysts which are exposed to isothermal conditions gave a flat profile of coke with pellet radius [Thakur and Thomas, 1985]. A mean pore diameter of at least 10–15 nm in macropores is recommended to minimize diffusion resistances.

5.4.2 Metal deposition on residue conversion catalysts

Metals accumulate more slowly on the catalyst surfaces (Figure 5–5), mainly because the inlet concentrations of metals are lower than for coke precursors. The accumulation of metals can be even greater than coke, for example the vanadium concentration can reach 30–50 wt% of the catalyst, on a fresh catalyst basis [Thakur and Thomas, 1985; Myers *et al.*, 1989]. Demetallation reactions can be considered autocatalytic, in the sense that once the surface of the catalyst is covered with metal sulfides, the catalyst remains quite active and continues to accumulate metal sulfides. The final loss of catalyst activity is usually associated with filling of pore mouths in the catalyst by metal sulfide deposits (Figure 5–5).

An example of deposition of metals inside catalyst pellets is illustrated in Figure 5–6 for Co/Mo on γ–alumina catalyst [Pereira *et al.*, 1990]. The catalysts were selected for large macropores in the γ–alumina; 5000 Å for Catalyst A and 1500–2500 Å for Catalyst B. The micropore diameters were 76 and 34 Å respectively. For comparison, the vanadium–bearing porphyrins have a diameter of ca. 25 Å. These molecules would move easily in the macropores, but with much more difficulty in the micropores. As the catalysts were used to process Lloydminster vacuum residue, the metal sulfides accumulated until about 10 wt% of the initial mass of catalyst had been deposited. These sulfides had a density of 4–4.5 g/mL. The profiles of Figure 5–6 show that the metals tend to deposit at the exterior of the catalyst pellets. The deposition was more even, using the interior volume more

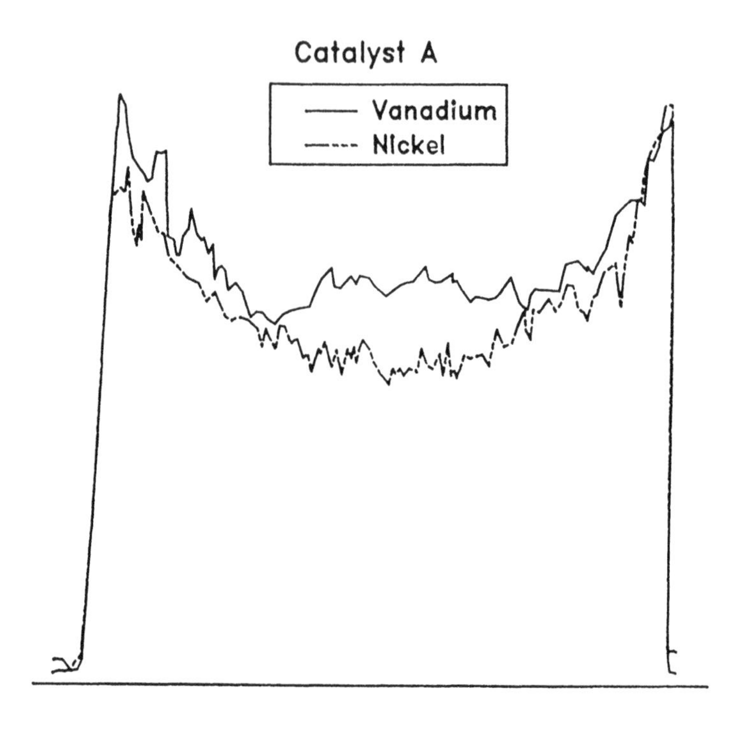

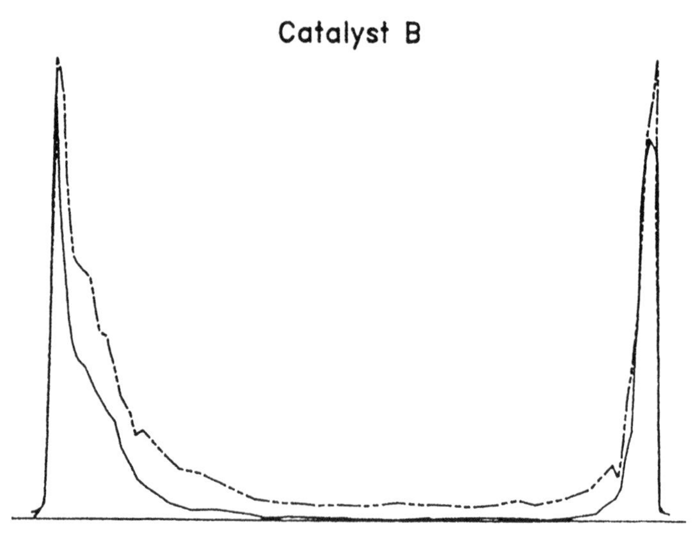

Figure 5–6

Typical nickel and vanadium profiles in catalyst pellets. Top – Catalyst A; Bottom – Catalyst B (Data from Pereira *et al.*, 1990).

efficiently, when the macropore volume was larger. Catalyst A had a macropore volume of 0.402 cm^3/g, compared to Catalyst B with 0.030 cm^3/g, giving much better performance. The micropore volumes were similar at 0.43—0.46 cm^3/g. Although the availability of macropores is significant to maintain access to the interior of the catalyst, the macropores were not significantly filled. Blockage at the entry to micropores is sufficient to explain the results of Figure 5—6. Catalyst A had a more extensive network of macropores so that more micropores were connected directly to macropores. Blockage of one micropore, therefore, would close off little of the catalyst volume. In Catalyst B fewer micropores would be connected directly to a macropore, so that blockage of one would disable proportionally more of the catalyst volume.

As in the case of coke deposition, larger catalyst pellets and higher operating temperatures increase the tendency for metal sulfides to deposit at the periphery of the pellet, as in Catalyst B. Higher hydrogen pressure also increases the rate of metal conversion, which increases the tendency toward high peripheral concentrations. The problem of maintaining good catalyst activity in the presence of metal sulfide deposits, therefore, is a challenging combination of chemistry and transport, and has led to the development of catalysts with improved pore networks to sustain deposition of metals for longer periods of time.

5.4.3 Kinetics of deactivation of demetallization catalyst

The main issue in the design of demetallation catalysts is not the active sites, but the access of the reactants to the interior of the pellet. Unlike coking, which tends to affect the number of active sites, accumulation of metals slows the rate of diffusion within the catalyst and reduces the accessible surface area. Pereira [1990] modeled catalyst activity for demetallization via the effectiveness factor, using the following expression for effective diffusivity:

$$D_e = D_b[\epsilon_M(1-(a/r_{p,M}))^4 + \epsilon_m(1-(a/r_{p,m}))^4]/\zeta \tag{5.1}$$

where D_b is the bulk diffusivity, ζ is the tortuosity factor, a the hydrodynamic radius of the diffusing molecule, ϵ porosity and r_p the radius of the pores. The subscript *M* indicates macropore properties, while *m* indicates micropore properties. The mean radius of the macropores was ca. 5000 Å, and the mean radius of the micropores was ca. 100 Å.

As the metal sulfides accumulate, the radius of the micropores will be most dramatically reduced, while the macropores will be relatively unchanged. Hence, the macropore radius and porosity are approximately constant with time, while the micropore radius decreases ($1 > r_m/r_m(t=0) > 0$).

The rate of demetallation is assumed to be first order in metal concentration in the liquid. The governing equation is

$$\frac{d}{dx}\left[D_e \frac{dC}{dx}\right] = -r = kA_pC = 2\left[\frac{\epsilon_M}{r_{p,M}} + \frac{\epsilon_m}{r_{p,m}}\right]kC \tag{5.2}$$

where $-r$ is the local volumetric rate of reaction, and A_p is the surface area available for reaction per unit volume. In this case, the reaction rate constant would include the effect of the initial coke deposition on catalyst activity. Assuming cylindrical pores, surface area $= 2\epsilon/r_p$. The boundary conditions are:

$$dC(0)/dx = 0 \tag{5.3}$$

$$C(L) = C_o \tag{5.4}$$

Equation (5.3) is a symmetry condition at the center of the particle ($C(0)$), while Equation (5.4) is the surface concentration. The rate of metal deposition in the pores with a first-order reaction is

$$dr_{p,i}/dt = -kC/\rho_{metal} \qquad i = m, M \tag{5.5}$$

The rate constant k includes the term for the partial pressure of hydrogen.

Pereira [1990] solved equation (5.2) to obtain values for $C(x)$ as a function of time and for the distribution of metal with x, subject to equations (5.1) and (5.5), by a regular perturbation approach. An example result is illustrated in Figure 5–7, which shows the nondimensionalized rate of reaction, r/Λ_p, as a function of nondimensionalized time (θ) and average micropore radius (r_m), where in this case the definition of the parameters is

$$\Lambda_p = D_b \epsilon_m^o C_o / \zeta L^2 = 1.9 \times 10^{-9} \text{ mol/cm}^3\text{–s} \tag{5.6a}$$

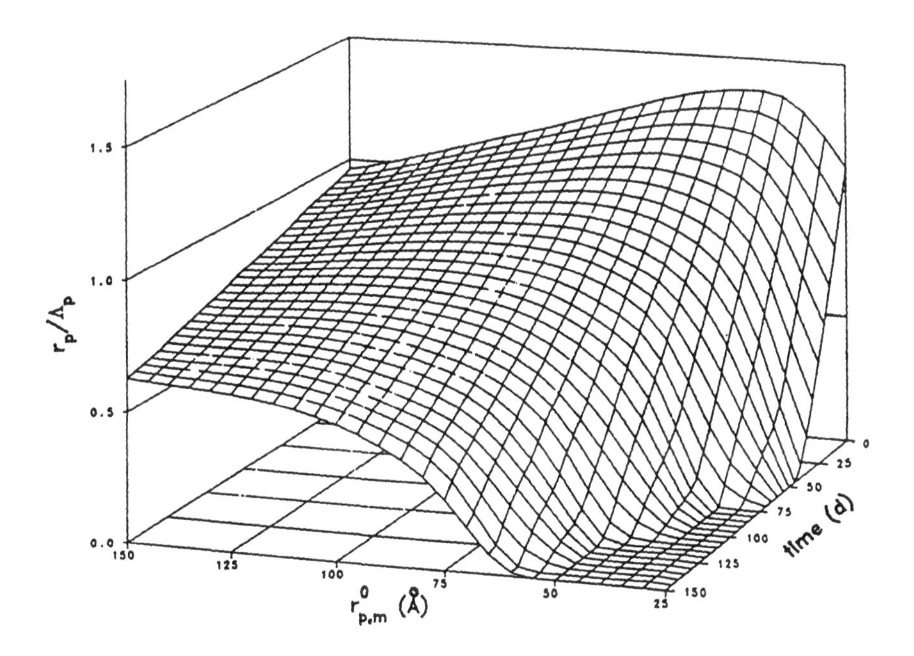

Figure 5–7

Rate of reaction (as r/Λ_p) for metals removal as a function of time and micropore radius ($r_{p,m}{}^0$) for a unimodal catalyst. $\beta = 0$, a = 10 Å, $\phi = 0.00152/\sqrt{r_{p,m}}$. *(Reprinted from C.J. Pereira, "Metal deposition in hydrotreating catalysts. 1. A regular perturbation solution approach", Ind. Eng. Chem. Res. vol 29, p. 517, 1990. Copyright by the American Chemical Society.)*

$$\beta = \epsilon^{o}_{M} / \epsilon^{o}_{m} \tag{5.6b}$$

$$\phi^2 = 2k\zeta L^2 / r^{o}_{m} D_b \tag{5.6c}$$

The parameter ϕ^2 in this case is a modified Thiele modulus, the ratio of the rate of reaction to the rate of bulk diffusion. The normal Thiele modulus is the ratio of the ratio of the rate of reaction to the effective rate of diffusion. The strength of Pereira's approach is that it specifically includes both macropores and micropores, thereby giving a tool that can predict demetallization performance.

5.4.4 Activity as a function of time–on–stream

The diffusion–reaction model of Pereira [1990] indicates the deposition of metals within a catalyst pellet. The more empirical approach to analyzing catalyst activity is to consider the activity with time–on–stream. As the catalyst ages, continued deposition of metals in the catalyst pellets gives a loss of activity. Typical activity profiles are illustrated in Figure 5–8 as a function of time on stream. The catalyst activity for residue conversion drops rapidly during the first 25–50 h (in this example) due to coking, then stays constant for hundreds of hours. This behavior is consistent with an initial deposition of coke on the catalyst, followed by conversion mainly via thermal (noncatalytic) reactions.

The density of the product liquids (Figure 5–8) was related to the overall hydrogenation activity of the catalysts. Catalyst action reduces density by removing metals and heteroatoms, and by hydrogenating aromatics. The activity of Catalyst A, for example, changed rapidly during the first 200 h, then stabilized until about 1200 h. After the jump in density to ca. 0.965, the performance of the reactor was equivalent to a run without catalyst, i.e. thermal conversion only. At 1200 h, therefore, metal deposition had essentially shut off all the catalyst pores. Catalyst D showed a similar jump in density at 900 h, signaling the end of its useful life.

One of the challenges in interpreting data for catalyst activity as a function of time, as in Figure 5–8, is to determine the contributions of coke and metal deposits to deactivation. Model compound studies using vanadium porphyrins suggest that vanadium accumulations in the range of 0–12 wt% have little impact on hydrogenation activity [Dejonghe *et al.*, 1990]. In this case the hydrogenation activity of the catalyst was based on toluene hydrogenation after the metals were deposited. When the catalyst was exposed to polyaromatic hydrocarbons, such as phenanthrene, then hydrogenation activity was decreased much more

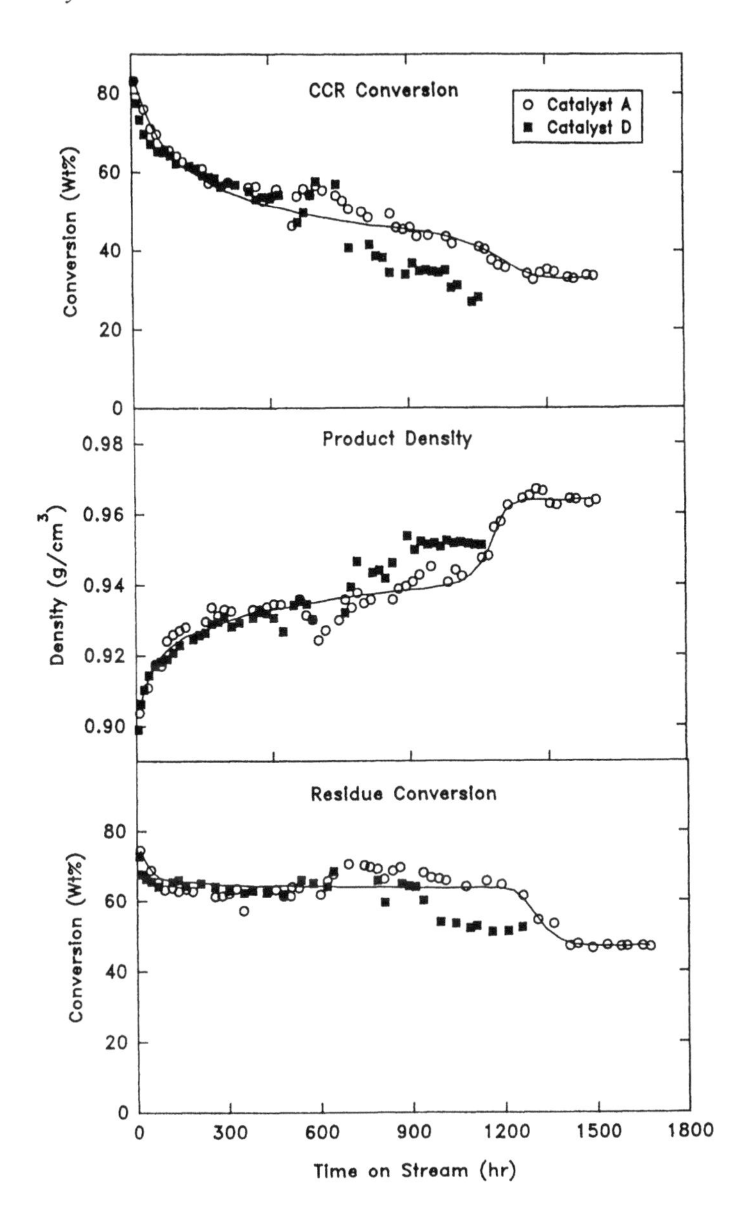

Figure 5–8

Catalyst activity with time in hydroconversion of residue in an agitated flow reactor. The feed was a blend of Lloydminster and Cold Lake atmospheric residues (399° C+) processed over a Ni–Mo on γ–alumina catalyst. Curves show the trend of activity for catalyst A. (Data from Oballa *et al.*, 1992) significantly. Bartholdy and Cooper [1993] attributed 60% of the deactivation to metals and 40% to coke in processing a Kuwait residue in a fixed–bed reactor. Although coke was deposited early in the experiment, they suggested that the coke characteristics changed with time and increasing operating temperature. Since fixed–bed reactors are operated for constant conversion, coke deposition can occur late in the catalyst life as the reactor conditions become more severe.

Myers *et al.* [1989] have suggested modeling deactivation of catalyst by simultaneous deposition of coke and metals by assuming that the deposits reduce activity by reducing pore volume. The catalyst activity, therefore, can be represented as a linear function of the remaining pore volume of the catalyst:

$$\alpha_c = c_1 \left[\frac{PV_0 - c_2[C] - c_3[M]}{PV_0} \right] + c_4 \tag{5.7}$$

where α is activity ratio (activity at time t divided by initial activity), PV_0 is the initial pore volume of the catalyst, [C] is the carbon concentration, [M] is the metals concentration, and c_1 to c_4 are empirical constants. The effective coke density in the catalyst pores was 0.8 g/cm^3, while the density of the metal sulfide was 3.4 g/cm^3.

The data in Figure 5–8 were obtained from a bench–scale Robinson–Mahoney type agitated catalytic reactor. These profiles could be used as a first approximation to estimate catalyst activity in an actual reactor. In the case of a fixed–bed unit, the time–on–stream data would correspond to the activity of the upper layer of the catalyst bed. By transforming the time axis into a metal–loading axis, based on a material balance, such data could be used to predict metal loading and reaction rate as a function of position within the reactor.

In the ebullated–bed systems, a portion of the catalyst is replaced daily so that the catalyst age distribution is given as follows:

$$dE(t) = \frac{e^{-t/\tau_c}}{\tau_c} dt \tag{5.8}$$

where dE(t) is the fraction of the catalyst with an age between t and t+dt, and τ_c is the mean age or mean residence time of the catalyst in the reactor. Equation (5.8) assumes that the catalyst is well–mixed in the reactor, following the study of McKnight and Nowlan [1993] who showed that catalyst was well mixed in an industrial ebullated bed reactor.

$$\tau_c = \text{Total catalyst in reactor/Rate of replacement} \tag{5.9}$$

This age distribution can then be used to calculate average activity as a function of catalyst replacement rate. In an ebullated bed the liquid and catalyst phases are well mixed, approximating a continuous stirred tank reactor. The overall rate of reaction i is given by the following equation:

$$r_i = X_i\, C_{i,0}/\tau \tag{5.10}$$

The same rate is also the sum of the rates for all of the catalyst pellets in the reactor at any time.

$$r_i = \int_0^\infty r(t)\, dE(t) = \int_0^\infty r(t) \frac{e^{-t/\tau_c}}{\tau_c}\, dt \tag{5.11}$$

Here r(t) is the rate of reaction as a function of time-on-stream, t. From equation (5.10), at any time the conversion in the bench-scale CSTR is linearly proportional to the rate of reaction, so that equation (5.11) can be rewritten in terms of conversion:

$$X_i = \int_0^\infty X(t) \frac{e^{-t/\tau_c}}{\tau_c}\, dt \tag{5.12}$$

where X(t) is the conversion after time t on stream, as in Figure 5-8. Integrating equation (5.12) for the CCR conversion data from Figure 5-8 gives the predicted overall conversion for the ebullated bed as a function of average catalyst, τ. The plot of overall conversion versus mean catalyst age (Figure 5-9) has a smaller slope than does the point measurement from the bench-scale reactor because the ebullated bed includes the contribution of catalyst with ages from zero to very long times.

5.4.5 Catalyst regeneration

Spent residue conversion catalyst is difficult to regenerate due to the presence of nickel and vanadium sulfide deposits from the oil feed. Coke deposits can be removed by calcining the catalyst at controlled temperature, but combustion leaves vanadium and nickel oxides.

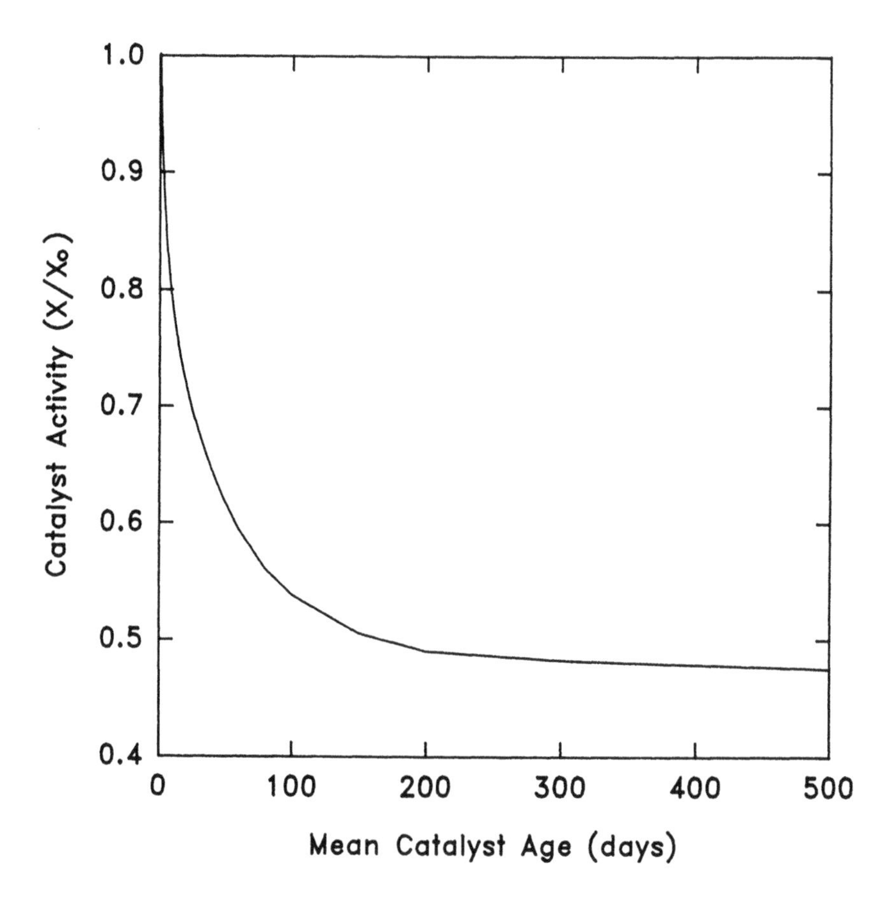

Figure 5–9
Catalyst activity in an ebullated bed reactor as a function of mean age of the catalyst

Leaching of catalyst in oxalic acid can give significant recovery in activity [Stanislaus *et al.*, 1993], provided that the leaching is controlled to avoid excessive leaching of molybdenum. Such regeneration, however, has not been widely used on a large scale. More severe acid leaching can completely remove the V, Ni, and Mo from the catalyst, leaving inert alumina [Jocker, 1993; Inoue *et al.*, 1993]. The extracted metals can then be separated and concentrated for reuse.

5.5 Additive–Based Processes

Additive–based processes offer an alternative to fixed– and ebullated–bed catalytic processing, and are founded on different objectives. Finely divided solids or soluble compounds are added to the feed. These additives are then carried through the reactor and exit the process with the unconverted residue fraction. The additives may contribute catalytic activity within the reactor, but they are not retained there.

5.5.1 Process description

Rather than attempting simultaneous HDS, HDN, HDM, and cracking reactions in a single reactor, the additive–based processes are normally designed with the objective of maximizing the conversion of residue. Downstream reactors are then used to treat the liquid products to obtain desirable characteristics of low sulfur and nitrogen content. The aim is to cut catalyst consumption in the primary reactor, at the expense of some addition to downstream processing load. Although some catalytic additives might be capable of giving HDS performance comparable to supported–metal catalysts, cost restricts the types of compounds that can be added.

The major emphasis on selecting the type and amount of additive is the control of coke formation in the reactor. A number of materials have been tested, including powdered coal impregnated with iron salts, iron oxide waste, and organometallic compounds such as metal naphthenates. At reactor conditions, naphthenates decompose to give metal sulfides. These sulfides are in a colloidal state, with particle sizes ranging from 20 to 250 nm [Kim *et al.*, 1989].

When finely powdered or dispersed catalyst is used, the reactive phase is a slurry or suspension, and high turbulence is not required to suspend the particles. These processes, therefore, use an upflow tubular reactor, where the liquid or liquid suspension of additive flows upward with the hydrogen gas. Recovery of the highly dispersed catalyst is usually not practical, and the use of iron–based additives is so inexpensive that the additive can be discarded.

An example of this approach is the CANMET process, developed in Canada (Figure 5–10). A number of processes of this type are either in development or in operation, including Veba Combi Cracking in Germany [Wenzel, 1992] and HDH in Venezuela [Guitian *et al.*, 1992]. The CANMET process uses powdered iron sulfate as an additive. The additive (or moderate activity catalyst depending on terminology) is added at the rate of 1–2% by weight of feed. The additive inhibits coke formation on the wall of the reactor, however, coke will tend to agglomerate on the particles. The additive remains in the unconverted residue fraction, forming a slurry which must then be burned or further processed. Table 5–4 gives the performance of the CANMET process in processing Cold Lake residue.

Table 5–4
Conversion of Cold Lake Residue by CANMET Process
[Pruden *et al.*, 1989]

Feedstock Properties

80% Cold Lake vacuum bottoms + 20% IPL vacuum bottoms, 3.3° API gravity

Asphaltenes	18.9 wt%
Solids	0.1 wt%
CCR	21.0 wt%
Nitrogen	0.56 wt%
Sulfur	4.9 wt%

Products	**Pitch Conversion, wt%**			
	70	77	79	86
C_1–C_4, wt%	5.7	6.9	7.3	8.6
Naphtha, vol%	18.8	22.6	23.7	27.6
Distillate, vol%	30.1	33.9	34.9	37.9
Gas Oil, vol%	33.1	33.1	33.1	33.1
Pitch (524° C+),vol%	25.6	19.1	17.4	11.3

The Veba Combi–Cracking process (Figure 5–11) uses a similar primary reactor (Liquid Phase Hydrogenation, LPH), followed by a secondary catalytic hydrotreating reactor (Gas Phase Hydrotreating, GPH) for treating the vapor product from the main reactor. Vapor phase processing offers higher conversion, in general, than liquid phase. A test of Cold Lake heavy oil achieved 94% conversion at 440–490° C and 15–25 MPa. The hydrogen consumption was approximately 305 m^3/t. Like the CANMET process, Veba uses a disposable additive in the liquid–phase reactor, and the unconverted bottom stream from the primary (LPH) reactor contains solids, both from the original additive

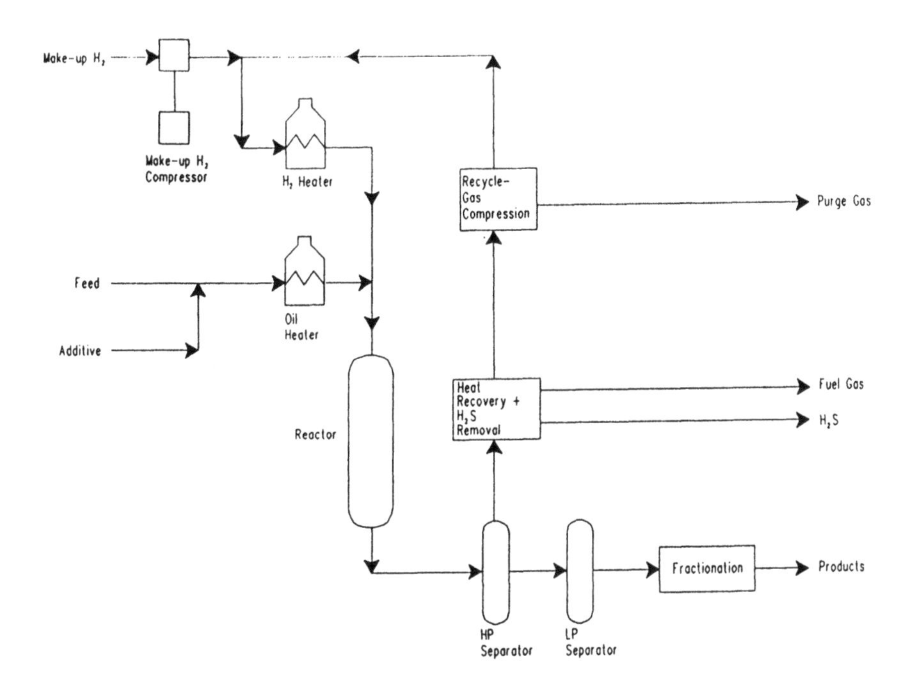

Figure 5—10
Process schematic for CANMET hydrocracking of residue feeds (From Pruden *et al.*, 1989)

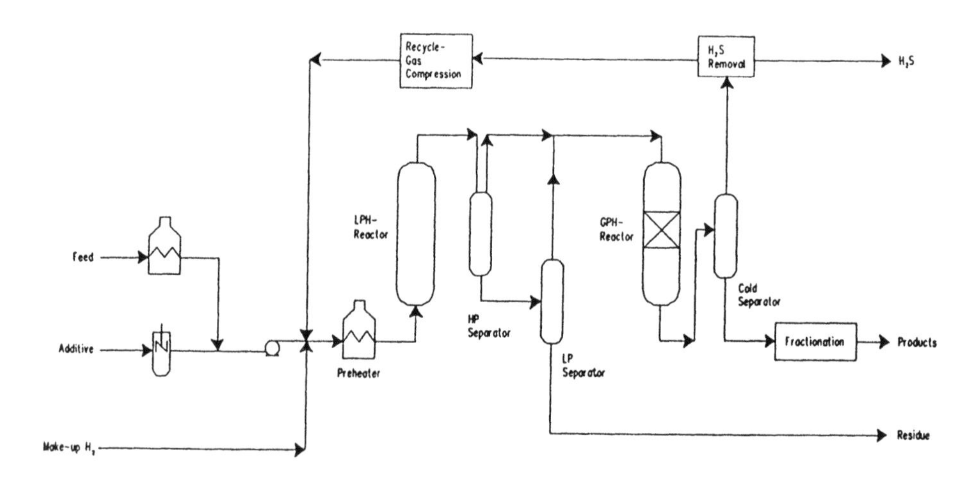

Figure 5–11
Veba Combi–Cracking Process (From Wenzel, 1992)

and from coke material formed in the reactor.

5.5.2 Reactor hydrodynamics

Unlike the ebullated–bed reactors, solids in the additive based processes are fine enough that the liquid/solid slurry can be treated as a homogeneous phase. Consequently, these reactors can be treated as conventional bubble reactors, with liquid feed from the bottom. Although bubble reactors have been studied extensively, almost all of the work has been with air/water or air/solution mixtures. Little work has been performed at elevated pressures, at the fluid densities appropriate to hydroconversion operations.

Özüm *et al.* [1989] used radioactive tracers to measure gas and liquid dispersion in a 25 mm x 670 mm tubular reactor, operating with Cold Lake vacuum bottoms feed at 455°C and 17 MPa. Using a dispersion model [Fogler, 1986] to fit data from gas and liquid tracers, they obtained Peclet numbers for the liquid in the range of 0.67 to 0.77, and for the gas 8.4 to 9.6. In this case, $Pe = UL_r/D_S$ where the reactor length $L_r = 0.67$ m and D_S was the dispersion coefficient. The small value of Pe for the liquid phase indicated almost perfect mixing, very close to the CSTR case. The larger Peclet number for the gas indicates less backmixing, but since the bubble diameter was unknown it is impossible to say if the results can be scaled up from a 25 mm diameter reactor. In general, industrial bubble columns exhibit much better axial mixing in both gas and liquid phases. Froment and Bischoff [1979] suggest that both phases be treated as well mixed (i.e. CSTR case).

As in the case of ebullated–bed reactors, pressure has a significant effect on gas holdup by suppressing coalescence of bubbles. De Bruijn *et al.* [1988] found that doubling the pressure to 13.9 MPa also doubled the gas holdup to as high as 0.35 (volume fraction of gas in reactor; Figure 5–12). Özüm *et al.* [1989] found gas holdups in the range of 0.27–0.28 over a range of gas and liquid flows. Both of these studies were in small reactors, wherein bubble diameter was almost certainly a significant fraction of reactor diameter. The applicability of these results at large scale is uncertain. Froment and Bischoff [1979] give equations for bubble size as a function of liquid properties and interfacial tension, but these methods have not been tested at high pressure.

5.6 Sediment Formation

The formation of solid sediments, or coke, during hydrocracking of bitumen and petroleum resids can be a major limitation on reactor performance and yield. For example, conversion of 95% of the residue fraction of Athabasca bitumen by the H–Oil process gave 50% solids in

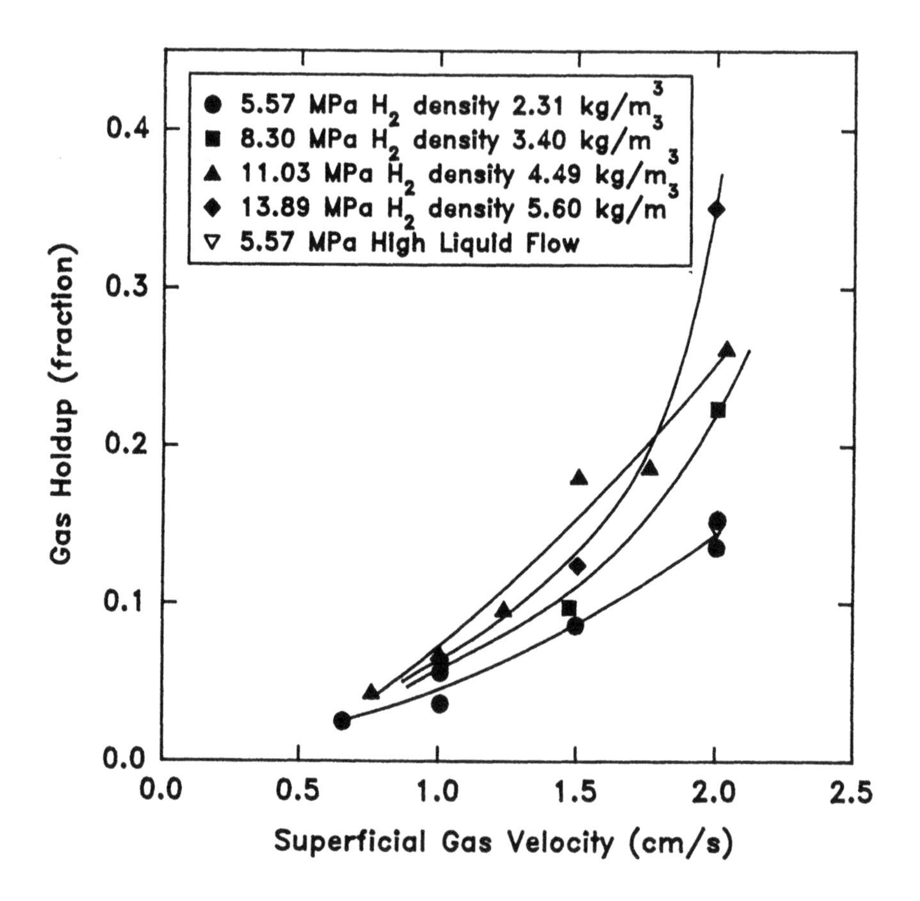

Figure 5–12
Gas holdup in a tubular reactor with Venezuelan residue at 300° C. The reactor dimension was 5.08 cm x 240 cm, and the liquid velocity was 0.04 cm/s (Data from De Bruijn *et al.*, 1988)

the +520° C product [Johnson *et al.*, 1977]. Mochida *et al.* [1989] observed both brown asphaltic particles and blue wax particles in the solids from catalytic hydroconversion of an Arabian vacuum residue.

The presence of these different types of solids shows that solubility controls solids formation. One view is to consider solids formation in terms of ternary phase behavior, as shown in Figure 5—13. The tendency for solid formation changes in response to the relative amounts of the light ends, middle distillates, and residues, and also to their changing chemical composition as reaction proceeds. A similar scheme was proposed by Takatsuka *et al.* [1988] with saturate, resin, and asphaltene fractions as components of the diagram (Figure 5—14). The limits indicated by Takutsaka *et al.* refer to the formation of sediment or solids in the process system. At high pressure, the light components be major components in the liquid phase and hence are included as a key pseudocomponent in Figure 5—13.

The tendency to form solids is normally considered as a sort of precipitation of asphaltenes from the oil as conversion shifts the solvent properties of the oil mixture. The asphaltenes, however, decompose or melt long before hydrocracking temperatures are achieved, even in the absence of any lighter oil fractions to serve as solvents. Hence the solids which precipitate from n—pentane or n—hexane at room temperature are not necessarily solids in the reactor at over 400° C. A much more reasonable model is that the liquids in the reactor split into a light, aliphatic phase, and a heavier, more aromatic phase. Recent work by Shaw and coworkers [Shaw *et al.*, 1988] with pyrene and tetralin suggests that liquid—liquid—vapor systems form at reactor conditions even with simple solvents. At 427° C and 12 MPa, for example, the two liquid phases in equilibrium contained 40 wt% pyrene and 70% pyrene (Figure 5—15). This phase splitting should also follow the ternary diagram (Figure 5—13) discussed above, depending on the chemical changes in each of the constituents.

The formation of a heavy, aromatic liquid would be the first step toward solid or coke formation. The hydrogen solubility in such a material would be much lower than in the light phase [Shaw, 1987], leading to poorer hydrogen transfer and an increase in condensation and aromatization reactions. Cooling this heavy phase at the reactor outlet could lead to the observed sediments or solids described for hydrocracking reactors. Continued reactions with a deficiency of hydrogen would lead to formation of particulate solids within the reactor, and deposition of coke on the reactor internals.

As hydrocracking proceeds, the heavy materials are progressively stripped of aliphatic side chains and become more aromatic in character [Gray *et al.*, 1989]. Therefore, a liquid phase that was initially

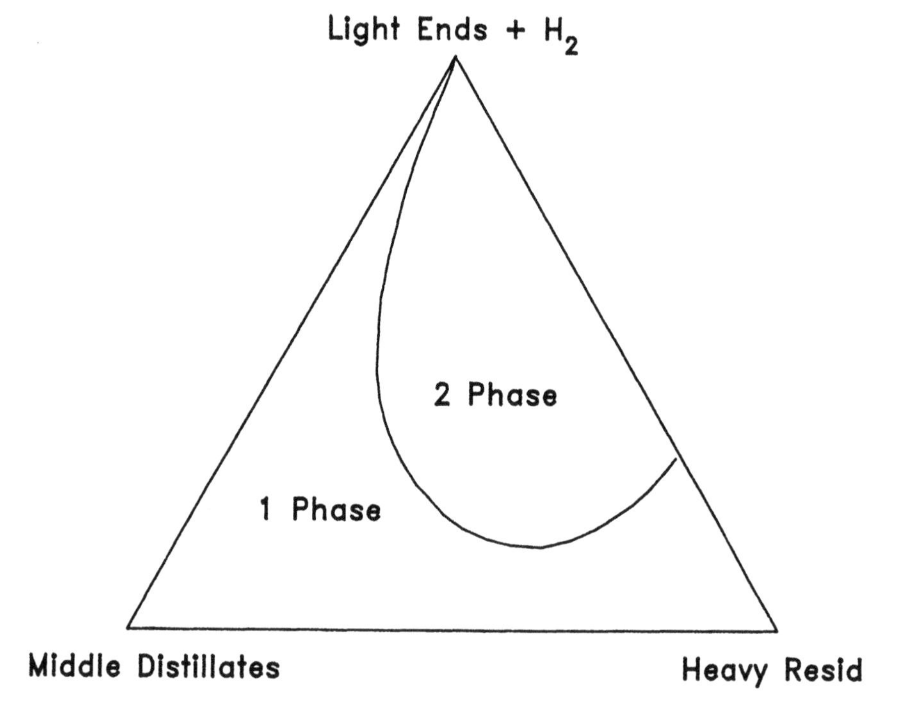

Figure 5—13
Hypothetical ternary diagram for liquid—liquid miscibility during hydroconversion

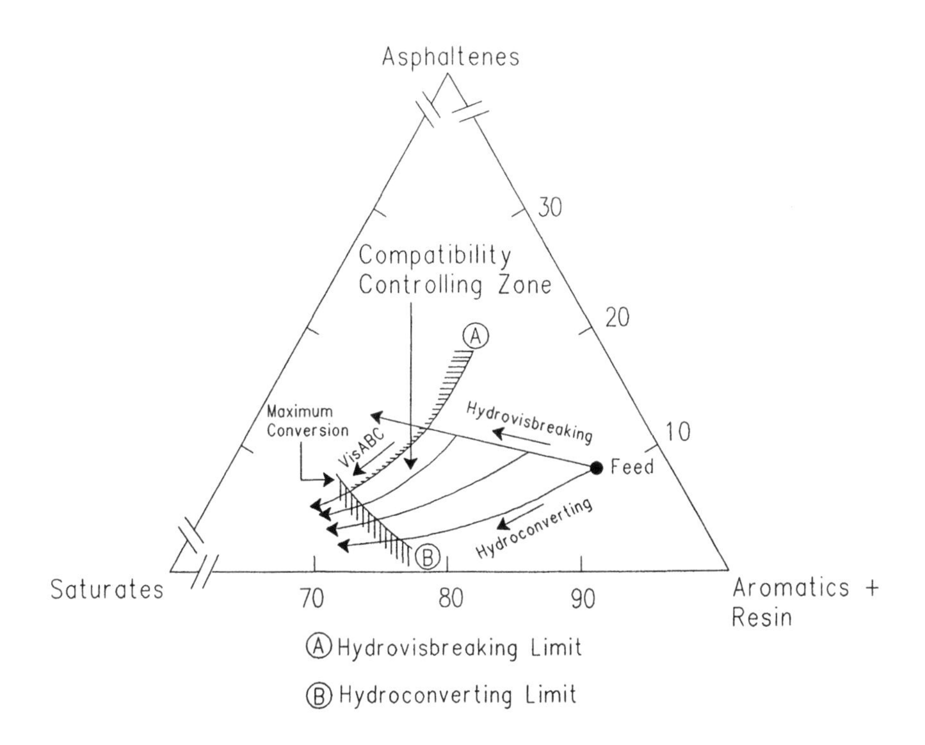

Figure 5—14
Phase diagram for hydrovisbreaking process, showing limits due to solids
(After Takatsuka *et al.*, 1988)

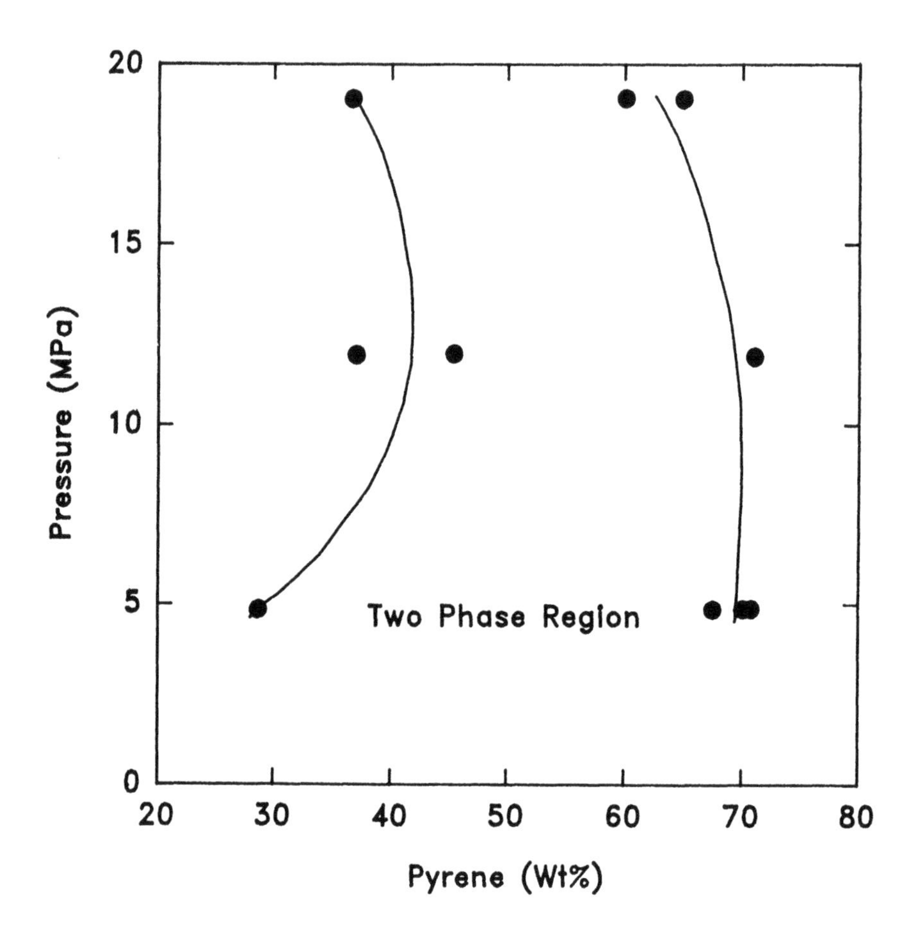

Figure 5–15
Liquid–liquid equilibrium in the pyrene–tetralin system (Data from Shaw *et al.*, 1988)

homogeneous at the reactor inlet could develop a second liquid phase as the reactions progressed. Phase splitting would be most likely to occur at high conversion levels, where some heavy fractions become chemically dissimilar from the bulk composition of the oil. Because these phenomena would occur at high pressure and temperature, they may not correspond to our expectations based on low pressure properties of bitumen.

A variety of approaches have been tested to control solids formation, typically by modifying the composition of the feed mixture or the recycle. The addition of hydrogenated middle distillates increases conversion and suppresses solids formation. Such methods have been known for many years (see for example Carlson *et al.*, 1958), and the success of this approach is normally attributed to the donation of hydrogen from the solvent to the bitumen. The additive-based processes use powdered additives to control the location of any deposition of solid material.

An alternate approach is the Eureka process, which controls solids formation at low pressure and 415—430° C by stripping the cracked products from the reactor with superheated steam [Aiba *et al.*, 1981; Sumida *et al.*, 1984]. The remaining pitch has reduced coking propensity because heavy aromatic compounds do not tend to separate in the reactor.

Catalytic processes, such as H—Oil, control coke formation by the hydrogenation activity of the catalyst, which mainly serves to transfer hydrogen to the heavy ends [Miki *et al.*, 1983]. The ebullating bed configuration also tends to remove the lighter cracked products, leaving a more homogeneous heavy material in the liquid phase to react. Solids formation can also be suppressed by adding an aromatic fraction to the feed to serve as a solvent for the solids [Beaton and Bertolacini, 1991].

A quite different approach is to use catalytic reactions to modify the least soluble components, and thereby suppress solids. In order to achieve significant hydrogenation of aromatics, a low—temperature hydrogenation ($\lesssim$ 390° C) is required prior to high severity cracking [Mochida *et al.*, 1990]. This concept will be discussed in more detail in section 5.8.

5.7 Kinetics of Hydroconversion Reactions

5.7.1 Residue conversion

Although conversion of high—boiling material is a central aim of hydroconversion, few studies of the kinetics of the reaction have been reported. A number of authors have presented models based on SARA analysis (e.g. Koseoglu and Phillips, 1988), but these models do not provide any insight into the key issue of conversion of nondistillable residue into distillates. Following on the kinetic theory presented in Chapter 4, two issues must be considered: first, what characteristics or components of the feed residue determine the observed rates of reaction and ultimate conversion? Even when residues have very similar composition, the conversion at identical conditions will differ due to the details of the chemical components. The second question is how can we analyze the kinetics given the insight that the residue is a complex mixture with a wide range of reactivity for cracking?

5.7.1.1 *Structure–reactivity relationships in residue conversion*

Given the limited data that chemical analysis can provide about a residue fraction, it is not surprising that only two studies have suggested quantitative structure–reactivity relationships. Gray *et al.* [1991] reacted four Alberta bitumens at identical conditions over a Ni/Mo on γ–alumina catalyst and found that the only variable that correlated the observed rate of residue conversion was the concentration of carbons α to an aromatic ring in the feeds (Figure 5–16). In a study of three Venezuelan vacuum residues, Guitian *et al.* [1992] found that residue conversion increased monotonically with sulfur content (Figure 5–17). Clearly, these studies of feeds from a single geological basin need to be extended based on data for a wider range of materials.

5.7.1.2 *Reaction order for residue conversion*

Recalling section 4.4 on kinetic analysis of reactions of mixtures, we would expect the apparent order of residue conversion to depend on the feed, the reactor type and the extent of conversion. High conversion in tubular reactors, without backmixing, should give the highest apparent order (approaching 2). Several studies have shown that residue conversion increases linearly with temperature, as illustrated in Figure 5–18 for two additive–based processes. The exact temperatures were not given, but Figure 5–18a for Veba Combi Cracking likely covers the quoted range of operation of the process from 425°C to 480°C. The HDH process (Figure 5–18b) would have a base temperature of T= 410 to 420°C.

If the residue is lumped as a single component, then the data in Figure 5–18b can be analyzed in terms of a simple kinetic model. Assuming first–order kinetics, constant liquid flow rate, and a fully backmixed liquid phase, the rate constant for reaction i is a function of residence time and conversion as follows:

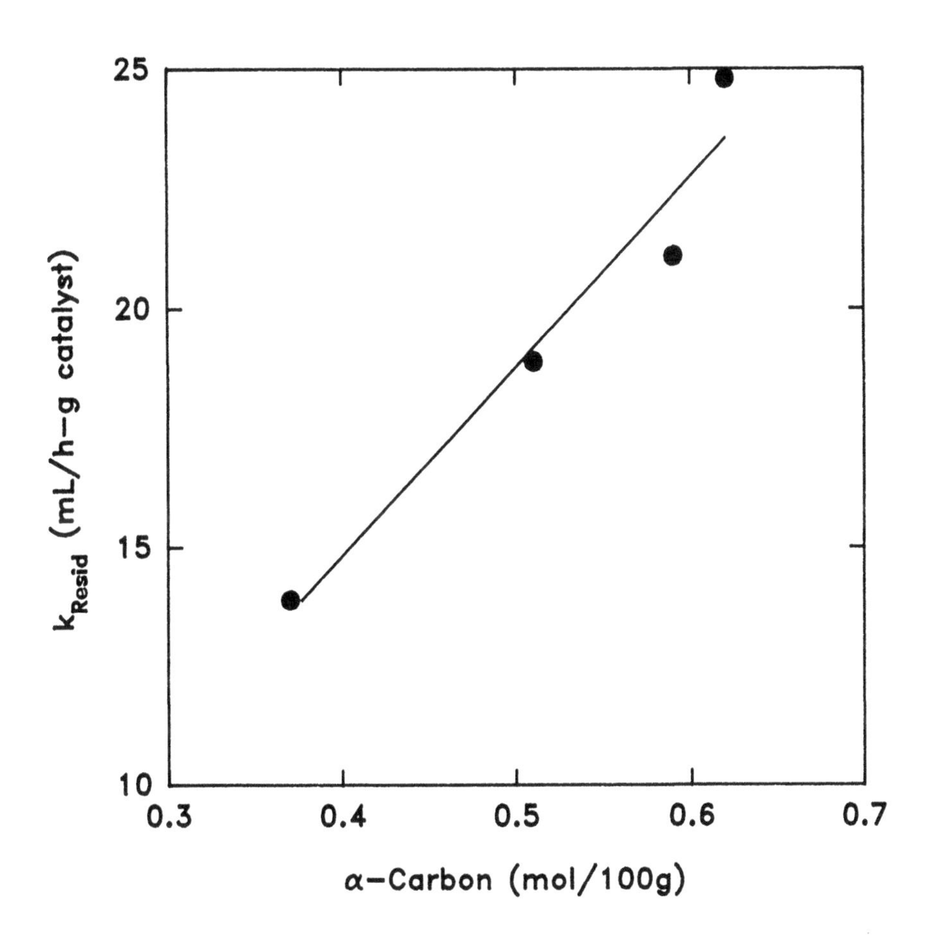

Figure 5–16
Reactivity of Alberta residues as a function of carbon α to aromatic rings (Data from Gray *et al.*, 1991)

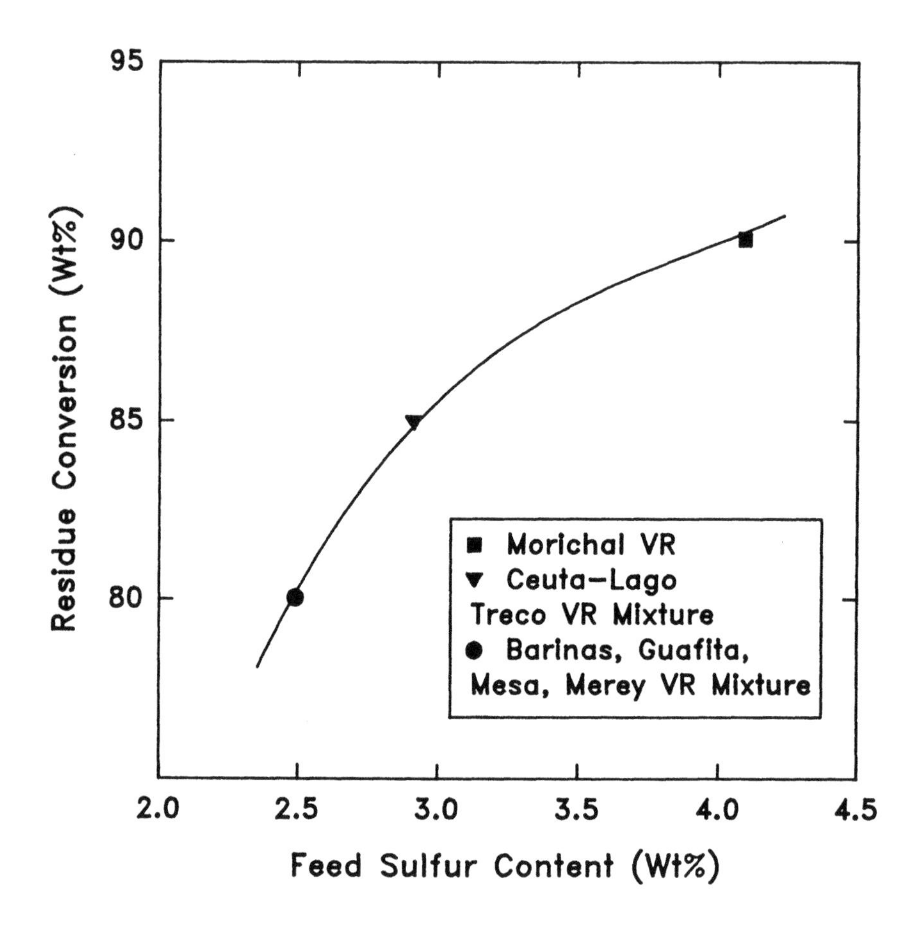

Figure 5—17
Reactivity of Venezuelan residues as a function of sulfur content (Data from Guitian *et al.*, 1992)

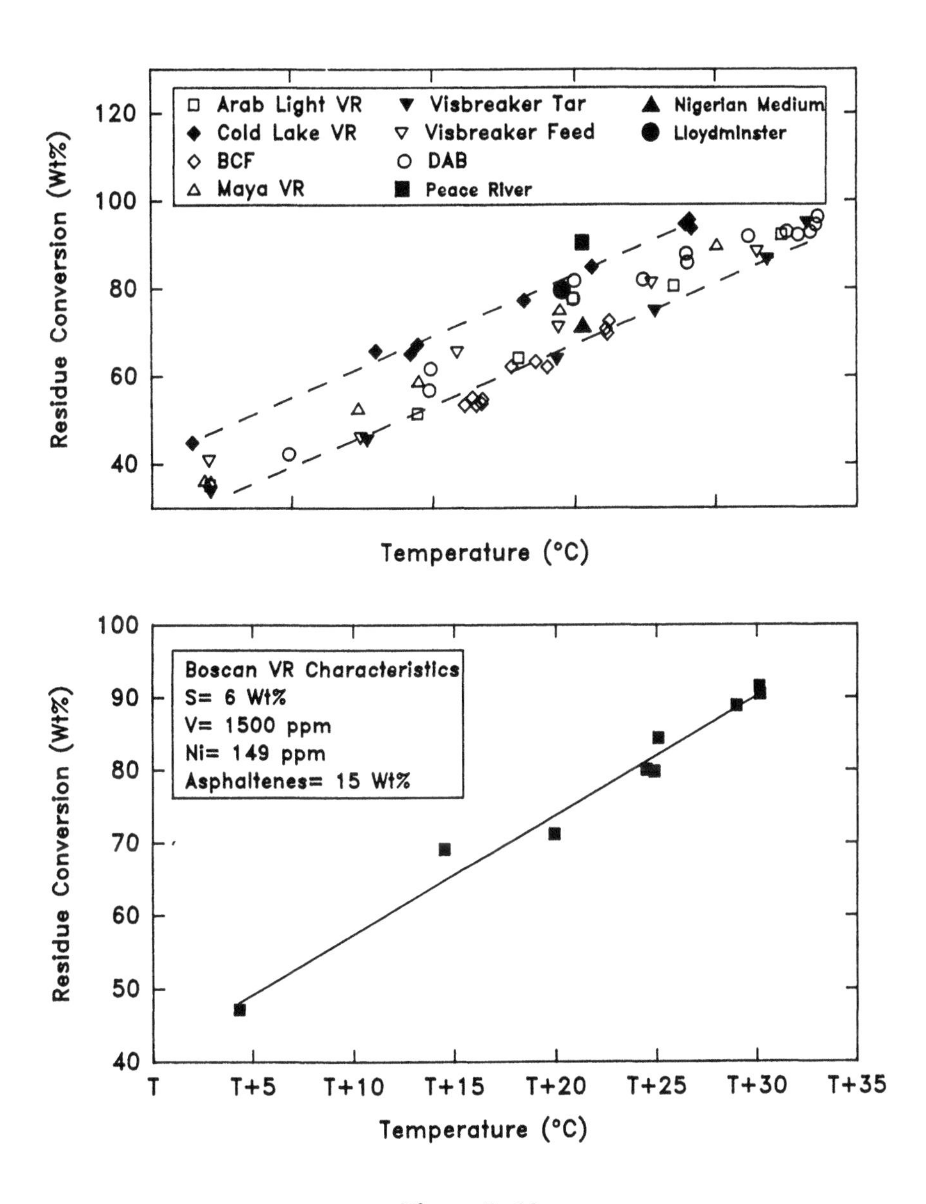

Figure 5—18

Residue conversion as a function of reaction temperature (Data in upper panel from Wenzel, 1992; Data in lower from Guitian *et al.* 1992)

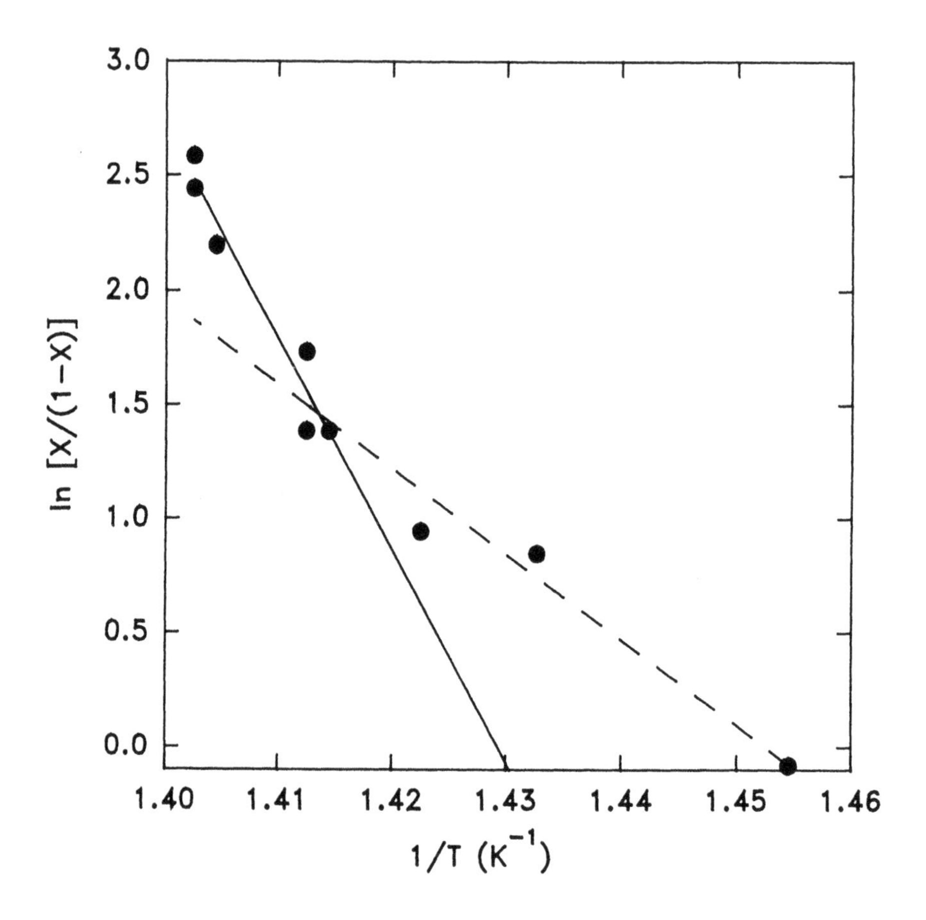

Figure 5—19
Arrhenius plot for data from Guitian *et al.* [1992] assuming a well mixed reactor and first—order kinetics

$$k_i = \frac{X_i}{\tau\ (1 - X_i)} \tag{5.13}$$

Assuming the Arrhenius form for the rate constant k_i, a plot of $\ln[X_i/(1-X)_i]$ versus 1/T should be linear at constant residence time, τ. The data from Figure 5–18b are plotted in this manner in Figure 5–19, and clearly fail to give a linear relationship. A zero–order reaction would give a better linear Arrhenius plot, i.e. a plot ln X versus 1/T from the corresponding equation for a zero–order reaction gave an r^2 of 0.97.

Before attempting to draw any significant inference from this result, the standard checks of reaction order must be used. Equation (5.13) suggests that a plot of τ versus $X_i/(1-X_i)$ at constant temperature should be linear if the reaction is first order. Alternately, the rate of reaction in the CSTR can be plotted against the outlet concentration of the reactant, C_i, at constant temperature:

$$r_i = \frac{C_{i,0} - C_i}{\tau} = k_i C_i \tag{5.14}$$

This classical approach usually gives the result that the residue conversion is first order overall at moderate conversion levels. For example, Oballa *et al.* [1992] found first–order kinetics for conversion of residue in a Cold Lake/Lloydminster blend, with apparent activation energies ranging from 200–300 kJ/mol for a series of Ni/Mo on γ–alumina catalysts. The same reaction order was observed by Gray *et al.* [1991] for Cold Lake feed at similar conditions.

At higher conversions, the apparent first–order reaction rate constant tends to decrease, so that the apparent reaction order becomes conversion–dependent [Beaton and Bertolacini, 1991]. As cracking proceeds, the nature of the remaining residue changes as illustrated in Figure 5–20. At low conversion, aliphatic side chains crack off easily, giving a reduction in molecular size and a gradual increase in aromatic carbon content in the remaining residue [Heck and DiGuiseppi, 1993]. As conversion continues, naphthenes are cracked, giving little change in molecular size but a more significant change in aromatic carbon. The general trend of increasing aromaticity in the unconverted residue is commonly observed, therefore, departures from linear kinetics should be expected in general at high conversion.

Mosby *et al.* [1986] developed a kinetic model for catalytic hydroconversion by assuming that thermal reactions determine the reduction in boiling point/molecular weight. The catalytic activity was

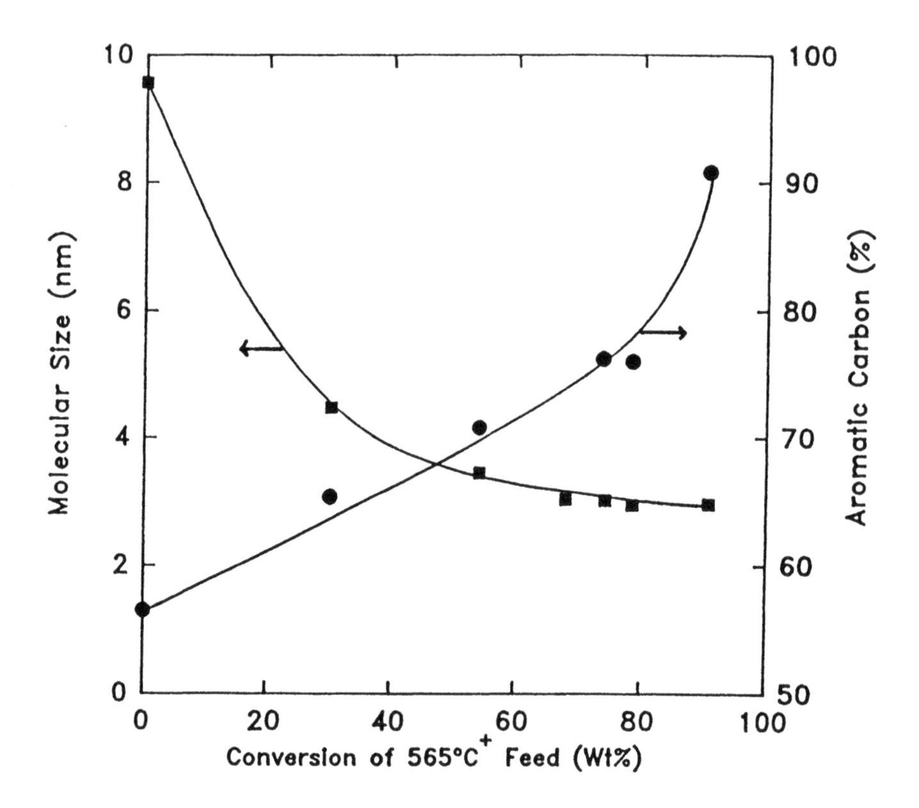

Figure 5–20

Molecular size and aromatic carbon content of asphaltenes as a function of residue conversion. Cracking of Maya 525° C+ residue with 650 ppm $Mo(CO)_6$ in a batch reactor and 13.8 MPa of flowing hydrogen. (Data from Heck and DiGuiseppi, 1993) assumed to act independently to remove sulfur and metals from the residue and the cracked products. The decrease in the rate constant for residue cracking with increasing conversion suggested an empirical separation of residue into "hard" and "easy" fractions. The resulting network is shown in Figure 5—21, along with typical relative rate constants (cracking of "hard" residue to gas oil taken as a basis). As reaction proceeds, more and more of the remaining residue is "hard", corresponding to the aromatic core groups that are difficult to crack apart or hydrogenate. The empirical model of Mosby *et al.* [1986] suffers from the drawback that the distinction between hard and soft residue is not based on any chemical information, nor can it be predicted *a priori.*

5.7.2 Kinetics of sulfur removal

The kinetics of sulfur removal from residues have been studied intensively, and similar problems arise as in the analysis of residue conversion. In the case of sulfur, thermal reactions can account for some portion of conversion because reaction of some residues in the absence of catalyst may result in 30—40% sulfur conversion. The sulfur removal beyond this level of conversion is driven by catalyst activity. This effect is illustrated in Figure 5—22 for sulfur in cracked products from a series of catalysts at identical temperature and pressure. The Ni/Mo on γ—alumina gave the lowest sulfur content in products, while the γ—alumina shows the sulfur when catalyst activity was absent.

Many studies have found fractional kinetics for sulfur removal, as summarized in Table 5—5. These results suggest that reaction order for HDS cannot be predicted a priori for a new feedstock.

Table 5—5
Kinetics of Residue Desulfurization in Laboratory CSTRs

Feed	Conversion %	Order	E_a kJ/mol
CL/LL[1]	80—90	1.5	140
Thasos[2]	10	1.9—2.3	132—145

1. Fraction boiling over 424° C, data from Oballa *et al.*, 1992.
2. Fraction boiling over 343° C (atmospheric residue), data from Ammus and Androutsopoulos, 1987.

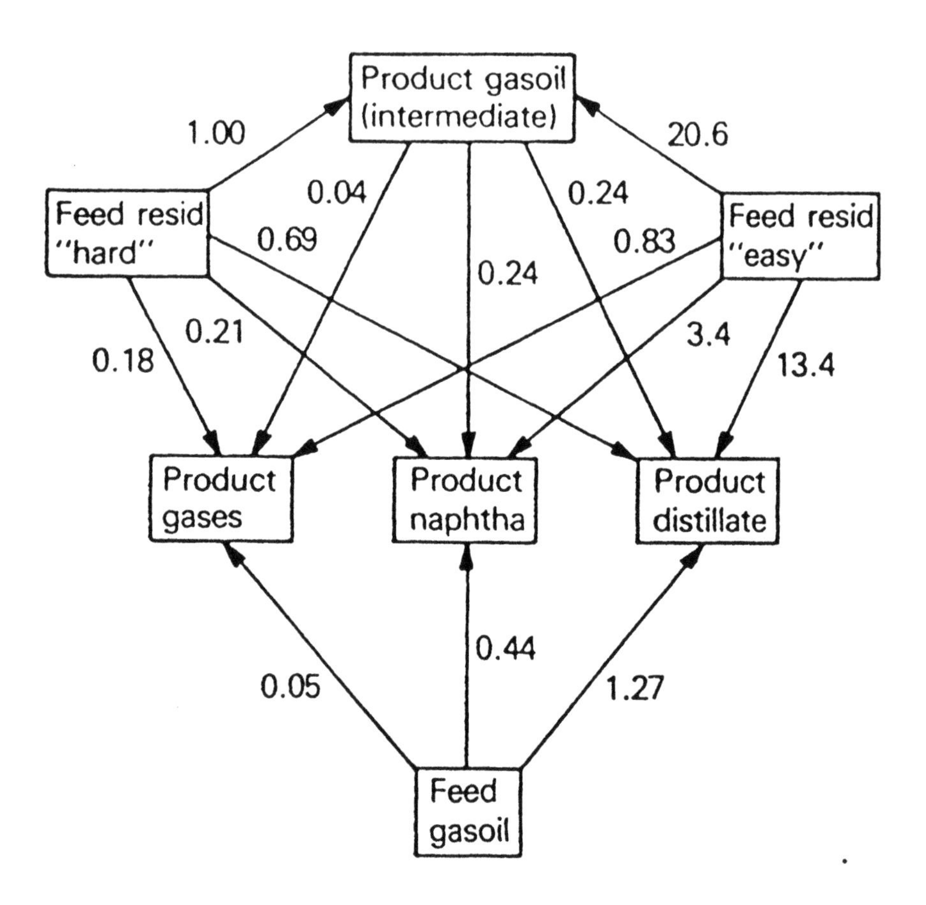

Figure 5–21

Reaction scheme for thermal reactions in residue hydroconversion. The numbers are first order reaction rate constants relative to cracking of "hard" residue to gas oil (k = 1.00). (From Mosby *et al.*, 1986)

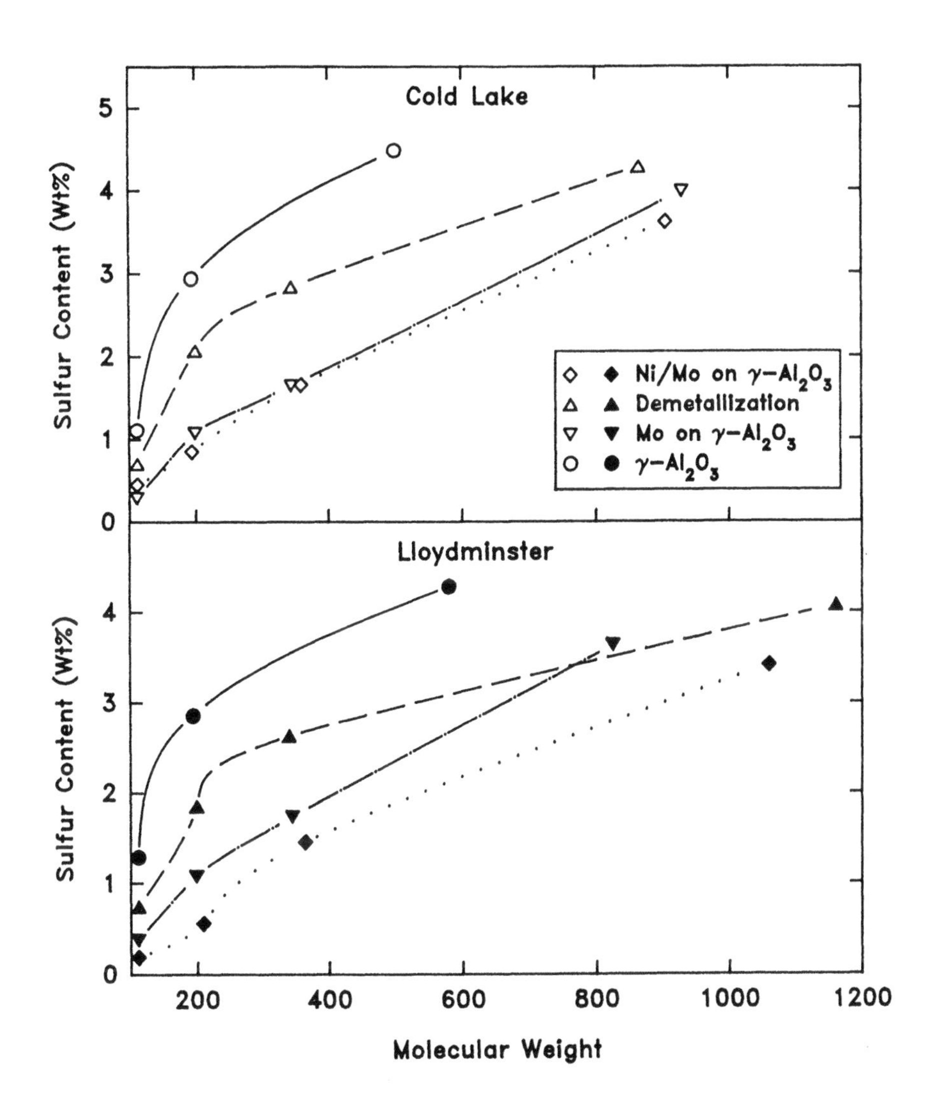

Figure 5–22

Sulfur content of product fractions from residue hydroconversion over different catalysts. From left to right the boiling fractions, plotted by molecular weight, are naphtha, middle distillate, gas oil, and residue (Data from Gray *et al.*, 1992)

Table 5–6
Catalyst Transport and Effectiveness for Residue HDS of Thasos Residue
(Data from Ammus and Androutsopoulos, 1987)

Sample	Pellet Diam, cm	Mean Pore Diam, Å	D_e cm^2/s	η
Cat. A	0.032	86	4.06E–7	0.46
Cat. A	0.058	82	2.6E–7	0.21
Cat. B	0.029	72	1.0E–7	0.26

5.7.3 Uptake of hydrogen

Beret and Reynolds [1985] developed a method for determining hydrogen uptake via the various simultaneous reactions that occur during catalytic hydroconversion of residue:

1. Hydrogenation/Dehydrogenation — Any decrease in the overall content of aromatic carbon (from ^{13}C–NMR) between the feed and the products requires hydrogen at a rate of one hydrogen atom per carbon atom reduced.

2. Desulfurization, Deoxygenation, and Denitrogenation — Sulfur was assumed to be thiophenic, consuming 2 hydrogen atoms per sulfur atom. Nitrogen was assumed to be pyrrolic and oxygen was assumed to be carboxylic and phenolic. Each heteroatom would require two atoms of hydrogen for removal.

3. Formation of C_1–C_3 Light Ends — These products each consume 2 hydrogen atoms per mol, one for each end of the broken C–C bond. This calculation assumes that no alkenes are formed, which is reasonable for catalytic hydrogenation conditions.

4. Hydrocracking and Condensation/Polymerization Reactions of Liquids — Breakage of a C–C bond requires two hydrogen atoms to cap the ends, again assuming that olefins are hydrogenated.

The cracking reactions are very important, but most difficult to estimate on a molar basis to assign hydrogen consumption. Beret and Reynolds [1985] calculated this portion of hydrogen consumption by difference:

$$\text{H for Cracking} = \text{Hydrogen consumption} - \text{H for aromatic carbon} - \text{H for O,N,S removal} - \text{H for } C_1\text{—}C_3 \tag{5.15}$$

A direct estimation method was proposed by Trytten and Gray [1990], using elemental analysis, molecular weight measurements, and NMR data. The method was useful for analyzing hydrogen consumption in laboratory reactors where measurements of total hydrogen consumption were inaccurate [Gray *et al.*, 1991]. Both Beret and Reynolds [1985] and Gray *et al.* [1991] found that determination of the distribution of hydrogen consumption was very useful in comparing reactions of different residue feeds.

Hydrogen is consumed by three different reactions; hydrogenation of aromatic carbon, hydrogenation of heteroatoms, and hydrocracking of C—C bonds to produce liquids and light ends. The kinetics of overall hydrogen uptake, therefore, are rarely determined. Papayannakos and Georgiou [1988] reported that hydrogen uptake by a Greek atmospheric residue followed second—order kinetics during hydrotreatment over a Co/Mo on γ—alumina catalyst. The activation energy was 86 kJ/mol. The reaction order for desulfurization in their experimental fixed—bed reactor was 2.5.

5.7.4 Kinetics of nitrogen and metals removal

The two other major reactions that have been studied are denitrogenation and demetallation. The extent of nitrogen removal is usually low in processing vacuum residues, and the kinetics are always first order with respect to nitrogen content. Demetallation reactions tend to be approximately first order also (see also section 5.4.3), although in some cases more complex series—reaction schemes have been proposed to help explain metal distributions within catalyst pellets [Thakur and Thomas, 1985]:

$$\text{Metal Bearing Precursor} \rightarrow \text{Hydrogenated Intermediate} \rightarrow \text{Metal Deposit} \tag{5.16}$$

Since the chemical form of roughly half the metals in the residue is not well understood (section 1.3.2.4), these reaction schemes are empirical.

5.7.5 Kinetics of heat release

Thermal cracking of C—C bonds is slightly endothermic, while the various hydrogenation reactions that occur during hydroconversion are strongly exothermic. The net result is an exothermic reaction, with only a negligible effect due to the endothermic reactions [Pruden and Denis,

1976]. The enthalpy of reaction for residue hydroconversion can be calculated from the hydrogen consumption; Galbreath and van Driesen [1967] estimated values of −49 kJ/mol H_2 consumed for hydrocracking and −56 kJ/mol of H_2 consumed for desulfurization.

Jaffe [1974] developed a method based a balance on the carbon bonds in the feed and products. He distinguished between three types of hydrogenation reactions, saturation of olefins (−113 to −125 kJ/mol H_2), saturation of aromatics (−60 to −69 kJ/mol H_2), and cracking of alkanes (−29 to −41 kJ/mol H_2). By summing the contributions of each type of reaction, the total enthalpy of reaction was calculated. Trytten and Gray [1990] showed how NMR data could be used to estimate carbon bond types in feed and product fractions. Combination of this approach with the method of Jaffe [1974] would give estimates of heat release according to each type of reaction.

5.8 Hydroconversion Preprocesses and Modifications

The process schemes presented in sections 5.2, 5.3, and 5.5 presumed that the only treatment that the feed has received was distillation to separate the residue fraction. A number of research studies have looked at pretreatments to either improve the yields of liquid products from hydroconversion, or to improve the product properties. These pretreatments are not widely used on a large scale either due to high cost or because they recover low–value portions of the residue feed.

5.8.1 Solvent deasphalting

One obvious method of cleaning the feed is to precipitate the asphaltenes using a solvent such as propane. Nelson [1958] gives detailed information on this well–established separation process. The deasphalted maltenes have less metals than the whole residue, but problems such as catalyst deactivation are not eliminated, merely slowed. These processes create a byproduct stream that is unattractive except as a raw material for asphalt. As such, deasphalting is often considered as an alternative to hydrocracking, but the economics of removing a portion of the feed stream and diverting it to asphalt production need careful consideration. At least one process design takes the opposite approach of separating asphaltenes from the hydrocracked product and recycling them to the reactor [Takeuchi *et al.*, 1983].

5.8.2 Low–temperature hydrogenation

Hydroconversion reactors for residue conversion are operated to achieve significant cracking of carbon–carbon bonds, which requires temperatures of at least 410–420° C. Lower temperatures are preferred, however, for catalytic hydrogenation of aromatics (section 3.3.2) and for minimizing coking of catalysts. Aromatic rings, therefore, either pass through unchanged or increase in concentration. The accumulation of aromatic rings in the unconverted residue becomes significant at high conversion, and definitely rate limiting (section 5.7.1). One method for decoupling the two reactions is to hydrogenate the residue at low temperatures, below 390° C, then crack the liquids using an additive-based process or an ebullated—bed reactor.

Beret and Reynolds [1990] found that the two step conversion of Maya residue (hydrogenation at temperatures below 390° C followed by cracking at T > 410° C) improved the efficiency of hydrogen usage. At equivalent levels of residue conversion, less hydrogen was consumed by light—end formation in the two stage process, and more hydrogen was incorporated in the products. Although the two—stage process consumed more hydrogen, it would reduce the load on downstream distillate hydrotreaters. The benefits of this two—stage approach, therefore, must be considered in the context of the overall hydrogen balance for the plant.

A particular benefit of low—temperature hydrogenation was the ability to hydrogenate the high—molecular weight components, which helps to convert precursors for CCR [Sanford and Chung, 1991], prevent coke formation in the reactor [Heck and diGuiseppi, 1993] and prevent precipitation of solids downstream of the reactor [Mochida *et al.*, 1990]. The protection against coking and precipitation was effective even at high conversion. These benefits, coupled with possible savings in catalyst consumption, suggest that two—stage processing may be attractive in some cases.

5.8.3 Coprocessing of residue and coal

Powdered coal has been used as an additive in hydroconversion processes to prevent coking, for example, the CANMET process. A number of studies have reported a boost in distillate yields when coal is added at a level of 5—10 wt% to residue feed. *Energy and Fuels* [vol 3(2), 1989] contains an excellent series of papers on the subject of coal—residue coprocessing. At low coal loadings, coprocessing can be considered an additive—based process technology. For example, Fouda *et al.* [1989] found that the addition of 2 wt% coal improved liquid yields from noncatalytic cracking of Cold Lake vacuum bottoms at 450° C and 13.9 MPa. Addition of 2% coal gave ca. 72% distillate yields, compared to 62% yield from residue alone or experiments with 10—30% coal. At concentrations ca. 2 wt%, the coal solids may provide catalytic activity as well as adsorption of coke precursors and interaction with free—radical reactions. The exact mechanism of the interaction is unknown. At low coal loadings, coal is an additive. Above 10 wt% in the feed, coal begins to contribute more to the products and the process becomes a hybrid of residue upgrading and coal liquefaction. The latter process is outside the scope of this book.

5.9 Environmental Aspects of Residue Hydroconversion

Hydroconversion of residue produces gas, liquid, and solid streams: hydrogen sulfide, purge gas, distillate products, unconverted residue, and spent catalyst. Hydrogen sulfide is acutely toxic to humans and animals. A detectable odor is apparent at concentrations ca. 0.03 ppm, the gas paralyzes the olfactory system at 150 ppm and causes eye and lung irritation, and exposure to concentrations over 500 ppm can cause death [Environment Canada, 1984; Alberta Community and Occupational Health, 1988]. Hydrogen sulfide from hydroconversion, as well as ammonia, are readily removed from gas streams by scrubbing solutions and are not emitted to the atmosphere. Recovered hydrogen sulfide is then converted to elemental sulfur by the modified Claus process [GPSA, 1987]. Safe handling of H_2S, therefore, is a serious concern on the plant site, but this byproduct is not released to the environment. Under emergency conditions, the H_2S would be flared with other vent gases.

The distillates from hydrotreating are also intermediates within the refinery operation. Very little research has been done on the toxicity of hydrocracked products, but their chemistry indicates that effects identical to crude oil would be expected. Hydrocracking does not produce unsubstituted fused–ring aromatics in significant concentrations, therefore, these intermediate streams would have mild mutagenicity and carcinogenicity comparable to conventional petroleum fractions. The mild chronic toxicity of these streams is mainly of concern for occupational health, and workers should avoid skin contact and inhalation [WHO, 1982].

Hydroconversion processes are designed to avoid aromatization reactions, and attendant coke formation. The unconverted residue fraction has a high molecular weight but is not significantly enriched in polyaromatic hydrocarbons relative to the distillate fractions. Although specific studies are lacking, this fraction is thought to be biologically quite inert, similar to the unprocessed bitumen (section 1.5).

Spent catalyst is the main environmental concern in the operation of hydroconversion processes. Whether catalyst is regenerated by acid leaching (section 5.4.5) or used only once (as in most ebullated–bed and additive based processes), spent catalyst is inevitably generated. In the past, these materials were sent to land fill, but this practice is being curtailed by tighter regulation [Habermehl, 1988]. Recycling of the metals content is technically feasible [Jocker, 1993; Inoue *et al.*, 1993], and is most attractive for catalysts containing molybdenum and cobalt. Recycling of iron–based additives is much less attractive, but it may be necessary if landfill disposal of iron sulfide waste is not available. Regardless of the catalyst or additive, recycling and disposal must be considered at the time that the hydroconversion process is selected.

Notation

a	hydrodynamic radius, m
A_p	area per unit volume, m^{-1}
c	empirical constant
C	concentration, $kmol/m^3$
[C]	carbon concentration, wt%
CSTR	continuous–flow stirred tank reactor
D_e	effective diffusivity in catalyst pellets, m^2/s
D_b	diffusivity in bulk liquid at infinite dilution, m^2/s
D_s	dispersion coefficient, m^2/s
E(t)	age fraction of catalyst
h	volumetric gas holdup
k	reaction rate constant, s^{-1} or $m^3 kmol^{-1} s^{-1}$
L	pellet diameter, m
L_r	reactor length, m
LHSV	liquid hourly space velocity
[M]	concentration of deposited metals, wt%
Pe	Peclet number, UL_r/D_s
PV	pore volume, m^3/kg
r	rate of reaction, kmol/s
r_p	pore radius, Å or m
t	time, s
T	temperature, K
U	velocity, m/s
x	linear distance, m
X	fractional conversion
α_c	catalyst activity
β	ratio of porosities in equation (5.6b)
ϵ	porosity
ζ	tortuosity
η	effectiveness factor
Λ_p	characterstic diffusion rate
ρ	density, kg/m^3
τ	mean residence time

ϕ	Thiele modulus

Subscripts

b	bulk fluid phase
c	catalyst
e	exit condition
HDS	hydrodesulfurization
i, j	counter variables
m	micropore
M	macropore
metal	metal sulfides deposited on catalyst
0	initial or inlet condition
p	pellet

Further Reading

J.H. Gary and G.E. Handwerk, *Petroleum Refining, Technology and Economics, 2nd Edition*, Marcel Dekker, New York, 1984.

Thakur, D.S.; Thomas, M.G. 1985. "Catalyst Deactivation in Heavy Hydrocarbon and Synthetic Crude Processing: A Review". *Appl. Catal.* 15, 197–225.

Problems

5.1 Mild Resid Hydrocracking (MRH) was proposed by Sue *et al.* [1988] as a method for upgrading heavy oils, according to Figure 5–23. The feed (Athabasca bitumen) is mixed with a fine powdered catalyst (not specified), heated, and fed to the reactor. Within this vessel the liquid is recirculated to promote mixing. The catalyst promotes hydrogenation, and adsorbs coke precursors. This adsorbed material, along with coke particles, is continuously removed from the reactor and reported as "coke" in the table below.

	Feed	Product
Specific gravity (15° C)	1.0207	
Sulfur, wt%	4.49	
Nitrogen, ppm by wt	3920	
Metals, ppm by wt	286	

	Feed	Product
Distillation Cuts		
C_5—171° C, wt%	—	15.6
Sulfur, ppm		7100
Nitrogen, ppm		550
171—343° C, wt%	5.6	35.5
Sulfur, wt%		2.2
Nitrogen, ppm		1400
343—525° C, wt%	37.4	30.8
Sulfur, wt%		2.7
Nitrogen, ppm		2900
Metal, ppm		0.5
525° C+, wt%	57.0	11.0
Coke	0.0	1.6
Hydrogen consumption, m^3 at NTP/m^3 feed oil		156

(a) The process gives light ends, liquid products, and hydrogen sulfide gas (2 wt% of the products). Calculate the conversion of sulfur from this process. If high—activity catalyst gives 80% conversion of sulfur, and thermal reactions alone give 30%, how would you rate the activity of the catalyst?

(b) What fraction of the hydrogen consumption goes to desulfurization?

(c) The process uses combustion in air to regenerate the catalyst, by burning off the coke deposits. Calculate the amount of sulfur burned in this way if the plant capacity is 500 t/d. State your assumptions. If environmental regulations cover sulfur sources of over 2 t/d (as SO_2), does this regenerator require regulatory approval?

5.2 Calculate the conversion of nitrogen in the LC—Fining process using Athabasca bitumen as feed, using the data from Tables 5—1 to 5—3.

5.3 Shell has developed a technology for processing heavy oil in a two-stage process [Robschlager *et al.*, 1992]. In the first stage bunker reactors are used for hydrodemetallization (HDM). Catalyst is fed to the top of these reactors, and removed at the bottom, so that a given spherical pellet of catalyst gradually progresses

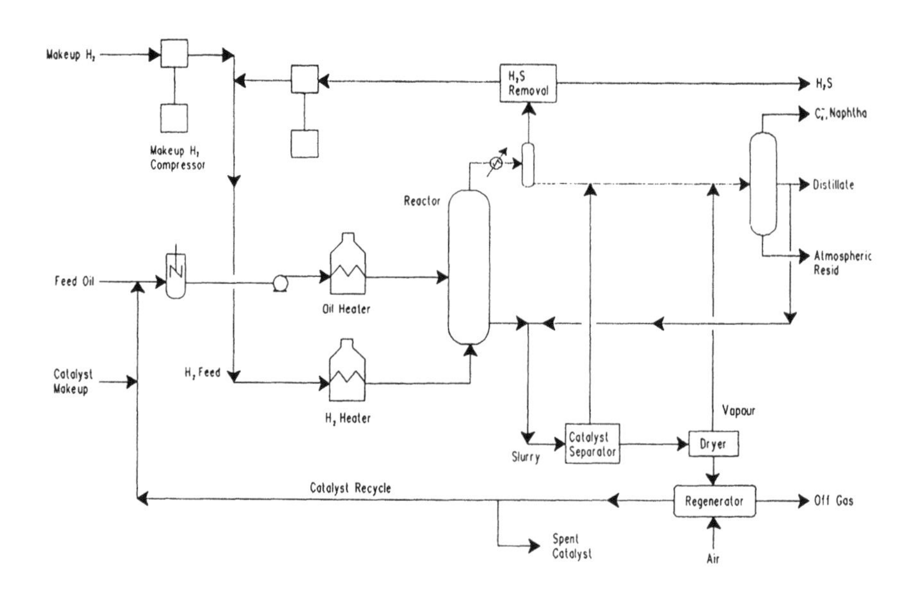

Figure 5–23
Process schematic for Mild residue hydrotreating (MRH), an additive-based hydrocracking technology (From Sue *et al.*, 1988)

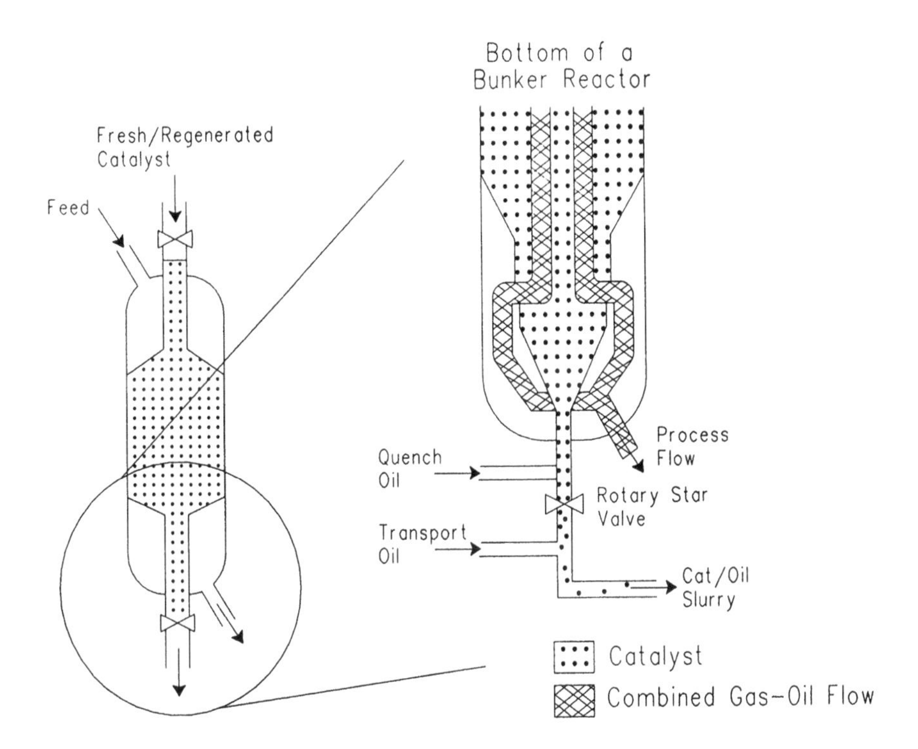

Figure 5–24
Schematic drawing of a bunker reactor (From Robschlager *et al.*, 1992)

downward through the reactor as part of a moving packed bed (Figure 5–24). The bunker reactor, therefore, is a method of changing catalyst in a trickle–bed reactor without shutting down the plant. The bunker reactors for HDM are followed by trickle-bed reactors for high–conversion desulfurization (HCON). These reactors use fixed–beds of catalyst. A test of the process was conducted with residue from Arabian heavy oil.

Feed:		Arabian heavy (580° C+)
	Rate	4400 t/d
	S, wt%	4.23
	N, wt%	0.26
	V+Ni, ppm	115

Performance		
	Desulfurization	92%
	Demetallization	95%
	Residue Conversion (to distillables < 525° C)	55%

(a) The silica catalyst can carry about 30% metals by weight. In steady state operation, how much fresh catalyst would be added to the bunker reactors each day to maintain conversion of metals?

(b) What is the minimum consumption of hydrogen for removal of heteroatoms based on the observed reactor performance? Assume that the nitrogen removal is 30%. What other reactions would also consume hydrogen?

(c) Could the unconverted residue from the process be used as Number 6 Fuel Oil for marine engines? This fuel type is a residual oil with less than 1% sulfur. Assume that all unconverted heteroatoms are concentrated in the unconverted residue.

CHAPTER

6

Thermal and Coking Processes

6.1 Severity of Thermal Processes

When petroleum fractions are heated to temperatures over ca. 410°C, then thermal or free–radical reactions start to give cracking of the mixture at significant rates (section 3.2). Thermal conversion does not require the addition of a catalyst, therefore, this approach is the oldest technology available for residue conversion. The severity of thermal processing determines the conversion and the product characteristics. Thermal treatment of residues ranges from mild treatment for reduction of viscosity to ultrapyrolysis for complete conversion to olefins and light ends (Figure 6–1). The higher the temperature, the shorter the time to achieve a given conversion. The severity of the process conditions is the combination of reaction time and temperature to achieve a given conversion.

If no side reactions occur, then very long times at low temperature should be equivalent to very short times at high temperature. Thermal reactions, however, can give rise to a variety of different reactions (section 3.2) so that selectivity for a given product changes with temperature and pressure. Table 6–1 summarizes the severity and conversion characteristics of thermal processes. The mild– and high–severity processes are frequently used for processing of residue fractions, while conditions similar to "ultrapyrolysis" (high temperature

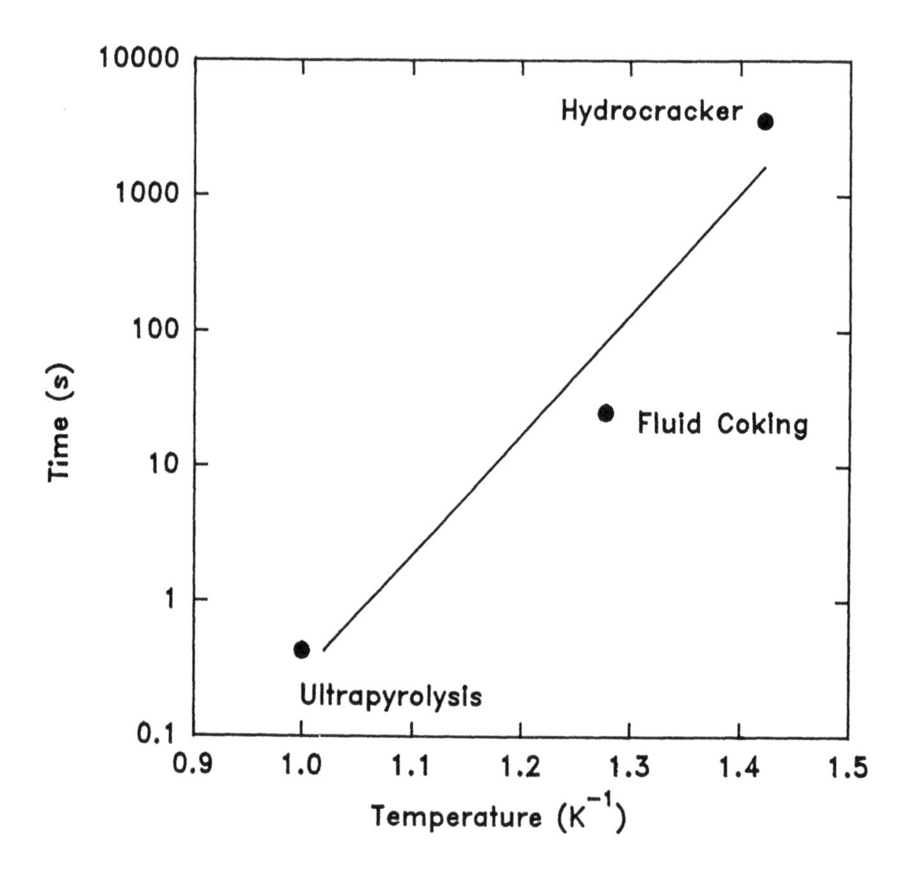

Figure 6–1
Reaction severity in terms of time and temperature

and very short residence time) are only used commercially for cracking ethane, propane, butane, and light distillate feeds to produce ethylene and higher olefins.

Table 6—1
Severity of Thermal Processing

Level of Severity	Process	Time s	Temp °C	Conversion
Mild	Visbreaking	90	425—500	Low
High	Delayed coking	(1)	435—480	High
	Hydrocracking	3600	420—440	Med—High
	Fluid coking	25	510—540	High
Extreme	Ultrapyrolysis	0.5		High

1. Semibatch process

6.2 Thermal Viscosity Reduction (Visbreaking)

This very mild thermal cracking process is aimed at reducing the viscosity of the residue or crude oil so that it can be pumped easily. Reduced viscosity is desirable for fuel oil applications and for pipeline transportation. Although the viscosity of residues drops dramatically with temperature (Figure 6—2 gives example data for Alberta bitumens), their viscosity is very high at ambient temperatures. The data of Figure 6—2 show that viscosity is not a concern within a process unit, where the temperature can be easily maintained above 100°C. Buried pipelines, however, experience temperatures which range from 10—15° C in the summer to 2—3° C in the winter (depending on the local climate). These temperatures require a viscosity of less than 2.5 x 10^{-4} m^2/s (250 cSt) at winter temperatures of 2—3° C, or about 1 x 10^{-4} m^2/s at 15° C.

The high viscosity is thought to be due to entanglement between the high—molecular weight components of the oil, and due to the formation of ordered structures in the liquid phase (Section 2.12.1 and Figure 1—4). Thermal cracking at low conversion can remove side chains from the asphaltenes and break bridging aliphatic linkages. A 5—10% conversion of atmospheric residue to naphtha is sufficient to reduce the entanglements and structures in the liquid phase, and give at least a 5—fold reduction in viscosity. Reduction in viscosity is also accompanied

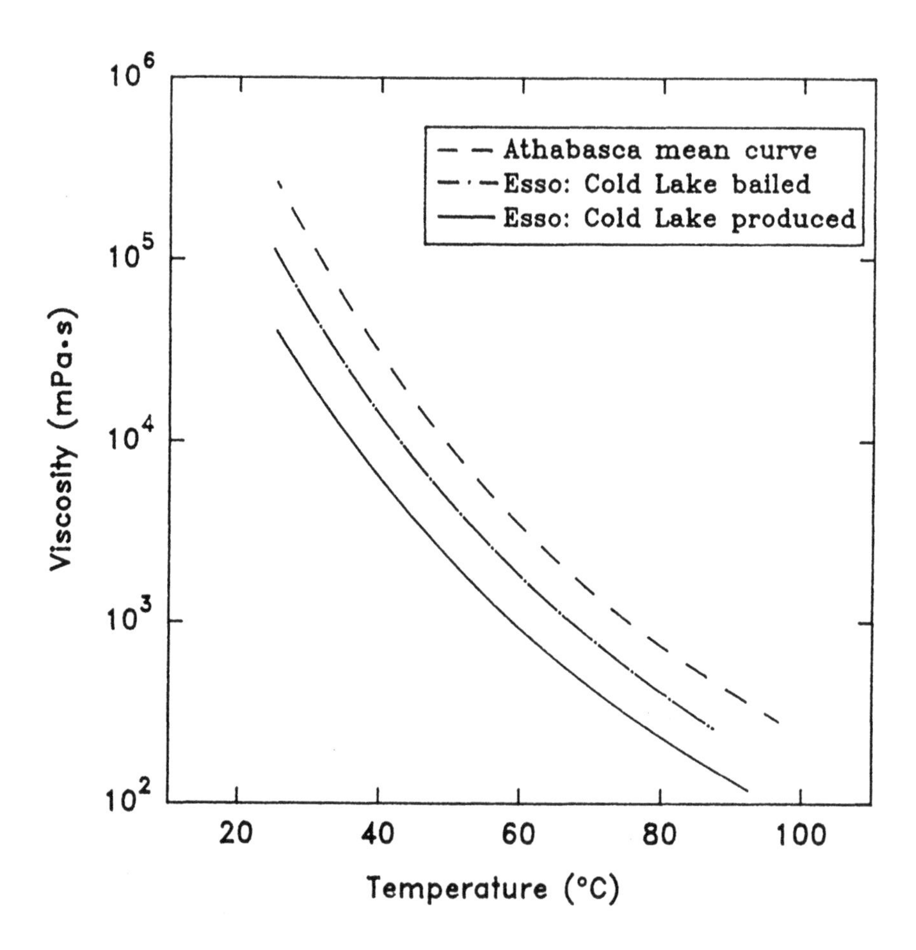

Figure 6–2

Viscosity of Athabasca and Cold Lake Oils. Data for Athabasca are mean results from several reports, Cold Lake bailed was from a well at formation temperature, as compared to Cold Lake produced which was from a steam injection project (Data from Seyer and Gyte, 1989) by a reduction in the pour point.

Thermal viscosity breaking, or visbreaking, uses this approach of mild thermal cracking as a relatively low cost and low severity approach to improving the viscosity characteristics of the residue without attempting significant conversion to distillates. Low residence times are required to avoid polymerization and coking reactions, although additives can help to suppress coke deposits on the tubes of the furnace [Allan *et al.*, 1983].

The visbreaking process is illustrated in Figure 6–3, and consists of a reaction furnace, followed by a quenching step with a recycled oil. The product mixture is then fractionated. All of the reaction in this process occurs as the oil flows through the tubes of the reaction furnace. The severity is controlled by the flow rate through the furnace and the temperature; typical conditions are 475–500°C at the furnace exit with a residence time of 1–3 min, with operation for 3–6 months before the furnace tubes must be decoked [Gary and Handwerk, 1984]. The operating pressure in the furnace tubes can range from 0.7 MPa to 5 MPa depending on the degree of vaporization and the residence time desired. For a given furnace–tube volume, a lower operating pressure will reduce the actual residence time of the liquid phase.

An alternative process design uses lower furnace temperatures and longer times, achieved by installing a soaking drum between the furnace and the fractionator. The disadvantage of this approach is the need to decoke the soaking drum.

The main limitation of straight thermal processing is that the products can be unstable. Thermal cracking at low pressure gives olefins, particularly in the naphtha fraction. These olefins give a very unstable product, which tends to undergo polymerization reactions to form tars and gums. The heavy fraction can form solids or sediments, which also limit the range of conversion. Sediment formation is a property of the feed composition that also determines the maximum conversion allowable, ranging from 12% for a South American crude through to 30% conversion of North Sea atmospheric residue [Allan *et al.*, 1983].

The reduction in viscosity of the unconverted residue tends to reach a limiting value with conversion, although the total product viscosity can continue to decrease as illustrated in Figure 6–4. Conversion of residue in visbreaking follows first–order reaction kinetics [Henderson and Weber, 1965]. The minimum viscosity of the unconverted residue can lie outside the range of allowable conversion if sediment begins to form [Rhoe and de Blignieres, 1979]. When pipelining of the visbreaker product is the process objective, addition of a diluent such as gas condensate can be used to achieve a further reduction in viscosity [Sankey and Wu, 1989].

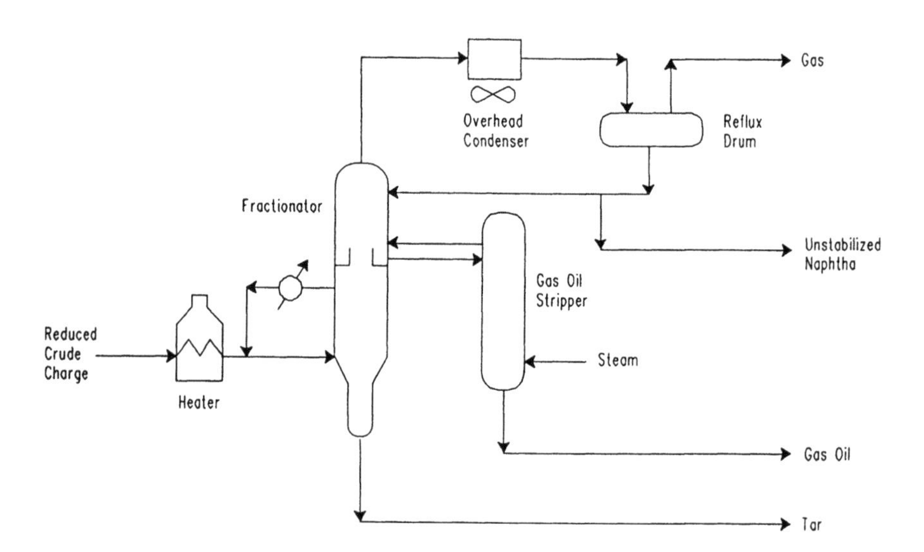

Figure 6–3
Process schematic of visbreaking process (From Allan *et al.*, 1983)

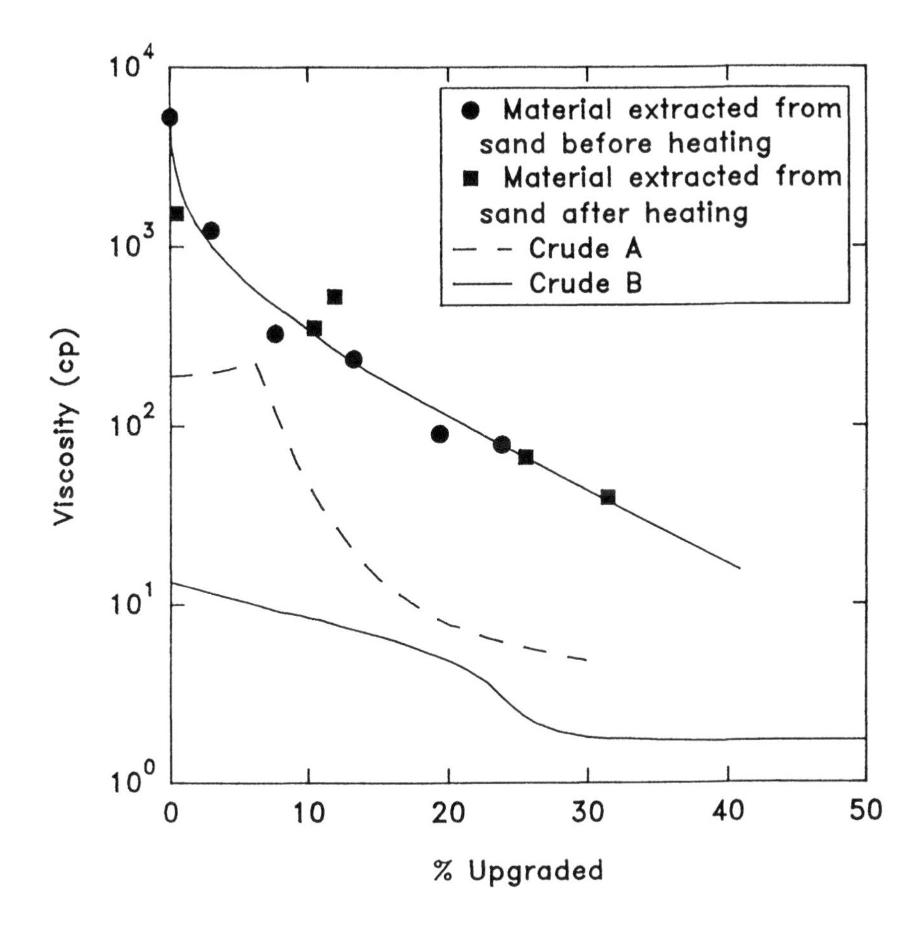

Figure 6–4
Reduction in viscosity with severity of visbreaking. Reaction at 371° C in a batch reactor (Data from Henderson and Weber, 1965)

6.3 Delayed Coking

The delayed coking process is widely used for treating heavy residues, and is particularly attractive when the green coke produced can be sold for anode or graphitic carbon manufacture (see section 6.5) or when there is no market for fuel oils, as in Western Canada. The process uses long reaction times in the liquid phase to convert the residue fraction of the feed to gases, distillates, and coke. The condensation reactions that give rise to the highly aromatic coke product also tend to retain sulfur, nitrogen and metals, so that the coke is enriched in these elements relative to the feed.

Table 6–2
Yields from Coking of Tia Juana Vacuum Residue
(Data from Edelman *et al.*, 1979)

Feed	525° C+ residue 2.9% S, 0.4% N, 22% CCR	
	Delayed Coking	Fluid Coking
Yields on Feed		
H_2S (wt%)	1.1	0.7
Light Ends (wt%)	11.1	11.6
Naphtha (vol%)	25.6	20.7
Middle Distillate (vol%)	26.4	15.8
Gas Oil (vol%)	13.8	32.5
Coke, wt%	33.0	20.0
Sulfur Content (wt%)		
Naphtha	0.45	0.6
Mid. Distillate	1.3	1.8
Gas Oil	1.9	2.5
Coke	3.7	4.1
Bromine Number		
Naphtha	69	107
Mid. Distillate	30	58
Gas Oil	22	40

An example schematic diagram for the delayed coking process is given in Figure 6–5. The conception of this semi–batch process is simple: the feed is heated to ca. 500° C, then accumulated in an

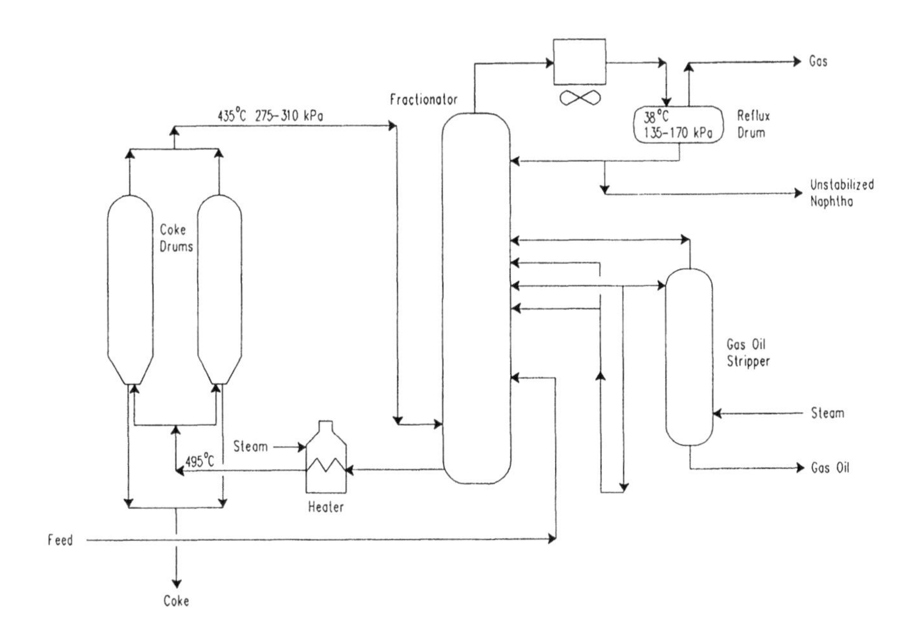

Figure 6–5
Process schematic for delayed coking process

insulated vessel, called a coke drum. The vapor products, consisting of gases and distillates, are drawn off the top of the coke drum at approximately 435° C and quenched by contact with colder oil. The quenching is often accomplished by feeding the vapors into the lower section of a fractionator. In the coke drum, meanwhile, the coke gradually accumulates until the drum is full. At this point, the feed is switched to the next coke drum and the coke product is recovered. A typical cycle spans 48 hours (Table 6–3), so that delayed coker plants are built with at least two coke drums.

Table 6–3
Typical Coke Drum Cycle
(From Gary and Handwerk, 1984)

	Time, h
Filling with coke	24.0
Switch and Steam out	3.0
Water cooling	3.0
Draining water	2.0
Hydraulic decoking	5.0
Head up and test	2.0
Warmup	7.0
Spare time	2.0
Total cycle	48.0

The accumulated coke must be drilled or broken out of the coke drum, and typical units use a hydraulic drill mounted on a gantry. The water from the drilling operation is skimmed to remove oil, filtered, and recycled (Figure 6–6).

6.4 Fluid Coking

The semi–batch coking process is most attractive for processing small volumes of residue, due to the effort involved in the decoking the drums at the end of each cycle. The yield of distillates from coking can be improved by reducing the residence time of the cracked vapors. In order to simplify handling of the coke product, and enhance product yields, Exxon developed fluidized–bed coking, or fluid coking, in the mid 1950's. In this continuous process, the feed is sprayed into a fluid bed made up of coke particles (Figure 6–7). The particles have a bulk density of 750–880 kg/m^3, with a particle density of 1440 kg/m^3 and a diameter of 100–600 μm. Coking occurs on the surface of these particles

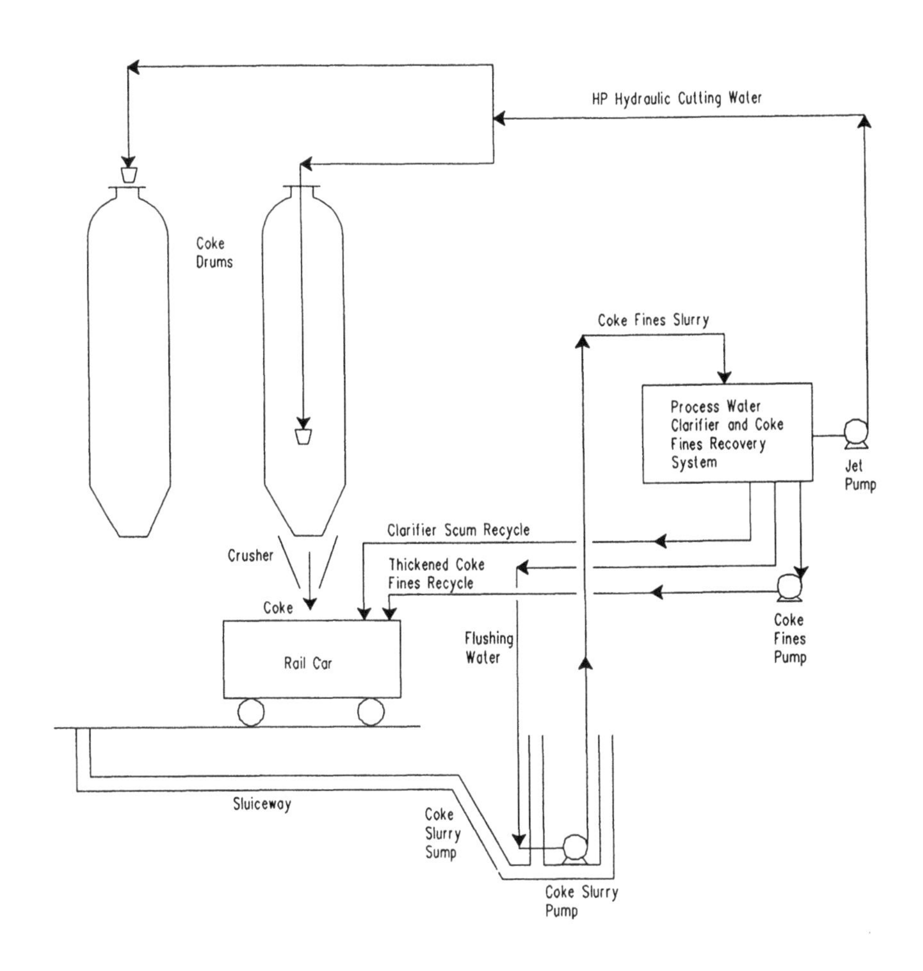

Figure 6–6
Typical Coke handling system for delayed cokers

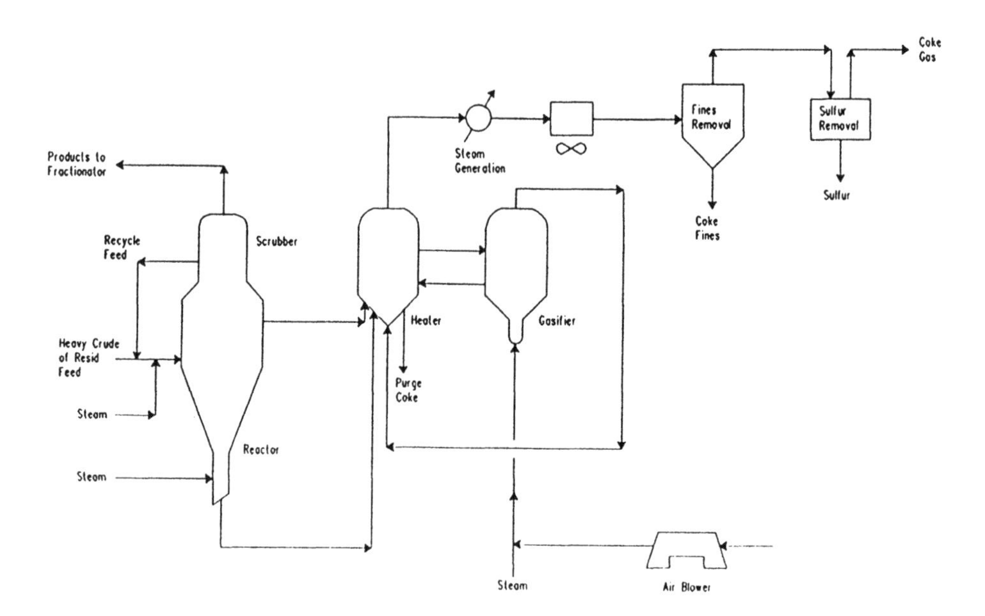

Figure 6–7
Process schematic of fluid coking process (From Edelman *et al.*, 1979)

at temperatures of 510—530° C. The cracked vapors rise to the top of the reactor, pass through cyclones to remove entrained particles of coke, and enter the scrubber in the top portion of the vessel. Here the vapors are quenched by contact with condensed liquid or fresh feed. After stripping of the coke with steam to remove liquids, the coke passes to the burner where a portion of the coke is burned to supply the heat requirements of the reactor. The product distribution from Athabasca bitumen is as follows:

Table 6—4
Yields From Fluid Coking of Athabasca Bitumen
(Design data from Syncrude Canada Ltd.)

Feed	72,900 bbl/d
Products	
Burner off—gas	8.7×10^3 m^3/d
Fuel gas	1.5×10^3 m^3/d
Butanes	2555 bbl/d
Naphtha	22,400 bbl/d
Gas oil	43,200 bbl/d
Fluid coke	1120 t/d

The yields of products are determined by the feed properties, the temperature of the fluid bed, and the residence time in the bed. The use of a fluidized bed reduces the residence time of the vapor—phase products in comparison to delayed coking, which in turn reduces polymerization and cracking reactions. Another important factor is that the excellent heat transfer in the fluid bed allows the reactor to operate at higher temperature, giving more cracking of volatiles from the coke. These factors give a lower yield of coke from fluid—bed operation than from delayed coking, and the yield of gas oil and is olefins increased (Table 6.2; olefin content is indicated by the bromine number).

The lower limit on operating temperature is set by the behavior of the fluidized coke particles. If the conversion to coke and light ends is too slow, then the coke particles become sticky and agglomerate in the reactor, giving a loss of fluidization of the particles.

The disadvantage of burning the coke to generate process heat (Figure 6—7) is that sulfur from the coke is liberated as sulfur dioxide. The gas stream from the coke burner also contains CO, CO_2 and N_2. An alternate approach is to use a coke gasifier to convert the carbonaceous solids to a mixture of CO, CO_2, and H_2. An example is Flexicoking from Exxon (Figure 6—8). Coke is converted to a low—heating value gas in a

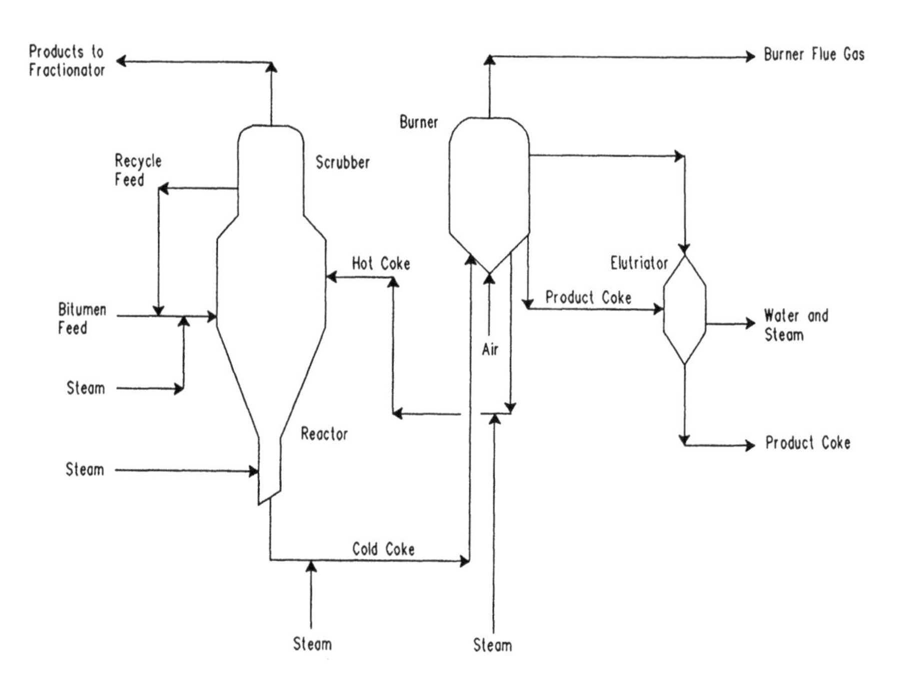

Figure 6–8
Schematic diagram for the Exxon Flexicoking process for coking of residue and gasification of coke (From Edelman *et al.*, 1979)

fluid bed gasifier with steam and air. The air is supplied to the gasifier to maintain temperatures of 830—1000° C, but is insufficient to burn all of the coke. Under these reducing conditions, the sulfur in the coke is converted to hydrogen sulfide, which can be scrubbed from the gas prior to combustion. A typical gas product, after removal of hydrogen sulfide, contains 18% CO, 10% CO_2, 15% H_2, 51% N_2, 5% H_2O and 1% CH_4 [Busch *et al.*, 1979]. The heater is located between the reactor and the gasifier, and it serves to transfer heat between the two vessels.

Yields of liquid products from Flexicoking are the same as from fluid coking, because the coking reactor is unaltered. The main drawback of gasification is the requirement for a large additional reactor, especially if high conversion of the coke is required. Units are designed to gasify 60—97% of the coke from the reactor. Even with the gasifier, the product coke will contain more sulfur than the feed, which limits the attractiveness of even the most advanced of coking processes.

6.5 Coke Yields and Properties

Coking processes have the virtue of eliminating the residue fraction of the feed, at the cost of forming a solid carbonaceous product. The yield of coke in a given coking process tends to be proportional to the carbon residue content of the feed (measured as CCR or MCR). The data of Table 6—5 illustrate how the yield of coke from delayed and fluid coking varies with CCR content of the feed.

Table 6—5
Yields of Coke from Delayed and Fluid Coking
(Data from Nelson, 1958)

Carbon Residue wt %	° API	Coke Yield, Weight% Delayed Coker	Fluid Coker
1		0	
5	26	8.5	3
10	16	18	11.5
15	10	27.5	17
20	6	35.5	23
25	3.5	42	29
30	2		34.5
40	−2.5		46

The formation of large quantities of coke is a severe drawback unless the coke can be put to use. Calcined petroleum coke can be used for making anodes for aluminum manufacture, and a variety of carbon or graphite products such as brushes for electrical equipment. These applications, however, require a coke which is low in mineral matter and sulfur. Table 6–6 lists typical specifications for anode–grade coke, and gives and example of an unsuitable coke product from Athabasca bitumen.

Table 6–6
Coke Composition and Specifications
(Data from Syncrude Canada and Reis, 1975)

Component	Athabasca Fluid Coke	Anode Grade Coke Specifications Maximum wt%
Carbon, wt%	79.99	
Hydrogen, wt %	1.66	
Sulfur, wt%	6.63	3
Nitrogen, wt%	1.9	
Ash, wt%	6.92	0.5
Volatiles	2.9	
Si + Fe, wt%		0.05
Vanadium, wt%		0.04

Athabasca bitumen is an example of a high sulfur, high ash feed that gives a coke that is unsuitable for anode use. The ash in this case is mainly composed of silicates, iron, and vanadium. Typical petroleum coke for anodes contains ca. 1.5% sulfur, which would result from a feed residue with about 0.9% sulfur [Reis, 1975]. As a target, therefore, residues for coking operations should contain less than 0.9% sulfur and proportionately low levels of vanadium.

If the residue feed produces a high sulfur, high ash, high vanadium coke, then the only other two options for use of the coke product are combustion of the coke to produce process steam (and large quantities of sulfur dioxide unless the coke is first gasified or the combustion gases are scrubbed), or stockpiling. Suncor at Fort McMurray, Alberta, uses the former option, while Syncrude Canada burns a regulated portion of the coke production in the burner vessel to drive the coker reactions, and stockpiles the rest.

For some residue feeds, particularly from Western Canada and Venezuela, the combination of poor coke properties for anode use, limits on sulfur dioxide emissions, and loss of liquid product volume have tended to relegate coking processes to a strictly secondary role to hydrogen addition processes in any new upgrading facilities.

6.6 Kinetics and Mechanism of Coke Formation

The available data suggest that coke formation is a complex process involving both chemical reactions and thermodynamic behavior. Like the asphaltenes, coke should be viewed as a solubility fraction. Its physical state at room temperature is solid, and it is insoluble in benzene or other solvents. The following two step mechanism has emerged from studies of whole oils and solubility fractions [Savage *et al.*, 1988; Wiehe, 1993]:

1. Thermal reactions result in the formation of high molecular weight, aromatic components in solution in the liquid phase.
   Reactions that contribute to this process are the cracking of side chains from aromatic groups, dehydrogenation of naphthenes to form aromatics, condensation of aliphatic structures to form aromatics, condensation of aromatics to form higher fused–ring aromatics, and dimerization or oligomerization reactions. Loss of side chains always accompanies thermal cracking, while dehydrogenation and condensation reactions are favored by hydrogen deficient conditions. Formation of oligomers is enhanced by the presence of olefins or diolefins, which themselves are products of cracking [Khorasheh, 1992]. The condensation and oligomerization reactions are also enhanced by the presence of Lewis acids, for example $AlCl_3$ [Mochida *et al.*, 1977].

2. Once the concentration of this material reaches a critical concentration, phase separation occurs giving a denser, aromatic liquid phase. Phase separation phenomena were discussed previously in sections 2.9 and 5.6.
   The importance of solvents in coking has been recognized for many years [e.g. Langer *et al.*, 1961], but their effects have often been ascribed to hydrogen–donor reactions rather than phase behavior. The separation of the phases depends on the solvent characteristics of the liquid. Addition of aromatic solvents will suppress phase separation (section 5.6), while paraffins will enhance separation. Microscopic examination of coke particles often shows evidence for mesophase; spherical domains that exhibit the anisotropic optical characteristics of a liquid crystal. This phenomenon is consistent with the formation of a second liquid phase; the mesophase liquid is denser than the rest of the hydrocarbon, has a higher surface tension, and likely wets metal surfaces better than the rest of the liquid phase. The mesophase characteristic of coke diminishes as the liquid phase becomes more compatible with the aromatic material [Lott and Cyr, 1992].

From this mechanism, we expect the following trends for coke yields:

1. Higher molecular weight fractions should give more coke (Figure 6—9).

2. More aromatic feeds need not yield more coke, depending on phase compatibility.

3. Acidic contaminants in a feed, such as clays, may promote coking.

4. Higher asphaltene content in a feed will, in general, correlate with higher coke yield [Banerjee *et al.*, 1986].

5. Coke may not form immediately if the solubility limit is not exceeded, so that an induction time is observed [Savage *et al.*, 1988; Wiehe, 1993], as illustrated in Figure 6—10.

6. The phase separation may be very sensitive to surface chemistry, hydrodynamics, and surface to volume ratio, similar to other processes that require nucleation. The data in Figures 6—9 and 6—10 provide striking examples of this effect. Banerjee *et al.* [1986] used quartz boats with 200 mg samples, pyrolyzed in flowing inert gas at 400° C, while Wiehe [1993] used an equivalent method at 400° C but with 2—3 g samples. In the former case, coking was completed in 10—15 min, while in the latter case an induction period of up to 90 min was observed before any coke was formed.

Wiehe [1993] has proposed a simple kinetic model to account for the data of Figure 6—10, using the two—step model.

$$M \xrightarrow{k_m} a\,A^* + (1 - a)\,V \tag{6.1}$$

$$A \xrightarrow{k_a} b\,A^* + c\,M^* + (1 - b - c)\,V \tag{6.2}$$

$$A^*_{max} = S_l\,(M + M^*) \tag{6.3}$$

$$A^*_{ex} = A^* - A^*_{max} \tag{6.4}$$

At long reaction times

$$A^*_{ex} \longrightarrow y\,C + (1 - y)\,M^* \tag{6.5}$$

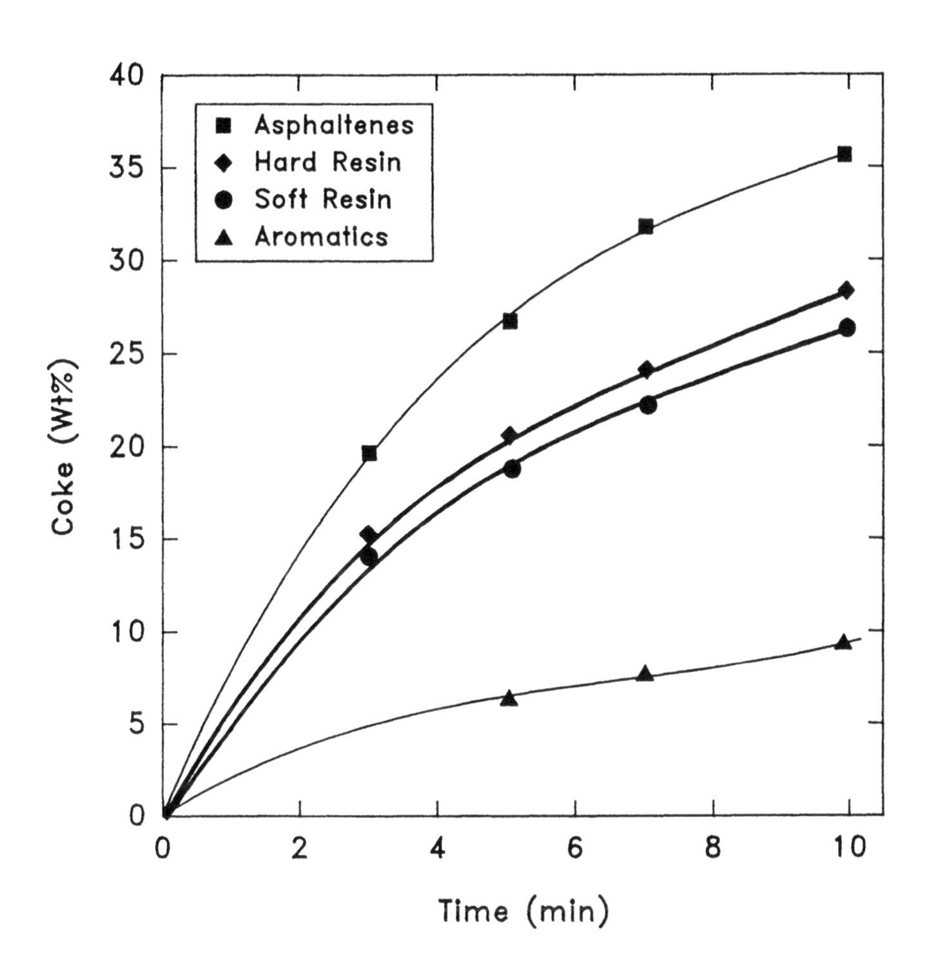

Figure 6–9

Yields of coke from fractions of Arabian vacuum bottoms at 400° C. Soft resin was separated from the maltenes by retention on Attapulgus clay and eluted by methyl ethyl ketone. Hard resin was eluted from the clay by tetrahydrofuran (Data from Banerjee *et al.*, 1986).

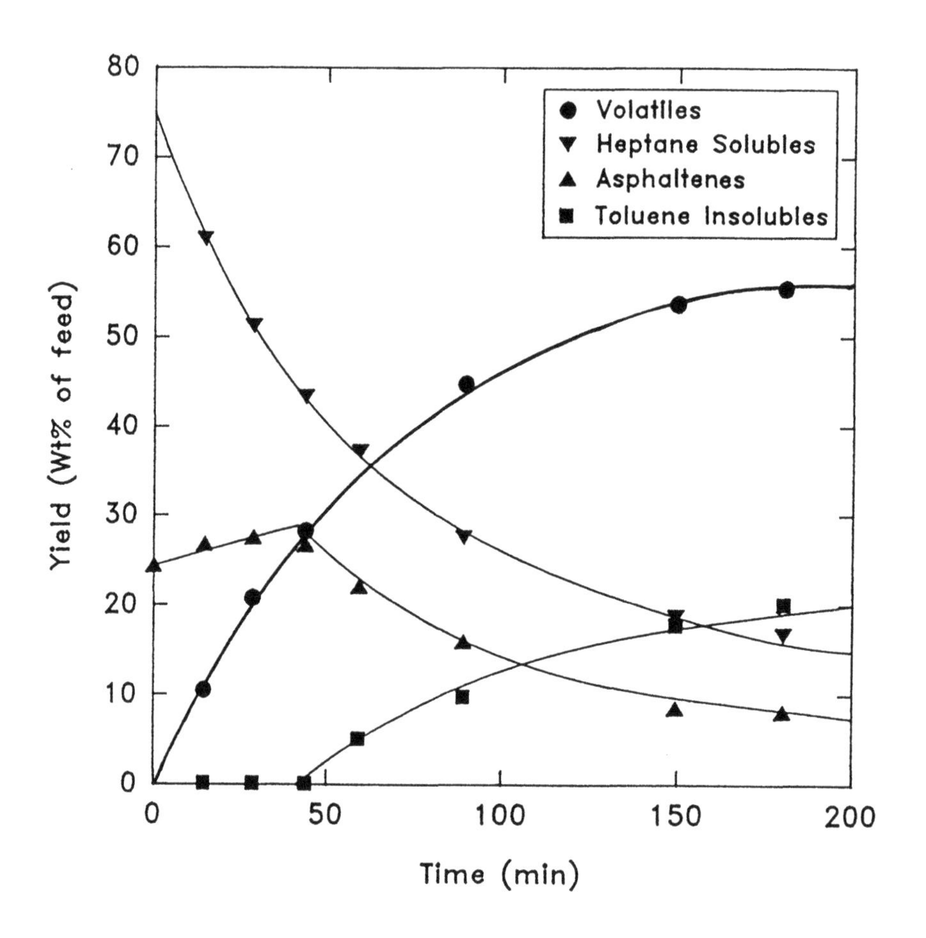

Figure 6–10
Yields of solubility fractions from pyrolysis of Cold Lake residue at 400° C. Data points are experimental values; curves are from equations (6.1) – (6.5) as discussed in the text. (Data from Wiehe, 1993)

where M and A are maltenes and asphaltenes in the feed, M* and A* are the corresponding products, V is volatiles, and C is coke. The parameters a, b, c, and y are stoichiometric coefficients. The solubility limit, S_l, is given as a fraction of total maltenes, and this limit determines the maximum cracked asphaltene concentration in solution, A^*_{max}. The excess asphaltene, A^*_{ex}, separates as a second phase and gives coke.

The cracking of the maltenes and asphaltenes (equations 6.1 and 6.2 respectively) was first–order, following previous work [e.g. Banerjee *et al.*, 1986]. The value of parameter k_m was 0.013 min^{-1} and k_a was 0.026 min^{-1} from experiments on the respective isolated fractions. The stoichiometric coefficients and solubility limit, S_l, were determined from data for pyrolysis of the maltenes and the whole residue. The values of the parameters were a = 0.221, b = 0.825, c = 0.02, y = 0.3 and S_l = 0.49 for whole residue and S_l = 0.61 when the maltenes were pyrolyzed separately. The curves shown in Figure 6–10 were from the equations (6.1) through (6.5) using these stoichiometric parameters and S_l = 0.49.

The ability of Wiehe's model to correlate data from pyrolysis of maltenes and whole residue with a single set of stoichiometric parameters suggests that this semiempirical approach has significant merit. The model was consistent with the observed induction behavior, and the maximum in asphaltene concentration. The solubility limit, S_l, depended on the initial composition as expected for a thermodynamic property of the reacting mixture. A more complete model will require prediction of the solubility limit based on the properties of the maltenes (solvent) and the reacted asphaltenes (solute), possibly using the solubility parameter (Section 2.9).

6.7 Integration of Coking and Hydroconversion Processes

Due to the favorable liquid yields from hydroconversion processes, these hydrogen–addition methods are very favorable for primary upgrading. Low hydrogen costs, and a lack of market for coke product are two factors that favor hydrogen addition. High conversion of residue requires recycling a large volume of liquid to the reactor inlet, which increases the size (and cost) of the reactor considerably (section 5.3.2). This effect of recycling is compounded by the low reactivity of the recycled material compared to the feed residue. To eliminate the unconverted residue as a by–product, it must be recycled to extinction.

An alternative is to couple a hydroconversion process with a coking unit. The unconverted residue from the hydrocracker is sent to the coker, where a portion is converted to distillates. This scheme, illustrated in Figure 6–11, combines the benefits of higher liquid yields from hydrogen

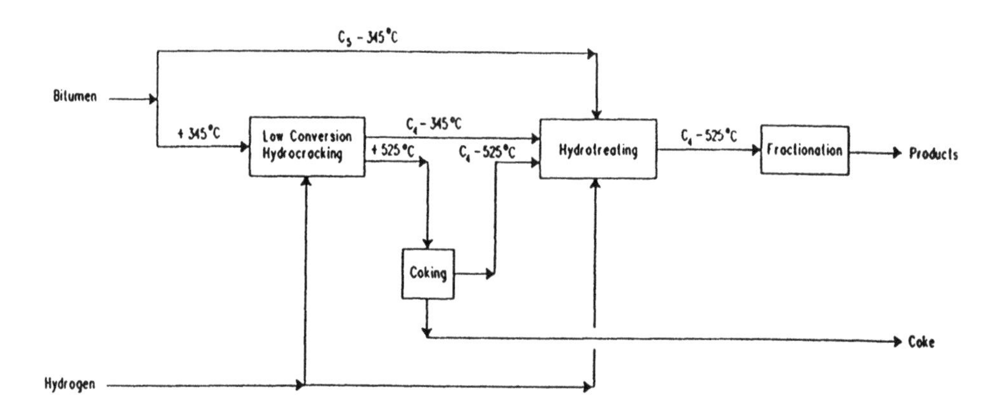

Figure 6–11
Coupling of hydroconversion and coking to eliminate residues

addition with the low cost of a coker. By feeding only unconverted residue to the coker, the overall yield of low–value coke is minimized.

6.8 Environmental Aspects of Thermal Processes

The major environmental concerns in coking processes have already been mentioned; sulfur dioxide emissions from combustion of coke that contains sulfur, and disposal of excess coke product. Sulfur dioxide has a wide range of effects on health and the environment, from bronchial irritation upon short term exposure to contributing to the acidification of lakes. Emissions of SO_2, therefore, are regulated in many countries.

The Flexicoking process offers one alternative to direct combustion of coke for process fuel; gasification to produce a mixture of CO, CO_2, H_2 and H_2S followed by treatment to remove the hydrogen sulfide. Maximizing the residue conversion and desulfurization of the residue in upstream hydroconversion units also maximizes the yield of H_2S relative to sulfur in the coke product. Current prices favor maximal residue conversion with minimum coke production, over gasification of coke [Menon and Mink, 1992].

Delayed coking requires the use of large volumes of water for hydraulic cleaning of the coke drum. This process water, however, can be recycled by skimming off the oil and filtering out suspended coke particles. If this water is used in a closed cycle, then the impact of delayed coking on water treatment facilities and the environment is minimized.

Thermal processing can significantly increase the concentration of polyaromatic hydrocarbons in the product liquids, because the low pressure, hydrogen deficient conditions favor aromatization of naphthenes and condensation of aromatics to form larger ring systems. To the extent that more compounds like benzo[a]pyrene are produced, the liquids from thermal processes will be more carcinogenic than unprocessed or hydroconversion products in the same boiling range. For example, Pasquini *et al.* [1989] found that a thermal petroleum pitch was significantly more mutagenic and carcinogenic than a petroleum asphalt. This biological activity was consistent with the higher concentration of polyaromatic hydrocarbons (PAHs), at 38.8 mg/g in the pitch compared to only 0.22 mg/g in the asphalt. Similarly, one would expect coker gas oils to contain more PAHs than unprocessed or hydroconverted distillates, and thereby give a higher potential for carcinogenic or mutagenic effects.

Subsequent hydroprocessing of the coker distillates would reduce the PAHs in the resulting product streams (Chapter 8), so that the only health concern outside the refinery itself is with high severity thermal products, such as pitches, which have not been hydrotreated. Coke solids would not pose a health hazard, and would have less environmental activity than unprocessed residue.

Further Reading

Gary, J.H.; Handwerk, G.E. 1984. *Petroleum Refining: Technology and Economics*, Marcel Dekker, New York.

Notation

a, b, c	stoichiometric parameters
A	reactant asphaltene
A*	cracked asphaltene
A^*_{max}	maximum solubility of cracked asphaltene
A^*_{ex}	excess of cracked asphaltene above soluble limit
C	coke product
k_a	rate constant for asphaltene decomposition, min^{-1}
k_m	rate constant for maltene decomposition, min^{-1}
M	reactant maltenes
M*	cracked maltenes
S_l	solubility limit for asphaltenes
V	volatiles
y	stoichiometric coefficient

Problems

6.1 An entrepreneur has approached your company trying to sell the rights to a secret additive which he claims will boost the yield of liquids from your coker, and at the same time desulfurize the liquid product. The asking price is $1.5 million, payable in advance. You must make a technical assessment of the proposal, and recommend whether or not the rights should be purchased. The following technical data are provided from batch experiments (conducted by the entrepreneur's assistant):

Feed: 1 kg. of Lloydminster bitumen

Analysis	Boiling Fractions	
3.38 wt% S	Naphtha (IBP—195)	7.1 vol%
0.25 wt% N	LGO (195—343)	28.3 vol%

0.99 wt% O	HGO (343—524)	21.6 vol%
84.33 wt% C	Resid (>525)	43 vol%
11.05 wt% H		

Reaction Conditions: Batch reactor, 560° C, 101.3 kPa, unknown residence time, secret additive

Products: Liquids: 806.8 mL, specific gravity 0.93
0.89 wt% S, 0.18 wt% N, 1.32 wt% O, 84.15 wt% C, 13.45 wt% H
Coke: finely divided solid, entrained in the liquid phase
26 g. yield, 7 wt% S, 0.5 wt% N, 89.5 wt% C, 3 wt% H
Gases: 0.19 m^3 (NTP), 40.7% H_2S, balance methane

All elemental and compositional analysis was conducted by a commercial laboratory.

CHAPTER

7

Fluid Catalytic Cracking of Residues

Fluid catalytic cracking (FCC) is one of the most commonly used refinery processes. FCC processes for conversion of gas oil to lighter distillates have been under continuous development for over 50 years, and they account for the majority of world–wide gasoline production and almost half the sales of catalysts to the petroleum industry [Occelli, 1988a]. Such a well established process has spawned a massive technical literature, therefore, the objective of this chapter is to describe briefly the fluid catalytic process, then discuss how it has been modified to handle feeds containing vacuum residue material. Adapting the FCC technology to handle more residue in the feed blend has received more attention over the past twenty years, and FCC technology is now available for feeds containing residues provided that total metals content is less than 30 ppm (Ni + V) and Conradson carbon residue (CCR) is less than ca. 8 weight% of the feed [Stripinis, 1991].

7.1 Process Description

7.1.1 Catalytic cracking chemistry

The objective of conventional FCC is to convert gas oil feed into naphtha and middle distillate. Heterogeneous catalysts bearing acid sites can accelerate the rate of cracking at a given temperature, and improve the selectivity toward stable products in the gasoline range as compared to non–catalytic cracking (thermal cracking). Acid sites can break aliphatic carbon–carbon bonds via a carbonium ion intermediate (Section 3.3.5). The reactions of the various hydrocarbon classes are summarized in the following table:

Table 7–1
Reactions in Catalytic Cracking
(From Decroocq, 1984)

Hydrocarbon	Reactions	Major Products
Paraffins	Cracking Isomerization	C_3+ isoparaffins and olefins
Naphthenes	Cleavage of rings and side chains Dehydrogenation	paraffins, olefins and aromatics
Hydro–aromatics	Opening of naphth–ring, cleavage of side chains Dehydrogenation	paraffins, olefins and aromatics
Aromatic rings	Negligible cracking	coke on catalyst
Aromatic side chains	Cleavage from ring	olefins and aromatics
Olefins	Cracking Hydrogen transfer	branched olefins, diolefins and paraffins
Diolefins	Oligomerization Cycloaddition	polyaromatics, coke

7.1.2 Cracking in riser reactors

Cracking of hydrocarbons over an acidic catalyst at low pressure inevitably leads to deposition of coke on the catalyst surface, which in turn deactivates the catalyst in 10–100 s. Such rapid deactivation led to the use of fluidized beds for catalytic cracking, which allowed pneumatic transport of powdered catalyst into the reactor and removal of the spent catalyst for regeneration. The spent catalyst was regenerated by burning off the coke, then it was returned to the reactor.

The development of newer high–activity catalysts allowed the development of riser reactors during the 1960's, as illustrated in Figure 7–1. The feed is mixed with the hot regenerated catalyst, which consists of particles of diameter ca. 70 μm. The vaporized feed travels upward with the catalyst at 470–510° C, through a dense bed of catalyst and through a series of cyclones to exit the reactor. A typical feed ratio is 5.7–9.5 kg catalyst/kg feed. Most of the cracking occurs as the feed and catalyst travel cocurrently up the riser. The catalyst deactivates very quickly, even with clean feed at low temperature, as illustrated in Figure 7–2 for a silica–alumina catalyst. The products are fractionated downstream, and a portion of the unconverted gas oil is recycled and blended with fresh feed entering the bottom of the riser. This recycle stream, often called FCC cycle oil, is much more aromatic the the fresh feed, and therefore much less reactive.

The catalyst passes downward from the dense bed through a steam–stripping zone, where hydrocarbons are removed, and into the regenerator. Combustion air is blown into the regenerator to maintain the catalyst in the fluidized state and burn the coke at 670–720° C [Decroocq, 1984]; the flow rate of air must be sufficient to provide 1–2% excess oxygen. Approximately 11–14 kg of air is required per kg of coke. The regenerated catalyst must have a low residual coke content to achieve high activity. In addition, low residual coke levels are desirable to avoid auto–catalysis of coking in the reactor. Typical residual levels are < 0.1 wt%. The combustion gases leave the regenerator through a series of cyclones (not shown in Figure 7–1). Additives such as Cr_2O_3, MnO_2, or Pt are used to enhance combustion of coke from the catalyst and give complete conversion of CO to CO_2 [Otterstedt *et al.*, 1986]. This complete oxidation maximizes the heat release from the regenerator, and can eliminate the need for a feed preheater and a flue gas CO boiler that were included in earlier designs.

The schematic shown in Figure 7–1 is representative of the riser–type units for processing gas oil and gas–oil/residue blends. Many other designs have been used [Murcia, 1992], but this diagram shows the basic pattern of catalyst use and regeneration. The catalyst suffers from attrition in all of this handling, and the fine particles of ground catalyst exit with the products from the reactor and with the combustion gases from the regenerator. Typical losses are 0.43–0.71 kg catalyst/m^3 feed, therefore, catalyst must be continuously added to the reactor from a feed hopper. This large consumption of catalyst helps to account for its domination of catalyst sales for petroleum processing.

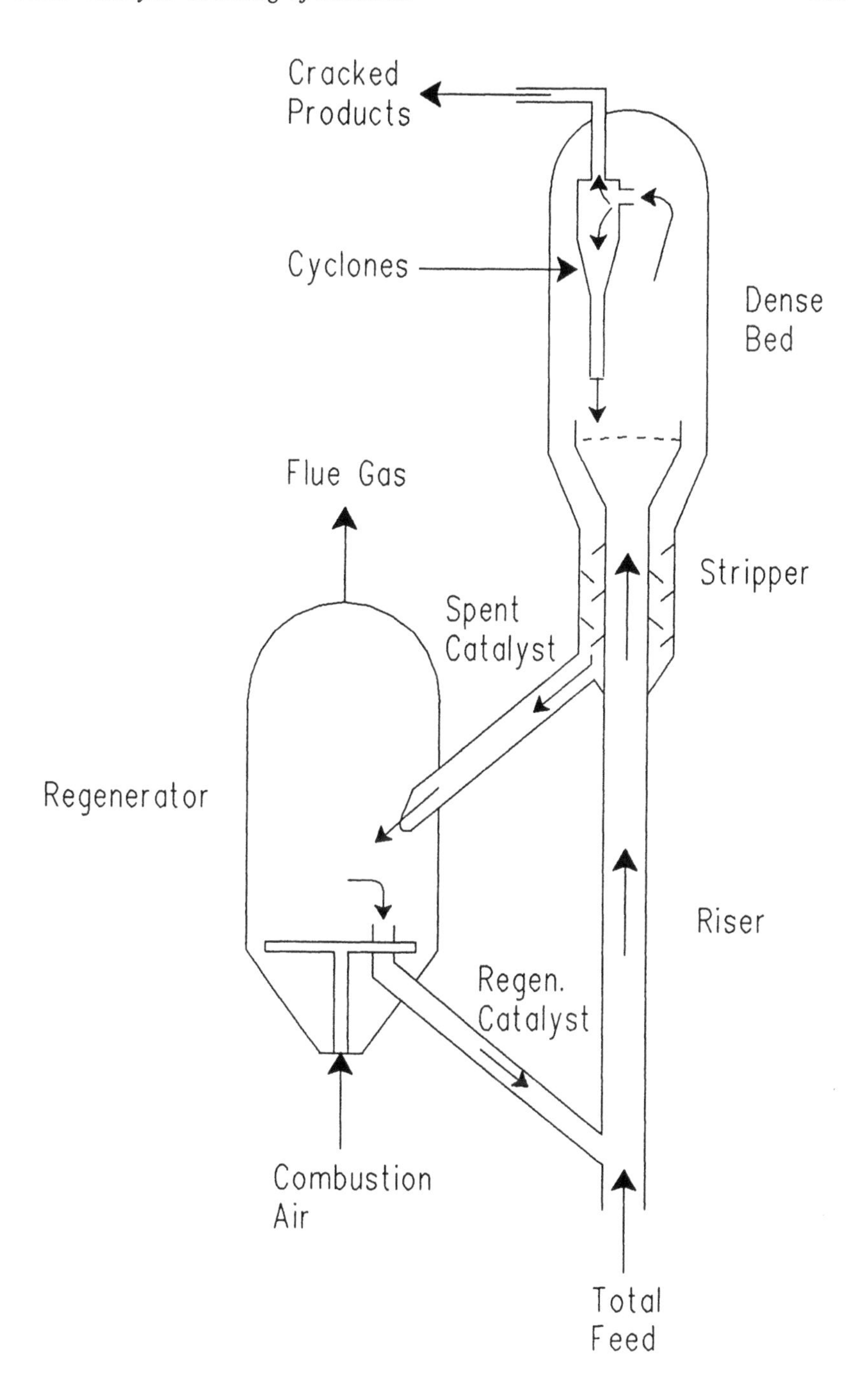

Figure 7—1
Representative fluid catalytic cracking unit (From Weekman, 1979)

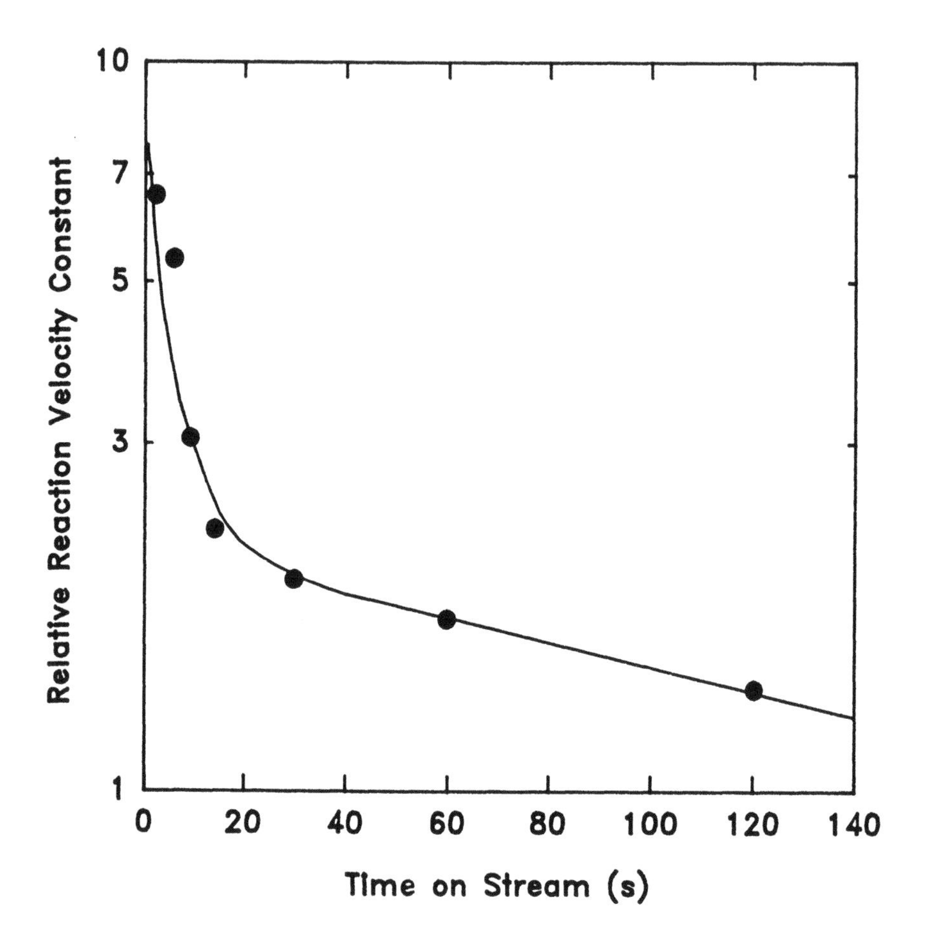

Figure 7—2
Deactivation of zeolite cracking catalyst with n—hexadecane feed at 480°C (Data from Weekman, 1979)

7.1.3 Catalyst composition

High yields of gasoline from the riser—type reactor are possible because of the development of high activity zeolite—based catalysts. Two types of acidic materials are combined in formulating FCC catalysts; 2—10 μm particles of zeolite are held in a matrix of amorphous silica-alumina. Typical catalysts contain 5—20% zeolite by weight. Zeolites are hydrated alkali—aluminum silicates; a common type for cracking of hydrocarbons is the synthetic sodium—aluminum silicate known as zeolite Y. The acidity is increased by replacing sodium with alkaline earths, rare earths, or hydrogen. The most common type of zeolite for catalytic cracking is ultrastable Y (USY), prepared from the zeolite Y form by exchanging the sodium for hydrogen, then treating with steam to reduce the alumina content for high—temperature stability and high acidity. The superacid sites on USY are able to protonate n—alkanes, thereby initiating carbonium ion reactions (Section 3.3.5).

The zeolite provides very high acidic activity, but has pores ca. 9 Å in diameter which are inaccessible to large hydrocarbon molecules. These species crack, therefore, on the exterior of the zeolite and in the silica-alumina matrix.

7.1.4 Enthalpies and kinetics of FCC units

The circulating catalyst carries not only coke, but also enthalpy because the regenerator operates at a higher temperature than the reactor. For steady—state operation, the following enthalpy balance must apply to the catalyst:

$$\text{Enthalpy gained in regenerator} = \text{enthalpy lost in reactor} \tag{7.1}$$

The enthalpy gained in the regenerator is related to the yield of coke in the reactor, y_C, as a weight fraction of the feed:

$$y_c(\text{Regenerator NHA}) = \text{Reaction Heat Required} \tag{7.2}$$

$$y_c = \text{Reaction Heat Required}/(\text{Regenerator NHA}) \tag{7.3}$$

$$\text{Reaction Heat Required} = \Delta H_r \cdot X + q_{loss}^{react} + \Sigma C_{p,i}(T_r - T_{i,in})w_i \tag{7.4}$$

$$\text{Regenerator Net Heat Available} = -\Delta H_c - q_{loss}^{regen} - \Sigma C_{p,j}(T_c - T_{j,in})w_j \tag{7.5}$$

where ΔH_r and ΔH_c are the enthalpies of reaction (endothermic) and combustion (exothermic) as kJ/kg of feed cracked and kJ/kg coke burned respectively, q is heat lost from vessels and transfer lines, and C_p are the heat capacities of the streams entering each vessel at an inlet temperature T_{in}. In the reactor, the feed, riser steam, stripping steam, feed diluent, and inert gases from the catalyst (with flows of w_i kg/kg feed) all must be heated to the reactor outlet temperature, T_r. In the regenerator, the combustion air and the coke on the catalyst (with flows w_j kg/kg coke) are heated to the combustion temperature, T_c. The terms in the Reaction heat required (equation 7.4) have units of kJ/kg of feed, while the terms in equation (7.5) for the regenerator have units of kJ/kg of coke on the catalyst.

Two methods are used to control the reactor temperature, T_r. The flow rate of regenerated catalyst into the reactor can be varied, using slide or plug valves. The relationship between catalyst flow rate and reactor temperature is given by:

$$m_{cat} = y_c \cdot F_{ao}(\text{Regenerator NHA})/[C_{p,cat}(T_c - T_r)] \tag{7.6}$$

where m_{cat} is the catalyst circulation rate and F_{ao} is the feed rate in kg/h. In this mode of operation, the coke yield ($m_{cat}y_c$) is almost constant, but the flow rate of catalyst varies to maintain T_r constant. The alternative mode of operation is to change the inlet temperature of the feed oil to maintain a constant reactor temperature.

The kinetics of the FCC reactors differ from the basic types discussed in Chapter 4, because the catalyst moves cocurrently with the feed up the riser. Following the derivation given by Fogler [1986], the design equation for the riser reactor is:

$$F_{ao}\, dX/dW = a\, W\, (-r_a) \tag{7.7}$$

where X is the conversion, W is the mass of catalyst in the riser, a is the catalyst activity, and r_a is the reaction rate. The feed to an FCC unit is a wide-boiling mixture, so following the discussion of section 4.4 it exhibits apparent second–order kinetics. Indeed, this observation of second–order behavior in pilot units for FCC prompted the early work on lumped kinetics [Weekman, 1979].

$$-r_a = kC_a^2 \tag{7.8}$$

where C_a is the concentration of reactant. The catalyst activity follows an exponential decay in activity with time of contact with the feed (Figure 7—2), which can be written:

$$-\,da/dW = k_d a/m_{cat} \tag{7.9}$$

and upon integrating over the length of the riser

$$a = \exp(-k_d W/m_{cat}) \tag{7.10}$$

Note that W/m_{cat} is the mean residence of time of catalyst in the riser, τ_{cat}. The relationship between concentration and conversion is given as

$$C_a = C_{ao}(1 - X) \tag{7.11}$$

Substitution of equations (7.8), (7.10) and (7.11) into equation (7.7), and integrating from the inlet to the outlet of the riser gives:

$$X/(1 - X) = \frac{k\ C_{ao}^2}{k_d} \cdot \frac{m_{cat}}{F_{ao}} [1 - \exp(-k_d \tau_{cat})] \tag{7.12}$$

Inspection of this equation shows the following proportionality between the conversion and the operating variables:

$$X/(1 - X) \approx A \cdot \frac{m_{cat}}{F_{ao}} \tag{7.13}$$

A plot of X/(1—X) versus the catalyst/oil feed ratio will, therefore, be linear.

7.2 Effects of Residue Feed on FCC Operation

Introduction of residue into the feed blend has several effects on the operation of an FCC unit, due to the differences in chemistry between gas oils and residues. The major effects are as follows:

1. Increase in CCR Content
   Residue can have a significant tendency toward coking, as measured by Conradson Carbon residue. When the feed is sprayed onto the catalyst at the base of the riser, some residue components will remain liquid and undergo coking reactions. Stokes and Mott [1989] give a rule of thumb that 75% of the CCR content of the feed is converted to coke on the catalyst.

2. Presence of metals
   The vanadium and nickel content of the feed will be deposited on the catalyst. The metals on the catalyst promotes dehydrogenation reactions, giving increases in the yield of coke and light ends. Nickel is generally more active for these reactions. At higher levels of accumulation, these metals can also block access to pores in the catalyst. This pore blockage reduces the surface area of the catalyst, and its activity (Figure 7—3).

3. Increase in sulfur content
   A typical residue fraction has more sulfur than a gas oil. A portion of this sulfur will be captured in the coke, and released as SO_2 from the regenerator. The remainder will exit the reactor as H_2S and sulfur in the distillates.

4. Increase in nitrogen content
   Nitrogen bases tend to poison the acid sites of the catalyst, which reduces the activity of the catalyst. These nitrogen compounds are subsequently removed in the regenerator, with attendant increases in the emissions of NO_x compounds.

The overall effects of adding residue to an FCC unit, either by blending or by feeding atmospheric residue, are to reduce performance relative to feeding gas oil alone. Residue increases catalyst coking and poisoning by nitrogen compounds, and deposits metals that reduce selectivity for desired products, so that the addition of residue incrementally reduces the yield from the gas oil diluent. Note that these comments are for typical residues only; addition of highly paraffinic residues which are devoid of heteroatoms and metals will actually give an incremental boost in gasoline yield [Decroocq, 1984].

Stokes and Mott [1989] defined Residue FCC as using a feed with more than 1% CCR content. They identified 31 units in North America operating with residue feed by this definition. Three sources of residue

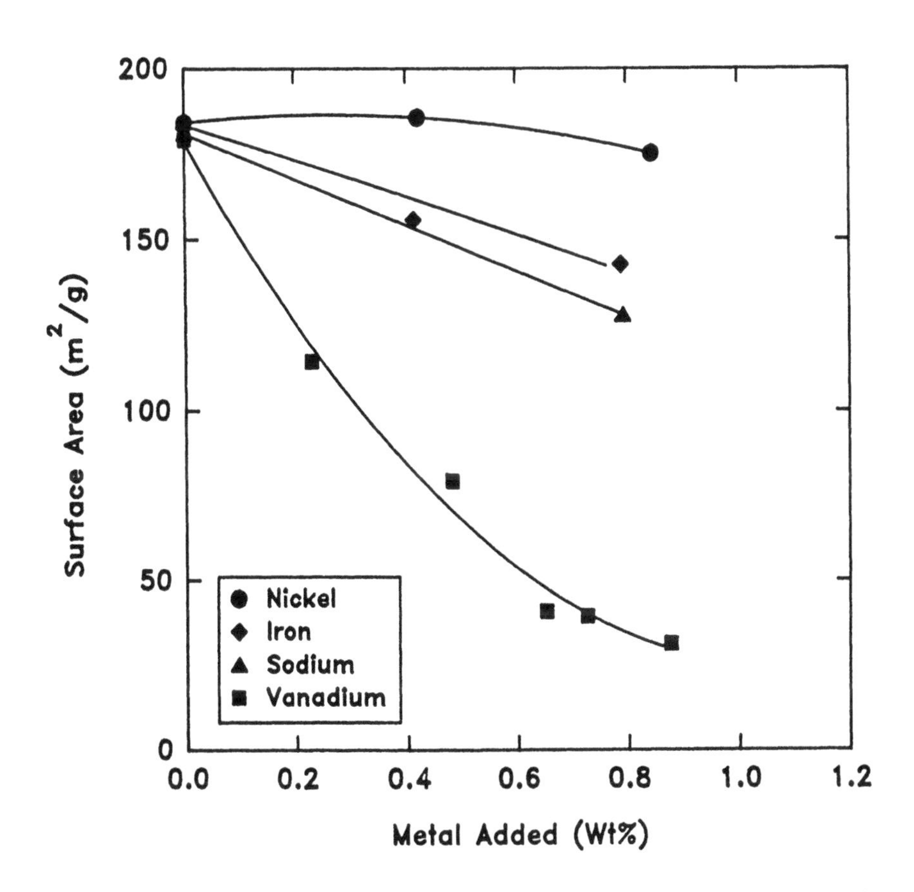

Figure 7–3
Loss of surface area due to addition of metals to FCC catalyst.(Data from Otterstedt *et al.*, 1986)

were identified: atmospheric tower bottoms, vacuum tower bottoms, and deasphalted oil. The fraction of residue in the feed averaged 43% when atmospheric tower bottoms were used as feed. When vacuum bottoms or deasphalted oil were blended with gas oil, the residue fraction averaged only 20% of the total feed.

Average operating conditions for these residue FCC units are given in Table 7—2. The regenerator temperatures were higher than the norm, as were the reactor temperatures, due to the increased deposition of coke on the catalyst and the heat balance relationship between the two operations (equation 7.2). The catalyst consumption was higher than typical FCC units (0.94 kg/m^3 compared to 0.43—0.71 kg/m^3) because of the deposition of metals on the catalyst. The equilibrium catalyst properties are given in Table 7—3.

Table 7—2
Average Operating Conditions and Yields for Residue FCC Units
(Data from Stokes and Mott, 1989)

Total feed rate, m^3/d	4770
Feed CCR content, wt%	2.09
Feed density, kg/m^3	904
Catalyst addition rate	
tonnes/d	4.5
kg/m^3	0.94
Reactor temperature, °C	523
Combined feed temperature, °C	249
Regenerator temperature, °C	716
Feed Conversion, vol%	76
Gasoline, vol% of fresh feed	59

Table 7—3
Average Equilibrium Catalyst Properties
(Data from Stokes and Mott, 1989)

	Residue Units	All Units
% Activity	66.3	67.3
Na, wt%	0.44	0.39
V, wppm	1950	1200
Ni, wppm	1670	880

The data of Table 7—3 show that the residue FCC units operate with higher levels of contaminants in the catalyst, so that the increase in metals in the feed is not exactly balanced by the increase in catalyst consumption. Such a buildup of metals reduces both conversion and selectivity for gasoline product. For example, in cracking a Mid Continent gas oil, increasing the metals content of a zeolite catalyst from 180 ppm to 3500 ppm reduced conversion from 79% to 71.5%, and reduced the yield of gasoline from 61 vol% of feed to 54% at constant 70% conversion [Decroocq, 1984].

Stokes and Mott [1989] suggest a useful kinetic relationship between conversion and coke deposition. Plots of X/(1—X) versus catalyst/feed ratio using data from a pilot plant were linear, following equation (7.13). For FCC units with slide valves, wherein catalyst/feed ratio is varied to control reactor temperature, the coke yield on the catalyst will be linearly proportional to the catalyst/feed ratio, as can be seen by rearranging equation (7.6):

$$y_c = (m_{cat}/F_{ao})[C_{p,cat}(T_c - T_r)]/(\text{Regenerator NHA}) \tag{7.14}$$

In this type of operation, the yield of coke and the catalyst/oil ratio will vary to maintain constant reactor temperature. Combining equations (7.13) and (7.14), we would expect $X/(1-X) \propto y_C$. For residue feeds, a plot of X/(1—X) versus y_C gave an intercept at ca. 0.75 times the CCR content of the feed. The minimum coke, therefore, was determined by CCR content, and thereafter the coke deposited on the catalyst increased with conversion. Stokes and Mott's operating curve follows equation (7.15):

$$X/(1 - X) = 0.75\ \text{CCR} + A'y_c \tag{7.15}$$

Operating lines of this type are illustrated in Figure 7—4 for four different feeds with CCR contents ranging from 1.25 to 6.9 wt%. According to Stokes and Mott [1989], the slopes of these curves depend on the metals in the catalyst, but no quantitative data were given. The general trend with CCR content, however, is quite clear in Figure 7—4. As the CCR content increases, the operating line shifts to the right and in two cases the slope of the line also decreases. The slope of the operating curve would decrease further if the metals, nitrogen, or aromatics content of the feed were to increase.

This discussion illustrates how FCC operation responds to residue components in the feed. The following sections deal with process modifications to enhance the performance of catalytic cracking at high levels of residue in the feed.

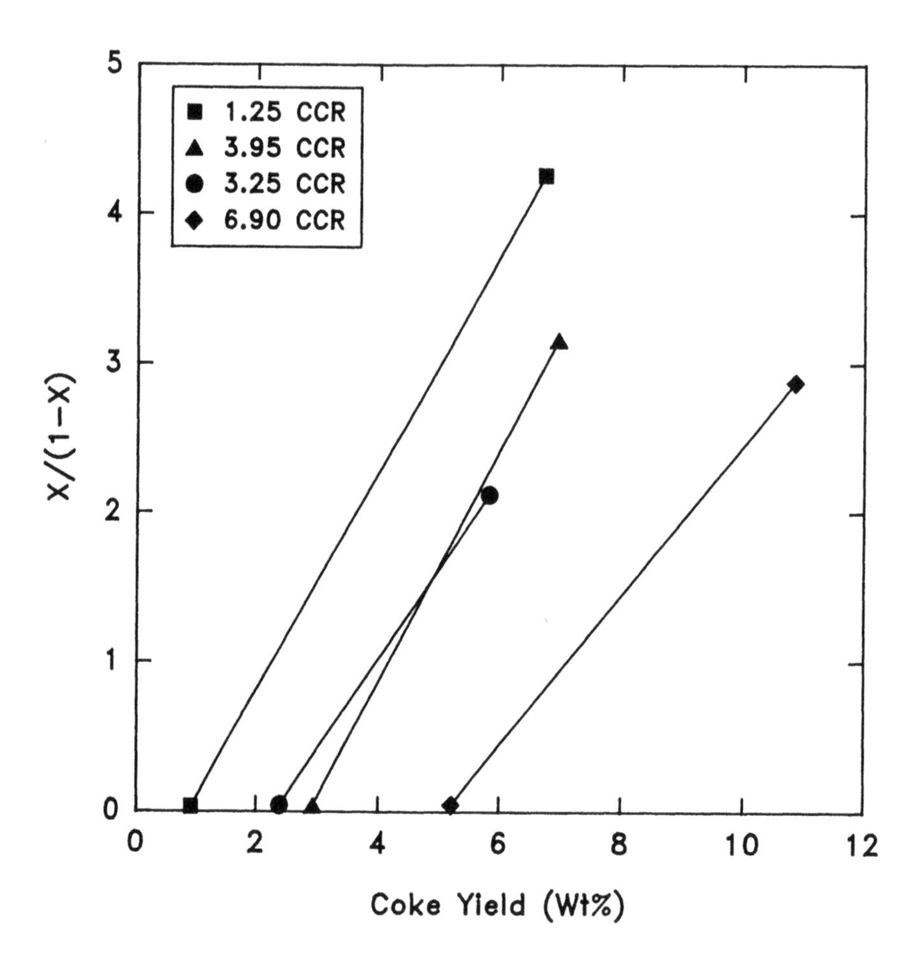

Figure 7—4
Performance lines for residue FCC (Data from Stokes and Mott, 1989)

7.3 Catalyst Modifications for Residue FCC

7.3.1 Metal–resistant catalysts

The metal contaminants in residue have three distinct effects on zeolite–based cracking catalysts; loss of surface area due to plugging (mainly attributed to V), destruction of the zeolite through reactions with the V, and catalysis of coking and hydrogen formation (mainly attributed to Ni). Sodium in the feed can cause deactivation by the active sites, while Cu and Fe can contribute to the effects noted above for Ni and V.

The main strategy for dealing with metals is trapping, whereby the metals react with scavenging compounds in the catalyst to form inert species. A wide variety of materials have been tested, and some examples are as follows:

Metal	Traps
Nickel	Sb, Sn, magnesium silicate [Occelli, 1988b]
Vanadium	magnesium silicate, alumina, coke

Addition of antimony to control the effect of nickel is a common strategy, and likely accounts for the successful operation of FCC catalysts with fairly high Ni content (Table 7–3). Commercial patents for the use of additives to control vanadium have been awarded to a number of companies [Otterstedt *et al.*, 1986]. The use of residual coke, described by Hettinger [1988], is an operating strategy that seeks to control the valence state of the vanadium and keep it below +5. By maintaining the vanadium as V_2O_4, migration in the catalyst was reduced and deactivation controlled.

An alternative strategy is to remove the contaminating metals by regeneration outside the FCC unit. Several processes are available to remove 30–40% of the vanadium and 80–90% of the nickel from zeolite catalysts [Elvin *et al.*, 1988]. Demetallization may be an attractive alternative to catalyst replacement, and may supplement the chemical trapping strategy.

7.3.2 Trapping of sulfur

Processing of residue feeds usually increases sulfur content. A significant fraction of the feed sulfur is deposited on the catalyst in the coke, then oxidized to SO_2 and SO_3 when the catalyst is regenerated, e.g.

$$S\ (\text{coke}) + O_2 \longrightarrow SO_2 \qquad (7.16)$$

$$SO_2 + 1/2\ O_2 \longrightarrow SO_3 \quad (7.17)$$

The deposition of sulfur in the coke depends on the feed chemistry and operating conditions; for feeds containing 0.21—0.34% sulfur, the coke contained 2.8 to 20% of the total sulfur in the products, so that 2.8—20% of the feed sulfur would be liberated as SO_2 from the regenerator.

By adding sulfur scavengers to the regenerator to compete with reactions (7.16) and (7.17), a portion of the sulfur oxides can be trapped.

$$MO + SO_2 + 1/2\ O_2 \longrightarrow MSO_4 \quad (7.18)$$

In this case, the scavenger is a metal oxide (MO) that reacts to form the sulfate. By selecting compounds that release H_2S under reducing conditions in the cracking reactor, a cycle can be established:

$$MSO_4 + 4\ H_2 \longrightarrow MO + H_2S + 3\ H_2O \quad (7.19)$$

A variety of metal oxides give sulfates at regenerator conditions and hydrogen sulfide at reactor conditions, including MgO, CeO, and Al_2O_3 [Otterstedt *et al.*, 1986; Hirschberg and Bertolacini, 1988]. Other metals are effective for promoting reaction (7.18), and include Pt and rare earths.

Two strategies can be used to introduce these sulfur traps into the regenerator: either incorporate these materials into the catalyst itself, or add the sulfur scavenger as a separate solid phase. In the latter case, the scavenger merely needs to have similar particle size and density to the catalyst. Addition of a scavenger independent of the catalyst has proven fairly popular with refiners, because they can adjust the operation of the FCC in response to changes in sulfur in the feed [Rheaume and Ritter, 1988]. By adjusting operating conditions, the commercially available sulfur traps can remove up to 80% of the sulfur from the regenerator [Occelli, 1988a]. More effective sulfur removal from the feed can only be achieved by flue gas desulfurization (very expensive), or by hydrotreating the feed prior to FCC.

7.3.3 High—boiling feed

When residue is added to a feed, components with boiling points over ca. 560° C are deposited on the catalyst as a liquid phase. If the catalyst lacks large pores, then this liquid will coke on the exterior of the catalyst giving poor conversion to gasoline and a possible reduction in catalyst activity. The importance of sufficient pore volume in residue FCC catalysts was discussed by Hettinger [1988], who suggested an optimal range of 0.4—0.55 mL/g, which is similar to Mo—alumina hydrocracking catalysts (Chapter 5). Achieving this pore volume, with an appropriate distribution of pore sizes, requires manipulation of the composition and manufacture of the Si—Al matrix of the catalyst. Hettinger [1988] describes the use of kaolin to aid in macropore formation, while Stokes and Mott [1989] discuss optimizing the ratio of zeolite to the matrix. Optimal design of the pore size distribution requires capacity for liquids, as well as good diffusive transport of high-molecular weight material to cracking sites on the Si/Al matrix, and resistance to pore—mouth plugging by deposited metals and coke.

7.4 Process Modifications for Residue FCC

7.4.1 Catalyst regeneration

The regenerator illustrated in Figure 7—1 uses air flow to fluidize the catalyst bed as coke is removed. This fluid bed is essentially well—mixed in the solid phase, so that the catalyst leaving the regenerator has a wide distribution of residence times and therefore a distribution of coke content. In addition, fluidization of the bed by bubbling air through it gives relatively poor gas/solid contacting and can give breakthrough of both CO and oxygen into the gas phase above the bed. Combustion of these gases heats the upper portion of the regenerator, but not the catalyst itself [Otterstedt *et al.*, 1986].

Two modified designs have been used to control these regeneration problems. One method is to use two stages of regeneration. Catalyst enters an upper bed, where coke is partly removed. The catalyst is then carried into a second regenerator chamber where it contacts fresh combustion air. In some designs the first stage is on top of the second and in others it is below, but in every case the two stages are within a single vessel [Murcia, 1982]. A second approach is to use a riser tube in the regenerator, to control the contacting between air and catalyst [Otterstedt *et al.*, 1986; Murcia, 1992]. In either case, the modified regenerator gives better control of the residence time of the catalyst in contact with combustion air.

The other aspect of residue processing is the heat balance. Operation with an extreme feed, with ca. 7—8 wt% CCR, at high conversion gives a large excess of coke over and above what is required to heat the reactor to optimal temperatures. The operating lines in Figure 7—4 illustrate the conversion—coke relationship. In a heat—balanced unit, the two alternatives are to run both vessels at higher temperatures or to run at lower conversion. The former case can give problems with catalyst deactivation and vessel metallurgy, due to high regenerator temperatures. The latter option results in a loss of valuable light product and more heavy gas oil.

The heat balance for the FCC unit can be made more favorable by inserting steam coils into the regenerator or catalyst transfer lines, which give the following version of equation (7.5):

$$\text{Regenerator Net Heat Available} = -\Delta H_c - q_{loss}^{regen} - q_{steam} - \Sigma C_{p,j}(T_c - T_{j,in})w_j \quad (7.20)$$

By reducing the net heat available from the regenerator, the unit can operate at a higher coke yield (y_C from equation (7.3)) without increasing the operating temperatures. Two approaches have been used in commercial units; steam coils in the regenerator vessel, and external catalyst coolers [Murcia, 1992].

7.4.2 Reactor modifications

As mentioned in section 7.3.3, the addition of residue to an FCC unit involves the addition of a liquid phase to the original gas/solid reactor design. Some of the changes in reactor design include [Otterstedt *et al.*, 1986]:

1. Using a vertical riser, as in Figure 7—1, to control contact time between catalyst and oil and minimize overcracking.

2. Installing a 90° bend at the top of the riser, followed by a cyclone to give rapid disengagement of the catalyst and the vapor stream.

3. Multiple feed injection points and addition of excess steam at the feed points to ensure good mixing between the feed and the hot catalyst. This mixing eliminates hot spots at the bottom of the riser tube.

7.4.3 Pretreatment of FCC feed

As in the case of hydroconversion (Chapter 5) and coking (Chapter 6), hydrogenation of the feed prior to cracking enhances the rate and selectivity of the cracking reactions. In the case of FCC, prehydrogenation has a double benefit because yield is improved by conversion of aromatics to naphthenes, and the deactivation of the catalyst is reduced because CCR content is lower, nitrogen content is reduced, and metals content is lower. A schematic process diagram is given in Figure 7—5. The impact of hydrogenating the feed to an FCC is illustrated by the data in Table 7—4:

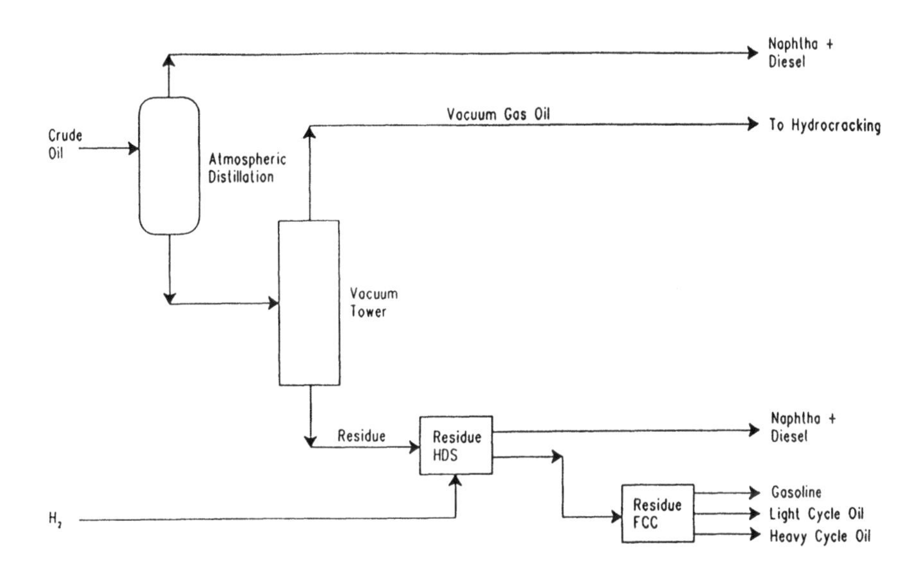

Figure 7–5
Combination of residue hydroconversion with FCC process

Table 7–4
Effect of Hydrogenation on FCC Yields from Arabian Light Atmospheric Residue (Data from Reynolds *et al.*, 1992)

	No Pretreat	Pre Hydro-treat
FCC Feed Properties		
Boiling point, °C	370+	370+
API	15.1	20.1
CCR, wt%	8.9	4.9
S, wt%	3.30	0.48
N, wt%	0.17	0.13
Ni + V, ppm	51	7
FCC Yields (liquid vol % except as noted)		
H_2S, wt%	1.7	0.2
C_2–C_4	24.8	25.3
Gasoline (C_5–221°C)	50.6	58.0
Light cycle oil (221°C – 360°C)	21.4	18.2
Bottoms (370°C+)	9.7	7.2
Coke, wt%	10.3	7.0
Catalyst use, kg/m^3	4.89	0.66
Catalyst cooler required	Yes	No

These data illustrate that significant hydrotreatment of the feed (in this case removal of ca. 85% of the sulfur and metals) can give a significant improvement in yields and reduce catalyst consumption. Note that hydrotreating reduced the metals content to 14% of its original value, and as we would expect the catalyst addition rate dropped to 14% of its original value (from 4.9 kg/m^3 to 0.66 kg/m^3). Using CCR as a criterion, hydrotreatment converted this feed from a material near or beyond the nominal limits for economical FCC operation (i.e. $<$ 8% CCR and 30 ppm metals [Stripinis, 1991]) to a more conventional feed within the normal residue FCC composition range discussed by Stokes and Mott [1989].

7.5 Environmental Aspects of Residue FCC

One of the environmental concerns for FCC processes has already been mentioned; sulfur dioxide emissions from combustion of coke on catalyst. Even with the addition of sulfur scavengers, the FCC process can be expected to emit 0.5 to 4% of the total sulfur in the feed oil as SO_2 from the regenerator. For example, Menon and Mink [1992] forecast that 0.8% of the feed sulfur to a deasphalting/FCC operation would be emitted as SO_2.

Related to emissions of sulfur are emissions of NO_x from the FCC regenerator. Nitrogen compounds tend to appear in the coke, just as sulfur does. This nitrogen can be released as NO_x during the combustion process. Some of the catalyst modifications discussed in section 7.3 can increase NO_x emissions, both from the coke and from the nitrogen in combustion air, including addition of combustion promoters and sulfur scavengers [Otterstedt *et al.*, 1986]. Control of CO and SO_x may, therefore, increase other emissions. Menon and Mink [1992] estimate that ca. 10% of the feed nitrogen would be emitted as NO_x (based on total moles NO_x emissions divided by total feed). The significance of this value must be assessed along with NO_x from other sources (heaters and compressors) on a given site.

The other emission of concern from FCC units is particulates from the regenerator. Attrition of the catalyst gives fine particles, which are carried in the flue gas. These dust emissions can be controlled by cyclones and scrubbers. Emission limits on the order of 50 mg/m^3 [Menon and Mink, 1992] can be met by these methods.

Like thermal processing, FCC can significantly increase the concentration of polyaromatic hydrocarbons in the recycle liquids. The same reactions that give coke on the catalyst can also give aromatization of naphthenes and condensation of aromatics to form larger ring systems. FCC cycle oils, therefore, can be expected to be more carcinogenic and mutagenic than the feed. Subsequent hydroprocessing or blending of cycle oils would reduce the PAHs in the resulting product streams.

Notation

a	catalyst activity
A, A'	parameters in equations 7.13 ans 7.15
C_p	heat capacity, kJ/(kg—K)
C_a	concentration of reactant, kg/m^3
F_{ao}	feed rate to reactor, kg/s
ΔH_c	enthalpy of combustion, kJ/kg coke
ΔH_r	enthalpy of cracking, kJ/kg feed cracked
k	reaction rate constant
m_{cat}	catalyst circulation rate, kg/s
NHA	net heat available
q_{loss}	heat losses, kJ/kg coke or kJ/kg feed
q_{steam}	heat removed for steam generation, kJ/kg coke

r_a	rate of cracking, kg/kg catalyst/s
T_c	regenerator temperature, ^oC
T_{in}	inlet temperature, ^oC
T_r	reactor temperature, ^oC
w	flow of stream as kg/kg feed or kg/kg coke
W	mass of catalyst in riser, kg
X	conversion of feed
y_c	yield of coke, as weight fraction of feed
τ_{cat}	mean residence time of catalyst, s

Subscripts

d	deactivation
i, j	counters
o	initial condition

Superscripts

react	reactor
regen	regenerator

Further Reading

Decroocq, D. 1984. *Catalytic cracking of heavy petroleum fractions.* Editions Technip, Paris, 123 pp.

Occelli, M.L. (ed) 1988. *Fluid Catalytic Cracking: Role in modern refining*, ACS Symp. Ser. 375, ACS, Washington DC, 353 pp.

Otterstedt, J.E.; Gevert, S.B.; Jaras, S.G.; Menon, P.G. 1986. "Fluid catalytic cracking of heavy (residual) oil fractions: A review". *Appl. Catal.* 22, 159–179.

Weekman, V.W. 1979. "Lumps, Models, and Kinetics in Practice", AIChE Monograph Series 11(75), 29 pp..

Problems

7.1 Use the data from Table 7.4 to construct operating lines for Arabian light residue in an FCC before and after hydrotreatment the FCC unit. How much coke would be expected at 80% conversion with each feed? Does the presence of metals on the catalyst change the slope of the operating line?

7.2 What is the equilibrium metals content of the catalyst in Table 7.4?

7.3 Assuming a sulfur capture efficiency of 70% (i.e. mol sulfate per mol metal), how much MgO must be added to the catalyst stream in the example of Table 7–4 to reduce SO_x emissions from the regenerator by 80%? Assume that the coke has the same sulfur content as the feed.

CHAPTER

8

Hydrotreating of Cracked Products

In Chapter 5, the term "hydroconversion" was used as a general term for processes that cracked hydrocarbons in high–pressure at temperatures in excess of 420°C. The use of catalysts in these processes ranged from heterogeneous supported–metal catalysts, such as Ni/Mo on alumina, to disposable additives such as iron sulfate. In contrast, hydrotreating is intended for selective removal of the heteroatoms from the feed with little attendant conversion of the hydrocarbons. Operating temperatures, therefore, are below 400—410°C, and an active supported-metal catalyst is required. The discussion in this chapter will focus on hydrotreating of distillates from primary cracking processes, including hydroconversion, coking, and FCC. When a feed has undergone a primary cracking step, the next stage of hydrogenation is sometimes referred to as "secondary upgrading". Other terms for this process include hydrofining and hydrorefining.

Direct hydrotreating of residues is desirable in some cases, either as a pretreatment for cracking processes or to reduce sulfur. Since residues usually contain some metals, such residue hydrotreatment would have characteristics intermediate between catalytic hydroconversion (Sections 5.2—5.4 and 5.7) and distillate processing because of deactivation of the catalyst.

8.1 Properties of Feeds for Hydrotreating

The liquid products from the primary upgrading of residues and bitumens have characteristics which are improved over the raw material, but fall short of the requirements of a conventional refinery. Typical specifications for synthetic crude oils are listed in Table 1—14. Even though the catalysts in the ebullated bed processes (LC—Fining or H—Oil) give some removal of sulfur and nitrogen, the resulting oil may still require secondary hydrotreating to reduce the heteroatom content to the levels of light crude. Products from coking or FCC processes may also require further hydrotreating to meet specifications, depending on the sulfur and nitrogen content of the feed. The following table lists some example products from upgrading processes:

Table 8—1
Examples of Product Properties from Primary Upgrading Processes

Property	LC—Fining[1]	H—Oil[2]	CANMET[3]	Fluid Coker[4]
S, Wt%	1.66			4.12
N, ppm.	332			3050
Naphtha (82—177° C)				
Wt%	17.4	16.9	18—28	0.5
N, ppm	300	430		
S, ppm	900	2000		
Mid—Distillate (177—343° C)				
Wt%	27.2	38.5	30—38	32.4
S, wt%	0.36	1.14		
N, wt%	0.14	0.086		
Gas Oil (343—525° C)				
Wt%	41.4	31.0	33	61.6
S, wt%	1.05	2.07		
N, wt%	0.45	0.38		
Residue (525° C+)				
Wt%	7.68	4.1	26—11	
S, wt%	3.0			
N, wt%	1.4			

1. Cold Lake bitumen, from Van Driesen *et al.*[1987]
2. Cold Lake bitumen, from Colyar *et al.*[1989]
3. 80% Cold Lake/20% vacuum bottoms, from Pruden *et al.*[1989]
4. Gas oil from Athabasca bitumen, from Rangwala *et al.* [1984].

All of these cracked oils have been processed thermally during the primary upgrading step, at temperatures of over 400° C, therefore, their composition can be fundamentally different from a virgin oil.

8.2 Characteristics of Cracked Distillates

Prior cracking by thermal or catalytic treatment gives a distribution of chemical types which is quite different from a virgin oil, even when the boiling distribution and heteroatom levels are comparable. This molecular difference between cracked and straight—run oils in turn gives rise to differences in the hydrotreating performance, such as rate of reaction, hydrogen consumption, and catalyst life.

In cracking of carbon—carbon bonds at high pressure, as in hydroconversion, the alkyl side chains are likely to cleave β or γ to aromatic or naphthenic rings, rather than decomposing to give 2—carbon products [Mushrush and Hazlett, 1984]. The naphtha and middle distillate, therefore, will contain paraffins in the C_5 to C_{25} range, and tend to be enriched in paraffinic carbon relative to the feed, as illustrated in Figure 8—1. Opening of naphthenic rings contributes additional paraffins and alkylaromatics. The thermal cracking reactions, however, leave the aromatic groups intact, as well as many naphthenics and naphthenoaromatics. These ring groups will migrate from the residue to the distillate products, so that any ring compounds identified in resins or asphaltenes can be expected to appear in the product from primary upgrading. Because large aromatic ring compounds, such as pyrene, cannot crack the residue distillate products of upgrading contain less aromatic carbon than the unconverted residue (Figure 8—2). If hydrogen is not available, as in FCC and coking, then condensation reactions will occur to build polynuclear aromatics in the heavy distillates.

8.3 Process Description

Hydrotreating of distillates is almost always accomplished in a reactor packed with pelleted catalyst. The hydrotreating catalyst is held in a fixed bed, and the liquid and hydrogen gas normally flow downward cocurrently. The boiling range of the feed and the operating temperature and pressure will determine the fraction of the feed in the vapor phase and the fraction in liquid. So long as liquid is present, then the reactor will operate in the trickling or pulse—flow regime, with liquid and vapor passing cocurrently through the catalyst bed.

An example process flow diagram is illustrated in Figure 8—3; numerous variations on this basic theme are found in the patent

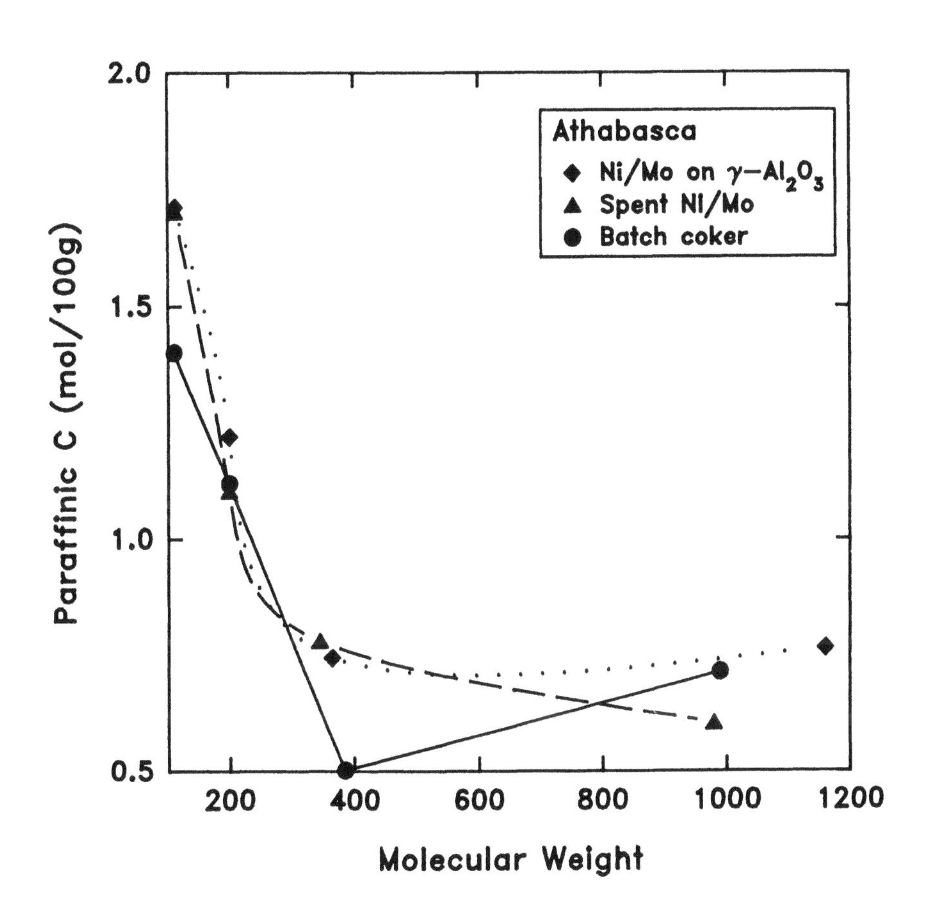

Figure 8—1
Paraffinic carbon content of hydroconversion fractions. CSTR reactor at 430° C and 13.9 MPa, LHSV=11.2 mL/g catalyst; Batch coker at 490° C and 120 kPa, 9 s gas residence time (Data from Gray *et al.*, 1992)

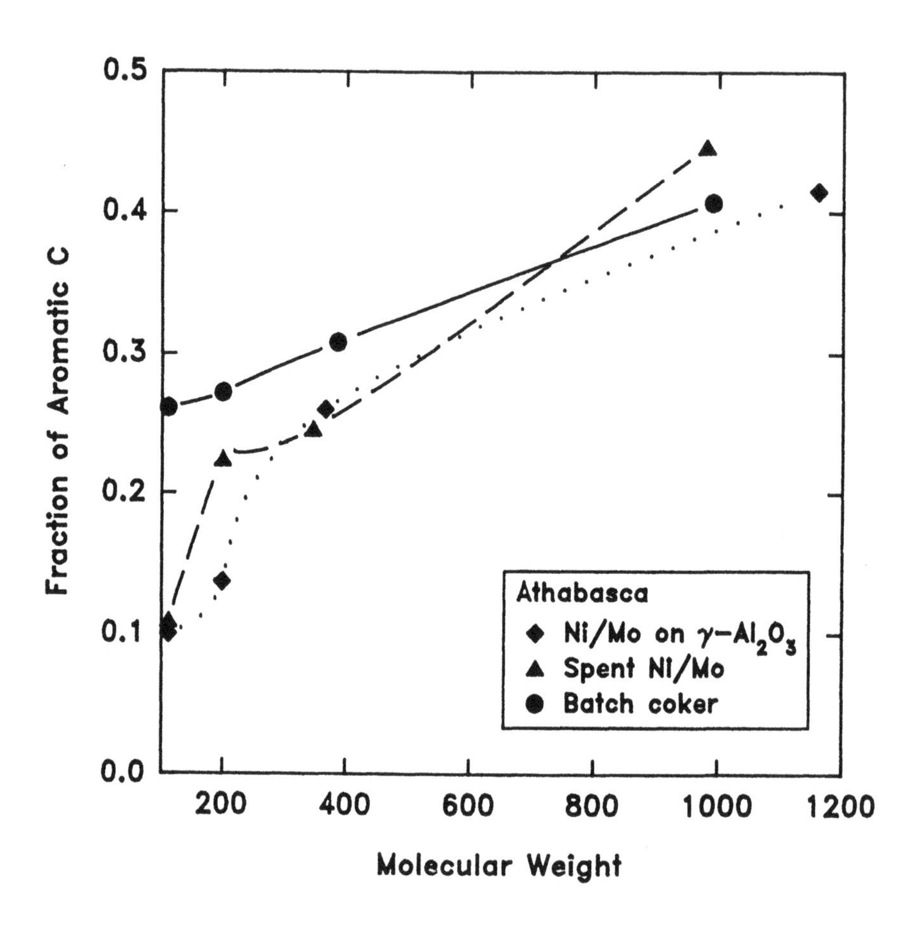

Figure 8–2
Aromatic carbon content of hydroconversion fractions. CSTR reactor at 430° C and 13.9 MPa, LHSV=11.2 mL/g catalyst; Batch coker at 490° C and 120 kPa, 9 s gas residence time (Data from Gray *et al.*, 1992)

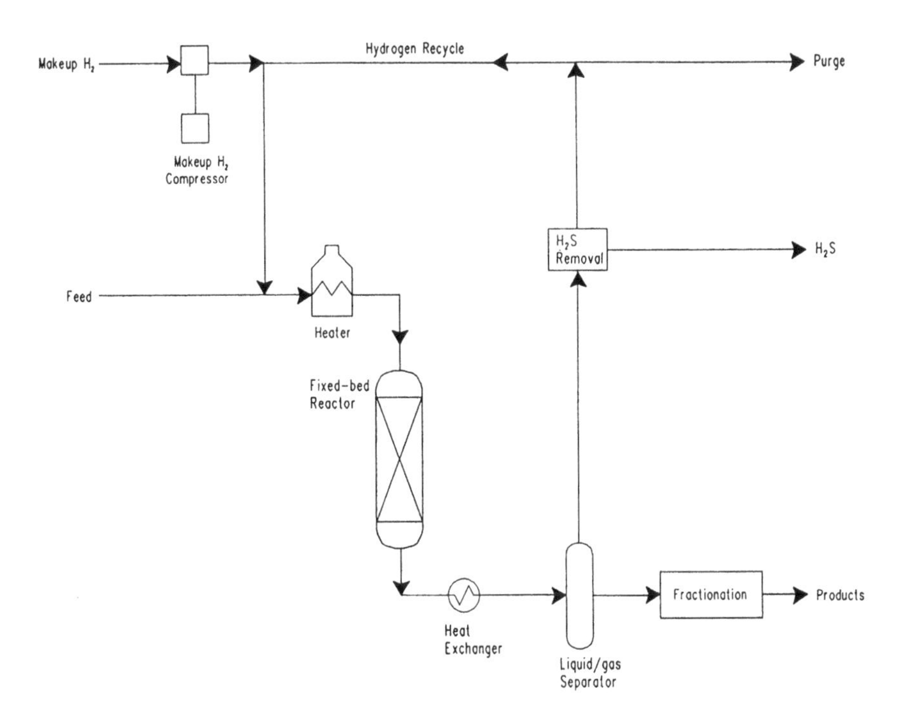

Figure 8–3
Typical Trickle–Bed Hydrotreating Process

literature [Speight, 1981]. The basic elements are the same in most processes; the feed is heated to the desired inlet temperature, and it passes through the reactor with the hydrogen stream. Hydrogen is supplied in excess of the reaction requirements in order ensure excess reactant and to limit the accumulation of hydrogen sulfide. Downstream of the reactor, the hydrogen sulfide is removed by alkanolamine absorption. The light gases that form due to cracking circulate with the hydrogen, and control the rate of purge. Fresh hydrogen make up is added to balance the losses in the purge stream and the consumption of hydrogen in reactions in the reactor. Typical operating conditions are listed in the following table:

Table 8—2
Operating Conditions for Distillate Hydrotreaters
(Data from Henry and Gilbert [1973] and Satterfield [1975])

Feed	Naphtha	Gas Oil
Pressure, MPa	2—3.5	3—7
Temperature, ^{o}C	350—380	380—425
H_2 in recycle stream, %	65	65
Superficial velocity of liquid, kg/(m^2—s)	8—25	0.8—8
Liquid hourly space velocity (LHSV), h^{-1}	0.5—8	1—10
H_2 flow, kg/kg feed	< 0.04	0.04—0.1
Catalyst diameter, mm	3—6	3—6
Catalyst bed depth, m	3—6	3—6

The catalyst in the reactor is arranged in beds with a depth of 3—6 m, with redistribution of liquid between each bed. This amount of catalyst is the practical limit due to the physical strength of the catalyst pellets and the need to ensure good distribution of the liquid. The hydrogenation reactions are exothermic, so that the exit temperature is always higher than the inlet. Cold hydrogen may be added between beds within the reactor to control the operating temperature.

The catalyst in the packed bed deactivates with time, so these processes are operated on a rising temperature profile. The reactor temperature is increased through the run to maintain a constant conversion level.

8.4 Reaction Chemistry

The catalytic hydrogenation of sulfur, nitrogen, and oxygen species was discussed in Section 3.3.3. This reaction chemistry is the same in both residue and distillate hydrogenation. The removal of heteroatomic compounds from cracked fractions (either from hydroconversion or thermal processes) is more difficult than in a corresponding virgin oil because of the prior thermal processing. The heteroatoms are more likely to be present in the less reactive heterocylic forms (e.g. thiophenes), as listed in Table 8—3 for the distillates boiling ranges, because easily converted compounds such as sulfide have already been removed by prior treatment. Thermal processing also enhances the migration of aromatic ring compounds, such as carbazole, into the distillate fractions by dealkylation reactions. Hence a variety of methyl and dimethyl quinoline and carbazole isomers which have been isolated from the asphaltenes of Athabasca bitumen [Strausz, 1989], have also been identified in coker gas oils [Schmitter *et al.*, 1984; Dorbon *et al.*, 1984].

8.4.1 Sulfur removal

A number of studies on hydrotreating of sulfur species have identified substituted dibenzothiophene (DBT) as the most resistant sulfur species in many distillates (e.g. Sapre *et al.* [1980]). Figure 8—4 shows the relative reactivities of various sulfur heterocycles, based on reaction of mixtures containing identified sulfur compounds. DBT has been assigned a reactivity of 1.0. From these data, the thiophenes are much more easily removed than the DBT's, but higher ring compounds are more reactive than expected. At least two trends contribute to these results; electronic effects on adsorption of the reactant onto the catalyst and subsequent reaction, and steric hindrance by substituents.

The high reactivity of benzonaphthothiophene was attributed to the rapid hydrogenation of the aromatic rings attached to the thiophene nucleus, compared to dibenzothiophene which mainly underwent direct removal of sulfur [Sapre *et al.*, 1980]. The larger ring compound would have a higher electron density at reactive carbons, and would be more easily adsorbed to the catalyst surface. The reduced reactivity of the 2—methyl dibenzothiophene (0.32 of DBT reactivity) can be attributed to steric hindrance, which would reduce access of the sulfur to the catalyst surface.

8.4.2 Nitrogen removal

The low reactivity of the nitrogen compounds can control the subsequent hydrotreating of the distillate products, particularly in naphthas that must meet a reformer—feed specification. Sulfur removal is

Table 8–3
Sulfur and Nitrogen Compounds in Distillates

	Substituent Aliphatic C on Heterocycles[1]		
	Naphtha 82–177° C	Mid–Dist 177–343° C	Gas Oil 343–525° C
Thiophene	 C_0–C_4	C_4–C_{11}	C_{11}–C_{19}
Benzothiophene	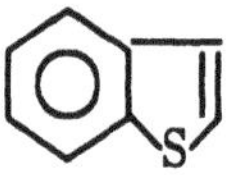	C_0–C_5	C_5–C_{13}
Dibenzothiophene	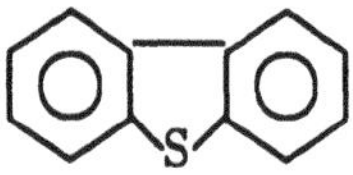	C_0	C_1–C_9
Pyrrole	 C_0–C_2	C_2–C_{10}	C_{10}–C_{20}
Indole	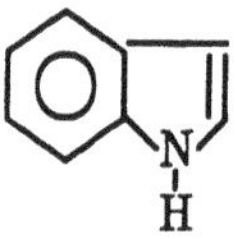	C_0–C_4	C_4–C_{11}
Carbazole			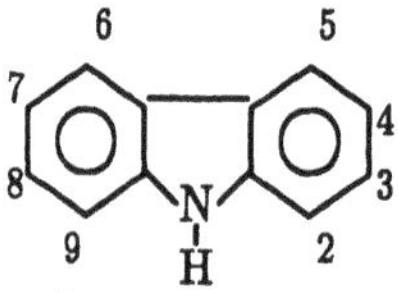 C_0–C_7
Pyridine	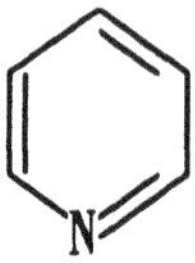 C_0–C_3	C_3–C_{10}	C_{11}–C_{19}
Quinoline	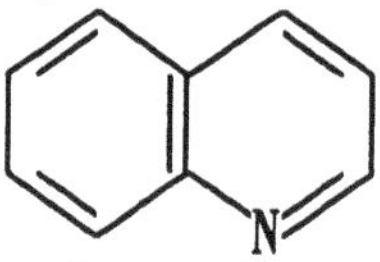	C_0–C_3	C_4–C_{11}
Acridine			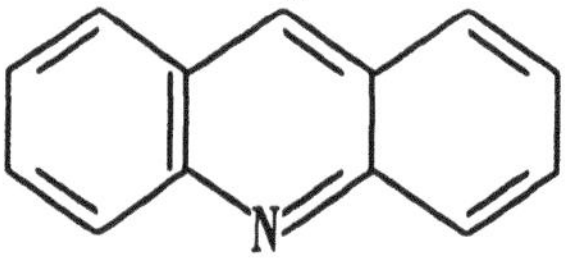 C_0–C_7

1. Boiling points of homologous series were estimated by adding Joback group contributions for CH_2 and CH_3 to the boiling point of the base ring compound [Reid *et al.*, 1987].

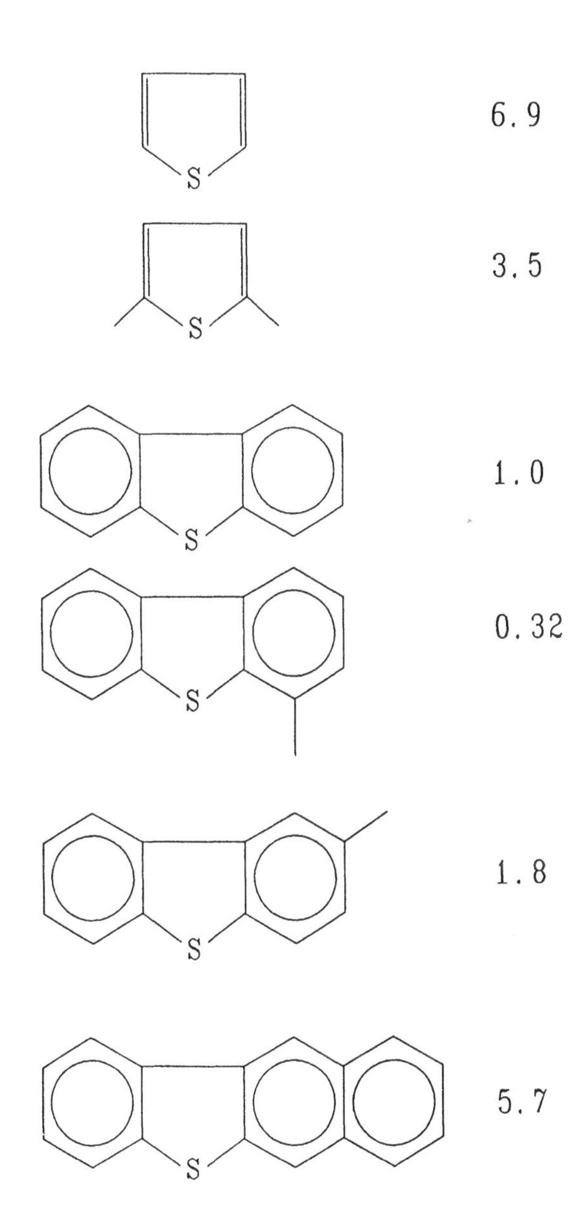

Figure 8–4

Relative reactivities of thiophenic sulfur compounds. The values shown are ratios of the overall first–order rate constants for desulfurization of the compounds to the value for dibenzothiophene at the same temperature (Data from Weisser and Landa, 1973, Sapre *et al.*, 1980, and Katti *et al.*, 1984) more often limiting in the heavier distillates. The nitrogen compounds which migrate to the distillates are resistant to conversion, as indicated by the data of the following tables:

Table 8–4
Removal of Pyrrole Benzologs from Coker Gas Oil
(Data from Dorbon *et al.*, 1984)

		Products			
	Feed	HDN1	HDN2	HDN3	HDN4
Temperature, °C		340	380	360	360
Hydrogen Press., MPa		5	5	7	10
Sulfur, ppm	3900	860	350	550	390
% Removal		78	91	86	90
Nitrogen, ppm	450	290	210	180	72
% Removal		35	54	60	84
Pyrroles, ppm		350	140	350	350

Table 8–5
Removal of Pyrrole Benzologs from Syncrude Coker Gas Oil
Hydrogen pressure 13.9 MPa (Data from Khorasheh *et al.*, 1989)

		Products		
	Feed	PR–1	PR–2	PR–3
Temperature, °C		400	400	400
LHSV, ml/h–g cat		12.5	16	23.5
N, mol/kg	0.217	0.108	0.123	0.131
% Removal		50	43	40
Pyrroles, mol/kg	0.053	0.038	0.043	0.051
% Removal		28	19	4

The nitrogen compounds are consistently more difficult to remove than the sulfur species. Both increasing temperature and increasing hydrogen pressure enhance the removal of total nitrogen (Figure 8–5), but some carbazole species can only be converted if the temperature is increased. Dorbon *et al.* [1984] found that benzocarbazoles were more readily removed than carbazoles, which is analogous to the high reactivity of benzo–naphthothiophene relative to dibenzothiophene (Figure 8–4), and acridine relative to quinoline [Moreau *et al.*, 1988].

Dorbon *et al.* [1984] also found that 1—methyl carbazole was converted more rapidly than carbazole, which is consistent with thermodynamic control of the conversion instead of kinetic control.

8.4.3 Oxygen Removal

Oxygen is present in distillates from bitumens and heavy oils, and has a bearing on product quality and catalyst life. The literature on oxygen removal from distillates is very scanty [Furimsky, 1983a], but Furimsky [1978] found that only 65% of the feed oxygen was removed by Ni/Mo catalyst at 400° C, 13.9 MPa, and LHSV = 2 h^{-1}. This result, and other data [Trytten, 1989] indicate that oxygen removal can be intermediate between sulfur and nitrogen depending on the feed.

8.4.4 Hydrogen consumption and aromatics conversion

Ni/Mo catalyst is more active than Co/Mo for both nitrogen removal and hydrogenation of aromatics, indicating nonspecific hydrogenation activity as opposed to a high affinity for the nitrogen compounds. This view is further supported by the hydrogen consumption, which is far above the theoretical requirements for removal of heteroatoms alone. For example, denitrogenation of a mixture of 2,4—dimethyl pyridine and 2—methyl napththalene consumed almost 20 times the minimum hydrogen requirement for nitrogen removal alone at an LHSV of 1 h^{-1} (Figure 8—6) [Ho, 1988]. Hydrotreating of Syncrude coker gas oil at 380—420° C and 13.9 MPa over a Ni/Mo catalyst hydrogenated 3 mol of aromatic carbon for every mol of heteroatoms (N, O, and S) removed [Gray, 1990]. A selective catalytic conversion of nitrogen compounds, by comparison, would remove the heteroatoms with little or no conversion of aromatic carbon.

8.4.5 Relative removal of heteroatomic compounds

Hydrotreating a cracked distillate involves a formidable combination of reactions which occur simultaneously: hydrodesulfurization (HDS), hydrodenitrogenation (HDN), hydrodeoxygenation (HDO), hydrogenation of aromatics, and thermal cracking. In addition the fate of basic and nonbasic nitrogen compounds must be distinguished because they have different effects on downstream processes, for example, pyrrolic or nonbasic nitrogen has been implicated in sludge formation and light instability of gas oils [Chmielowiec *et al.*, 1987]. Temperature, pressure, residence time, and catalyst also vary from study to study. Given this number of variables, contradictions are bound to occur. For example, Dorbon *et al.* [1984] found that carbazoles were more resistant than quinolines to hydrogenation, while at higher nitrogen content and temperature both basic and nonbasic forms of nitrogen were persistent

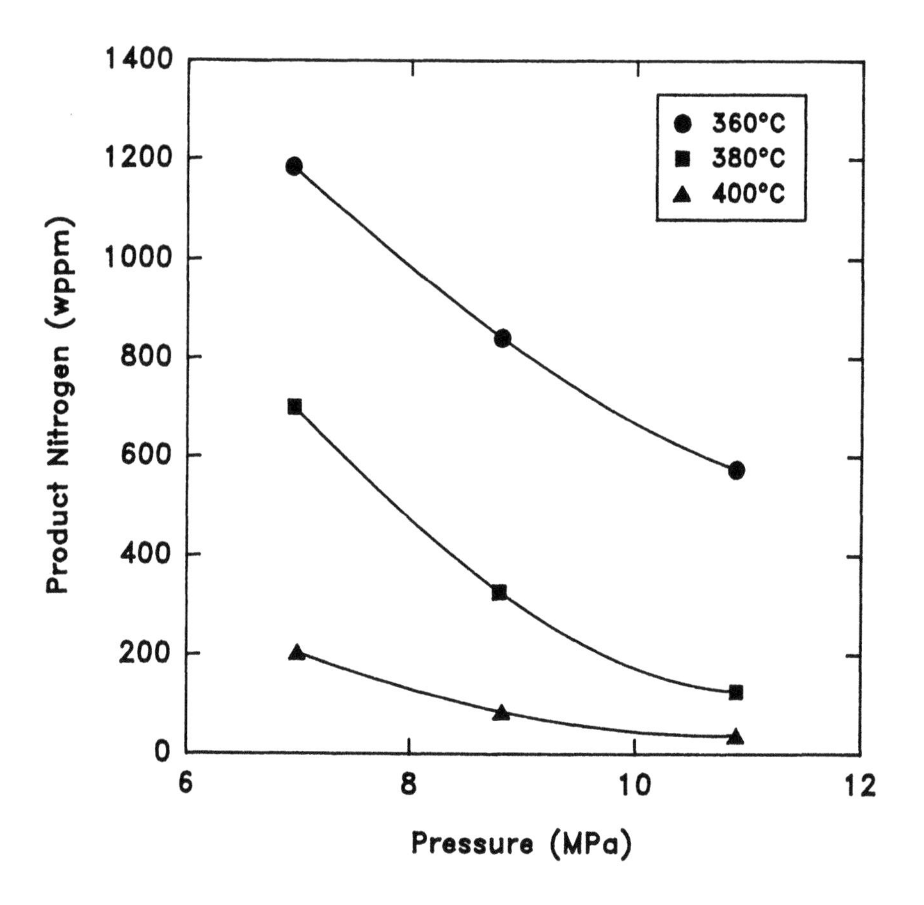

Figure 8–5
Effect of temperature and pressure on nitrogen conversion in a gas oil feed (Data from Yui, 1989)

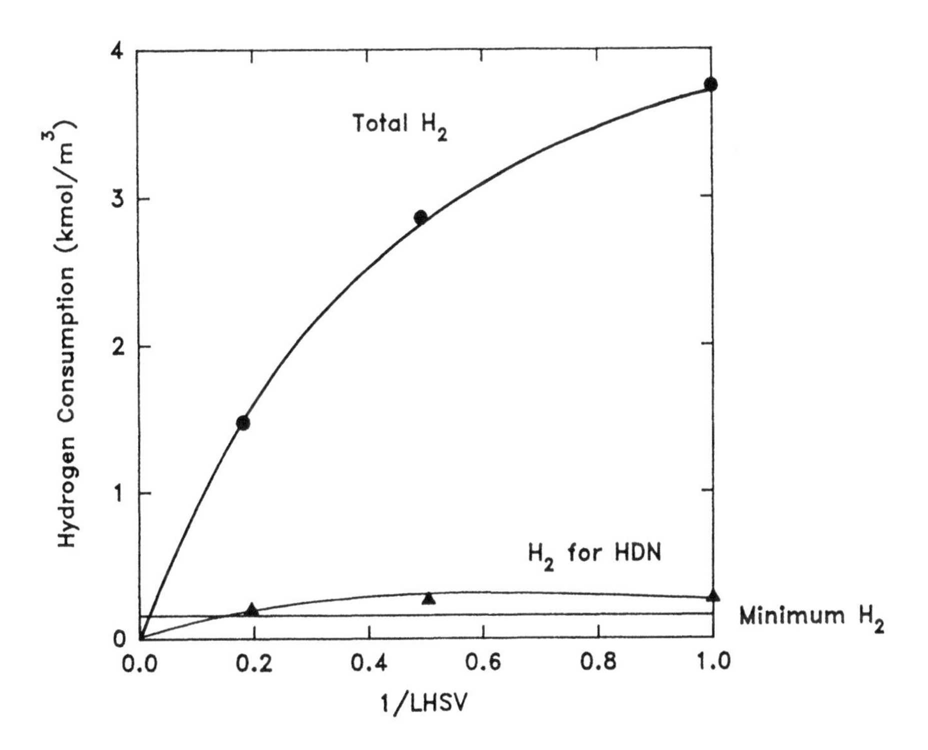

Figure 8–6
Hydrogen consumption for simultaneous hydrodenitrogenation of 2,4–dimethyl pyridine and hydrogenation of 2–methyl naphthalene. The units of LHSV are h^{-1} (Data from Ho, 1988)

[Furimsky *et al.*, 1978; Table 8–4].

Trytten *et al.* [1990] found that reaction rates for HDN and HDS were both strongly dependent on the molecular weight and chemical characteristics of the feed oil (Figure 4–14). The effectiveness of the catalyst actually went through a maximum as the boiling point of the feed was increased, due to the interaction between diffusion effects and the reduction in intrinsic reactivity (Figure 8–7). At reaction temperatures of 400 and 425° C, nitrogen species in low–boiling fractions were reduced to low levels, but the the basic and pyrrolic compounds in the high–boiling fractions were persistent even at 425° C (Figure 8–8). The explanation for the difference in reactivity an increase in the number and size of alkyl substituents on the heterocycles, and in more refractory benzologs in higher–boiling fractions (Table 8–3). The relative ease of removal of the two types of nitrogen, therefore, depends on feed and reactor temperature and hydrogen pressure.

8.4.6 Thermal reactions

Thermal reactions also occur during hydrotreating at *ca.* 400° C. Khorasheh *et al.* [1989] showed that the formation of C_1–C_3 hydrocarbons was independent of the presence of a catalyst (Figure 3–6). The yields of C_4 and C_6 were enhanced by catalyst, consistent with hydrogenation of aromatics followed by cracking and ring–opening reactions. At temperatures over 430° C, the rate of HDN remained constant, whereas HDS continued to increase with increases in temperature. This result indicated one limit to the use of higher temperatures to force HDN reactions to completion. Thermal cracking reactions become very active at temperatures over 410–420° C, and these reactions will compete for the available hydrogen. The rate of HDN is much more dependent on hydrogen pressure than is HDS, so at high temperature HDN was controlled by the availability of hydrogen.

8.5 Kinetics and Hydrodynamics of Hydrotreating

8.5.1 Apparent reaction order and kinetics

The kinetics of heteroatom removal, particularly HDS, have received a great deal of attention. HDS reactions give an order ranging from 1 to 2 depending on the feedstock and the extent of conversion, whereas pure compounds follow first–order kinetics (section 4.4). This apparent change in order is consistent with the analysis by Ho and Aris [1987]. The work of Trytten *et al.* [1990] showed that a gas oil contains compounds with a very wide range of reactivities, and that reaction rates can decline by an order of magnitude when molecular weight is doubled (Figure 4–14). The most common approach to analyzing the kinetics of

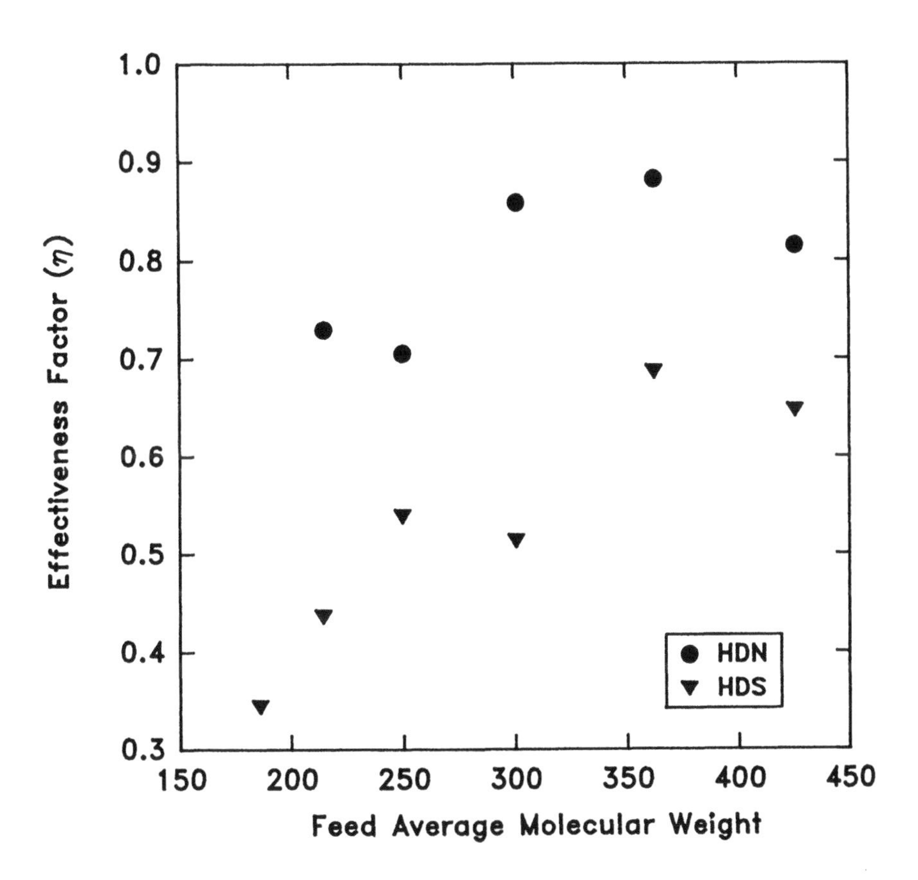

Figure 8–7
Effectiveness factors for hydrotreating of fractions of a coker gas oil derived from Athabasca bitumen (Data from Trytten *et al.*, 1990)

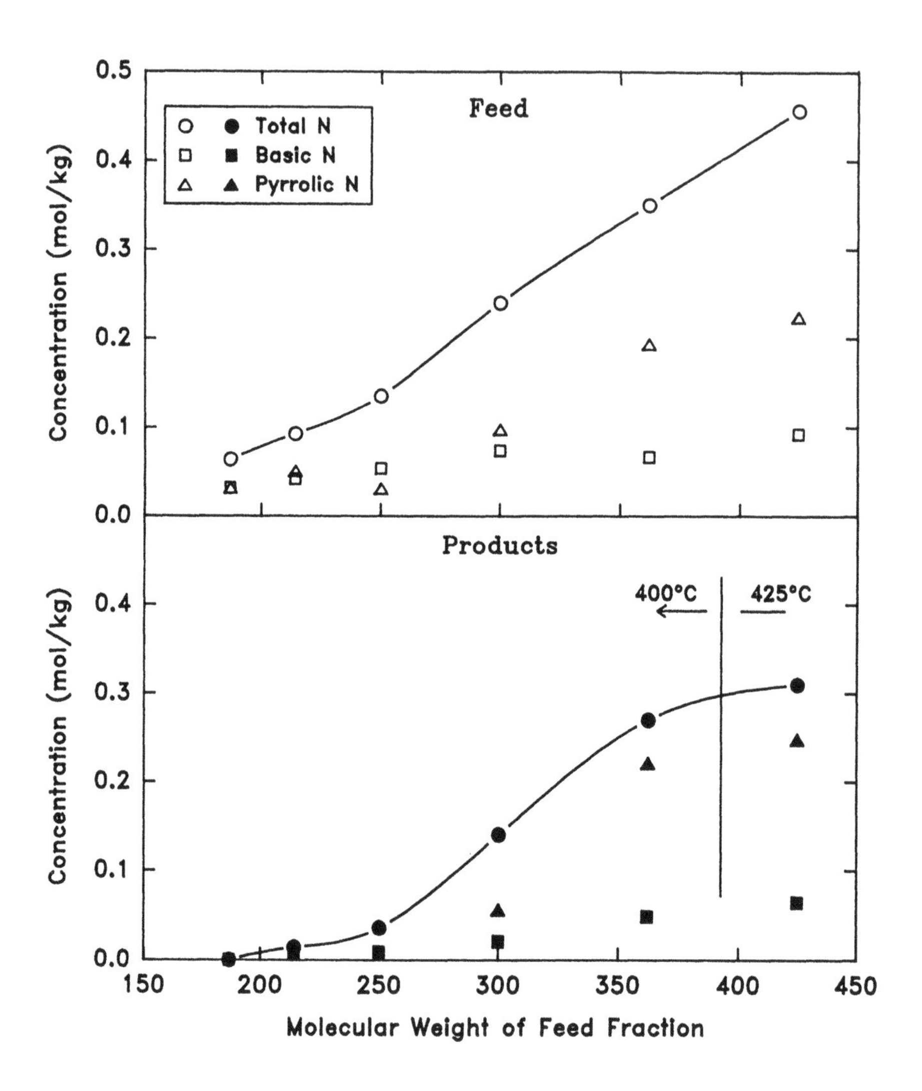

Figure 8—8

Conversion of Nitrogen Species in Coker Gas Oil Fractions. Rate constants at 13.9 MPa hydrogen pressure, 8 g Ni/Mo on γ—alumina catalyst in a 150 mL continuous reactor (Data from Trytten, 1989) hydrotreating is to use a power law expression:

$$\text{Rate} = -k\,C^{n}\,p_{H_2}^{\beta} \tag{8.1}$$

where n and β are empirical constants. For example, Yui [1989] found that n=1.5; β=0.86 for HDS, and n=1.0; β=1.39 for HDN in hydrotreating of coker gas oil from Athabasca bitumen. The power–law form is a convenient expression, even though the actual kinetics for the individual components would follow a Langmuir–Hinshelwood form [Sundaram *et al.*, 1988]. Reaction orders with respect to sulfur have been reported in the range 1.2 to 3. The reaction order for HDN is almost always 1 with respect to nitrogen content [Ho, 1988; Christensen and Cooper, 1990]. As the analysis of section 4.4 showed, the power law approach is useful for interpolation, but it cannot give intrinsic kinetics or be used with confidence to extrapolate reactor performance.

The trickle–bed reactor also gives problems in measuring phenomenological (as opposed to empirical) kinetics of mixtures because the rate expression is integrated over the entire length of the reactor. The apparent reaction kinetics, therefore, are an average over a range of conversion values, possibly for series of compounds with very different reactivities. Data presented by Christensen and Cooper [1990] illustrate the problem very well (Figure 8–9). A variety of gas oils were reacted in a trickle–bed reactor, and the HDS kinetics seemed to follow 1.7 order with respect to sulfur content. Further analysis showed that the apparent order was changing with sulfur conversion.

The kinetics of other hydrotreating reactions have been reported less frequently. Yui and Sanford [1991] considered the kinetics of hydrogenation of aromatics, using equation (8.1) along with a reversible rate constant (as in Section 4.2.3). Yui [1989] and Yui and Sanford [1989] analyzed the kinetics of mild hydrocracking, defined as conversion of material boiling over 343°C. The power law in equation (8.1) was used, with n=1.0 and β=0.323 giving a best fit to the data. The activation energy for mild hydrocracking was only 82 kJ/mol, which is much lower than one would expect for thermal reactions. The conversion of the gas oil to middle distillate, however, would be due to a combination of thermal and catalytic reactions [Trytten and Gray, 1990].

8.5.2 Reactor hydrodynamics

The trickle bed operation involves downward flow of gas and liquid over the catalyst pellets. Depending on the superficial velocities of the two flowing phases, a variety of complex flow regimes are observed. At low liquid flow rate, the liquid phase trickles over the solid as a

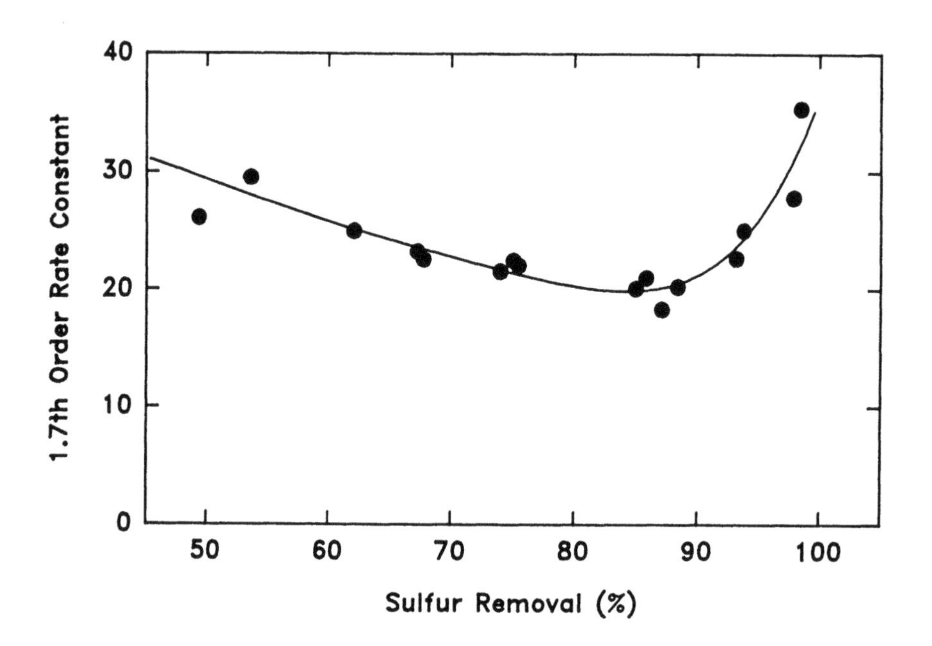

Figure 8—9
Change in apparent reaction rate constant for HDS of vacuum gas oil with conversion (Data from Christensen and Cooper, 1990)

continuous film or in rivulets, and the gas flows continuously through the voids in the bed. At higher liquid flow rates, rippling or slugging flow can be observed, where the liquid volume fraction at a given point in the reactor fluctuates with time. These fluctuations can give rise to hysteresis, where changes in the liquid or gas flow to the reactor can give unpredictable liquid hold up in the reactor. These complex flow problems can occur in any type of trickle bed reactor, not just hydrotreaters. Satterfield [1975] provides an excellent review on trickle bed reactors, and recent studies by Funk *et al.* [1990] and Sundaresan [1993] illustrate the continuing work on understanding these complex reactors.

The main kinetic effects of the complex hydrodynamics are on catalyst wetting and liquid hold up. At low liquid velocities, the catalyst is not fully wetted by the liquid phase, and a loss of catalyst effectiveness results. At higher liquid superficial velocities, the catalyst is wetted consistently and it is fully utilized [Henry and Gilbert, 1973]. At these higher velocities, the liquid hold up may be more difficult to predict due to time varying flows.

Commercial–scale trickle bed reactors for hydrotreating are intended operate in a regime wherein the catalyst is fully utilized due to the high liquid velocities (Table 8–2; Henry and Gilbert [1973]). Under these conditions, the trickle–bed reactor approximates an ideal plug–flow reactor for kinetic analysis (see section 4.2.2). Allowing for the fact that the liquid does not occupy the entire void space of the reactor, equation (4.50) gives the following for first–order kinetics [Henry and Gilbert, 1973]:

$$h/LHSV = -1/k \ln[1-X_e] \qquad (8.2)$$

where $LHSV = Q_0/V$ (volumetric flow rate of liquid feed over total reactor volume) and h is the fractional liquid volumetric hold–up in the reactor (m^3 liquid/m^3 reactor volume). In this case, LHSV/h is the true space velocity of liquid. So long as h is constant, then the apparent rate constant will be independent of flow rate. In laboratory reactors, however, both liquid hold–up and catalyst wetting can be sensitive to flow and apparent reaction orders greater than one can be observed [Henry and Gilbert, 1973]. Hydrodynamics, therefore, can have similar effects on the apparent reaction order as lumping of kinetics.

Liquid hold up can be correlated to the ratio of the Reynolds number to the Froude number, which gives the following empirical relationship [Henry and Gilbert, 1973]:

$$\ln(1-X_e) \propto L^{1/3}(LHSV)^{-2/3} d_p^{-2/3} \nu^{1/3} \qquad (8.3)$$

An alternate equation was suggested by Mears [1974] based on studies of catalyst wetting:

$$\ln(1-X_e) \propto h^{0.32}(LHSV)^{-0.68} d_p^{0.18} \nu^{-0.05} \tag{8.4}$$

Both equations give an equivalent dependence of conversion on LHSV. For scale–up from the laboratory reactor, the rate constant would be determined from equation (8.3) or (8.4), then used to predict conversion on the basis of equation (8.2). An alternative approach is to treat the exponent of LHSV as an adjustable parameter in fitting kinetic data for a specific reactor. For example, Yui [1989] found that exponents of –0.38 and –0.63 for LHSV gave the best fit for HDS of a gas oil over two different catalysts. The same catalysts in the same pilot–scale reactor gave exponents of –0.76 and –0.72 for LHSV in fitting the data for HDN. For more detailed modelling of wetting phenomena, see Funk *et al.*, [1990] and the references therein.

8.5.3 Stoichiometry of hydrotreating

The hydrogen used in a hydrotreater is in excess of the theoretical requirement for heteroatom removal alone (e.g. Figure 8–6), mainly due to the attendant hydrogenation of aromatic carbon. The minimum stoichiometric requirement is one mol of H_2 per mol of sulfur removed, with similar ratios for oxygen and nitrogen depending on their chemical type (e.g. phenol versus furan).

Gray [1990] studied the kinetics and stoichiometry of the hydrogenation and cracking reactions that accompany hydrotreating of gas oils to remove sulfur and nitrogen. Data from NMR spectroscopy were used to calculate the concentrations of carbon types in feed and product samples from hydrotreating of coker gas oil. The carbon types were then treated as pseudocomponents in a lumped kinetic analysis. When two groups were related by a common reaction, their rates of reaction were in constant proportion according to a stoichiometric coefficient:

$$\sigma_{ij} = r_i/r_j \tag{8.5}$$

where r is the rate of reaction. In catalytic hydrotreating, all of the significant changes in carbon types were driven by the catalytic removal of heteroatoms (O, N, S), so that the rates of reaction were proportional to the rate of heteroatom removal. This proportionality was independent of temperature and residence time in a stirred–tank reactor, as illustrated in Figure 8–10. The data of Table 8–6 show the stoichiometric ratios; the negative coefficient for naphthenic carbon shows the formation of this carbon type due to the conversion of aromatic carbon.

Table 8–6
Stoichiometric Ratios from Rates of Reaction
(Data from Gray, 1990)

Carbon Group Type	$r_i/r_{O,N,S}$ [1]	Coeff of Variance
Aromatic C (Total C_{ar})	3.0	14
Aromatic C bound to O,N,S (C_{ar}–(ONS))	1.8	8.3
Methyl β to Aromatic Ring (β–CH_3)	0.57	17
Naphthenic C	–4.2	10.7

1. Ratio of rate of reaction i (r_i) to total rate of heteroatom removal

A variety of carbon types, such as paraffins, did not give a significant change upon hydrotreating. The power of this kinetic analysis based on carbon types is that the changes in the oil wrought by sulfur, nitrogen, and oxygen removal can now be analyzed systematically. Characteristics of hydrotreating such as hydrogen consumption and product quality (i.e. cetane number) could be determined as a function of heteroatom removal based on the conversion of carbon types. This analysis also underscores the lack of selectivity toward heteroatoms, given the significant increase in naphthenic carbon due to ring hydrogenation.

8.6 Catalyst Properties

8.6.1 Catalyst selection

The catalysts for hydrotreating are similar to the materials for catalytic hydroconversion (section 5.2), except that the provision of macropores in the alumina matrix is of less concern. The preferred catalyst for hydrotreating distillates that contain nitrogen compounds is Ni/Mo sulfide on γ–alumina. If nitrogen content is low, then Co/Mo

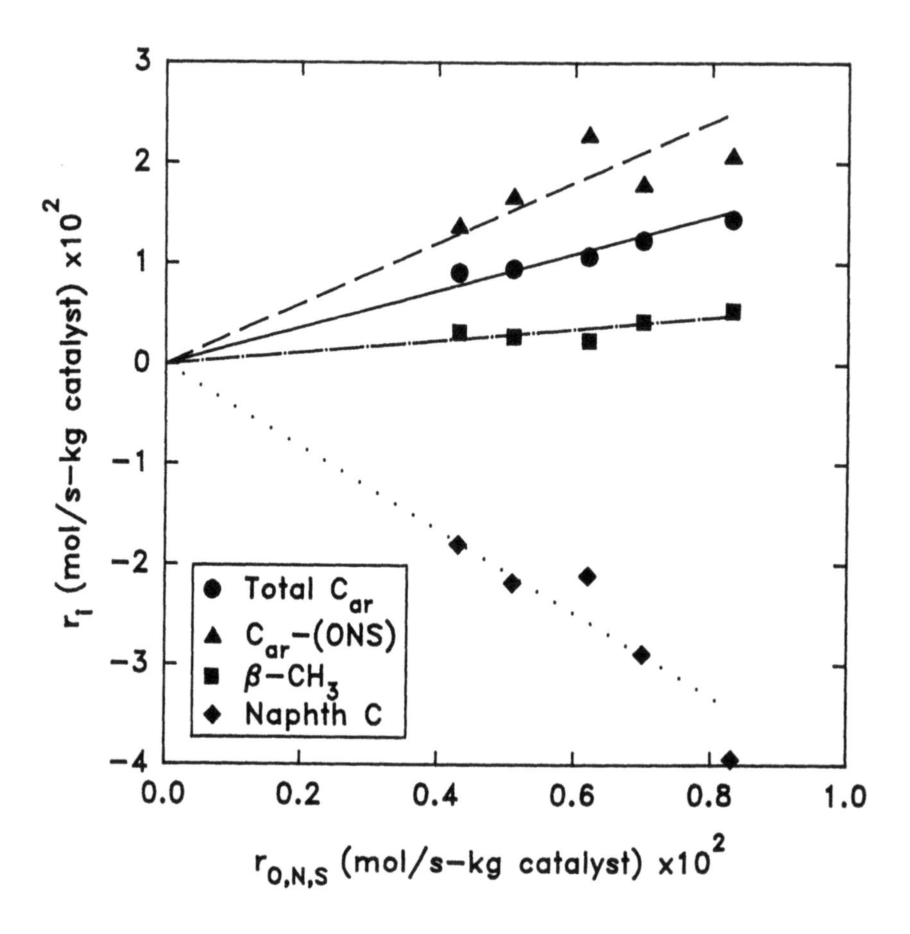

Figure 8—10
Stoichiometry of carbon types in hydrotreating of coker gas oil. Rate constants at 380—420° C, 12.5—23 mL/h—g catalyst with 8 g Ni/Mo on γ—alumina in a 150 mL continuous reactor (Data from Gray, 1990)

sulfide is commonly selected. Catalysts based on tungsten are also active (e.g. Ni/W sulfide on γ—alumina; Weisser and Landa [1973]), but are less commonly used because of the high price of tungsten.

Under hydrotreating conditions, the active metals are in the sulfide form, with Ni acting as a promoter on crystallites of MoS_2 [Ho, 1988; Prins *et al.*, 1989]. Molybdenum contents are typically 10—14% of the catalyst by weight (on a metal basis), while the promoter Ni or Co is typically added to a content of 2—4% by weight. Common additives in distillate hydrotreating catalysts are phosphorus, which tends to enhance nitrogen removal at the expense of some loss in HDS activity [Christensen and Cooper, 1990], and fluorine to enhance the acidity [Weisser and Landa, 1973].

Given the complexity of the feed and the number of related reactions, development of a single optimal catalyst and reactor environment is unlikely. Several catalyst vendors have recommended staged beds of catalysts to overcome this problem by using optimized catalysts for different reactions. One proposed combination is Ni/Mo catalyst, to remove N, followed by Co/Mo for higher sulfur removal. Use of staged beds would also allow combination of much more specific catalysts than are in current use, without sacrificing overall performance. For example, Ho [1988] suggested use of catalysts that were highly specific to breakage of C—N bonds, with HDS handled by other catalysts in the reactor.

Kirchen and Sanford [1989] tested a layered bed, consisting of Ni/Mo followed by Co/Mo catalyst (both on γ—alumina), for hydrotreating a coker gas oil. They found no significant change in desulfurization, but a loss of denitrogenation and mild hydrocracking performance. No difference in deactivation characteristics was observed. Clearly, any proposed combinations of catalysts must be tested with representative feeds.

8.6.2 Catalyst deactivation

A major concern in hydrotreating of distillates is not merely the initial activity of the catalyst, but the activity and regenerability of the catalyst over an extended service life. The metal sulfide catalysts used to treat distillates can be deactivated by at least two mechanisms: selective poisoning of active sites, and fouling of the catalyst surface by carbonaceous deposits [Furimsky, 1979]. Heavy gas oils contain only low concentrations of metals (Ni, V, and Fe) which are easily removed by a guard bed, so that accumulation of metals is not a primary mechanism for deactivation in treating distillates [Sanford, 1986].

Metal sulfide catalysts are resistant to poisoning by components of the feed, and have a useful service life of a year or more before regeneration is required. One demanding service is naphtha hydrotreating, because of the 1 ppm nitrogen limit in the product. Another is deep desulfurization of middle distillates to meet stricter sulfur limits on transportation fuels. As hydrotreating catalysts deactivate, the reactor temperature is increased to compensate. Metal sulfides supported on γ—alumina are often referred to as bifunctional; hydrogenation activity in the metal crystallites and cracking activity on the alumina surface. Basic nitrogen compounds are present in syncrude feeds, however, and these compounds have been recognized as poisons for acidic cracking sites for over 40 years [Mills *et al.*, 1950]. Even the nonbasic pyrrolic types can give rise to basic intermediates when they are partly hydrogenated. Consequently, strongly acidic sites will be poisoned in actual service, and any cracking will be due to a mixture of thermal reactions and weak acid sites. This view is consistent with the composition of the light ends, which are not enriched in butanes (from cracking via carbocation intermediates) and indeed are formed in equal amounts whether catalyst is present or not [Khorasheh *et al.*, 1989].

Loss of HDN and HDS activity with time, therefore, is due to accumulation of carbonaceous deposits, or coke. This material is thought to foul the surface, blocking pores within the catalyst and access to the active surface [Thakur and Thomas, 1985]. The carbon content of the catalyst rises rapidly to an initial level, then changes little during the service life. The amount of carbon on the catalyst tends to increase down the length of the reactor, due to the exothermic temperature and the drop in hydrogen fugacity [Wukasch and Rase, 1982; Sanford, 1986; Gosselink *et al.*, 1987; Egiebor *et al.*, 1989].

8.6.2.1 Chemistry of carbonaceous deposits

The composition of the carbonaceous deposits gives important information on the deactivation processes. The deposits from gas oil and naphtha hydrotreating catalysts are dramatically enriched in nitrogen and oxygen, relative to the feed oil [Furimsky, 1978b; Wukasch and Rase, 1982; Choi and Gray, 1988]. For example, methanol extracts from a naphtha catalyst contained 3.7—4% N, compared to 0.3% in the coker naphtha feed to the reactor [Choi and Gray, 1988]. Apparently the adsorption and subsequent reaction of nitrogen and oxygen compounds is an important factor in catalyst deactivation. Analysis of solvent extracts from spent catalysts showed the presence of amides and pyrroles [Choi and Gray, 1988]. Stronger bases (i.e. quinolines) were only minor constituents of the extract. Wukasch and Rase [1982] showed that the exterior of spent gas oil catalyst was higher in carbon, while the interior was higher in nitrogen.

Although the main focus in deactivation has been on carbon

content, the role of nitrogen and oxygen compounds is significant. Furimsky [1978b] suggested that polymerization of nitrogen and oxygen heterocycles on the surface by Lewis acid sites would account for the observed enrichment. Experiments by Stohl and Stephens [1986] support this view. Ni/Mo catalyst was contacted with different classes of compounds isolated from coal liquids, then the residual hydrogenation activity was assayed. Residual activity was 59% after reaction with polyaromatic hydrocarbons, compared to only 19% with phenolic compounds, and only 5% after reaction with a nitrogenous extract.

Although the carbon deposits have been called coke, recent work with ^{13}C—NMR suggests that a wide range of deposits can occur on catalysts used to hydrotreat distillates [Egiebor *et al.*, 1989]. Catalysts near the reactor contained deposits with a high content of aliphatic carbon, ranging from 30—50% of the total, as compared to 60—70% in the feed. The high molecular weight methanol extract from a naphtha catalyst contained even more aliphatic carbon, at 78% [Choi and Gray, 1988]. Samples from lower in the reactor had a higher carbon content and a higher aromaticity, as much as 99% in a naphtha hydrotreating catalyst [Egiebor *et al.*, 1989]. These data indicate that two modes of fouling may occur: one due to heteroatomic species near the inlet, giving a N and O—rich deposit, and a second lower in the reactor due to higher temperature and lower hydrogen pressure.

8.6.2.2 Deactivation of high acidity catalysts

A significant area of interest in hydrotreating gas oils is to increase the formation of naphtha by mild hydrocracking. Because the catalyst tends to deactivate by accumulating coke, this cracking cannot be achieved by merely increasing the acidity of the alumina support in Ni/Mo catalysts. One option is to increase the operating temperature to enhance thermal reactions, and to offset the increased coking tendency by using a less acidic support. This strategy has been proposed by several groups for hydrocracking of residues, and research has identified promising nonacidic supports that still give good hydrogenation activity, for example carbon [Groot *et al.*, 1986] and titania/alumina mixtures [Nishijima *et al.*, 1987].

Use of a more active cracking catalyst, such as Si/Al or zeolite, requires protection from nitrogen compounds. For example, Gosselink *et al.* [1987] tested catalysts in series, with Ni/Mo first followed by a cracking catalyst to get better yields of naphtha from a gas oil. This combination was more attractive than Ni/Mo alone at higher temperatures, over 395° C. The deactivation process, however, was very complex because breakthrough of nitrogen compounds from the upper bed gave rapid poisoning of the cracking catalyst.

A fundamental problem of this approach, however, is that the components of the feed are unlikely to cooperate. As discussed above, the nitrogen compounds are concentrated in the heavy gas oil, and give the lowest intrinsic reactivity. Cracking of these compounds would enhance reactivity for HDN, but the reverse sequence of nitrogen removal followed by cracking would be preferred. This result could be achieved if the first stage were high—temperature removal of nitrogen on a nonacidic catalyst support, followed by cracking. Given the complex nature of the feed, combinations of more selective catalysts may prove to be very attractive in the future. In any case, the main limitation on such schemes is the rate of deactivation due to deposition of organic material on the catalyst.

8.6.2.3 Kinetics of catalyst deactivation

The carbon content of hydrotreating catalysts rapidly reaches a stable level, then remains constant through most of the operating life similar to residue hydroconversion catalysts (Figure 5—5). The stable level of carbon depends on the feed composition, temperature, and hydrogen partial pressure. Even though a catalyst has a constant level of coke, based on carbon, it will continue to deactivate. The physical structure (surface area and pore volume) changes relatively little during this deactivation process. Hence, a coke—deposition approach to deactivation is not in itself capable of describing the deactivation of hydrotreating catalysts. Similarly, a time—on—stream equation may empirically describe activity, but it cannot be used with any confidence to predict performance of other reactor types.

Several studies have implicated heteroatomic species containing N and O in the formation of coke on hydrotreating catalysts. Experiments consistently show a much higher concentration of N and O on the catalyst surface in comparison to the feedstock or the hydrotreated product [Furimsky, 1978b; Choi and Gray, 1988]. The carbon content on the surface is not a simple function of the aromaticity of the feed or the presence of asphaltenes. Analysis of extracts from spent catalysts suggests that polymerization of nitrogen and oxygen compounds is important in forming coke deposits, probably along with side reactions from the hydrogenation of aromatics.

Another major factor is the relationship between the surface deposits and the reaction temperature and partial pressure of hydrogen. Figure 8—11 shows representative data for the rate of deactivation of benzene hydrogenation over Ni on SiO_2/Al_2O_3 catalyst due to poisoning by thiophene [Zrncevic *et al.*, 1990]. Because a a larger value of $-\ln k_d$ indicates a smaller value for k_d, the data of Figure 8—11 indicate that the rate of deactivation was decreased at higher hydrogen partial

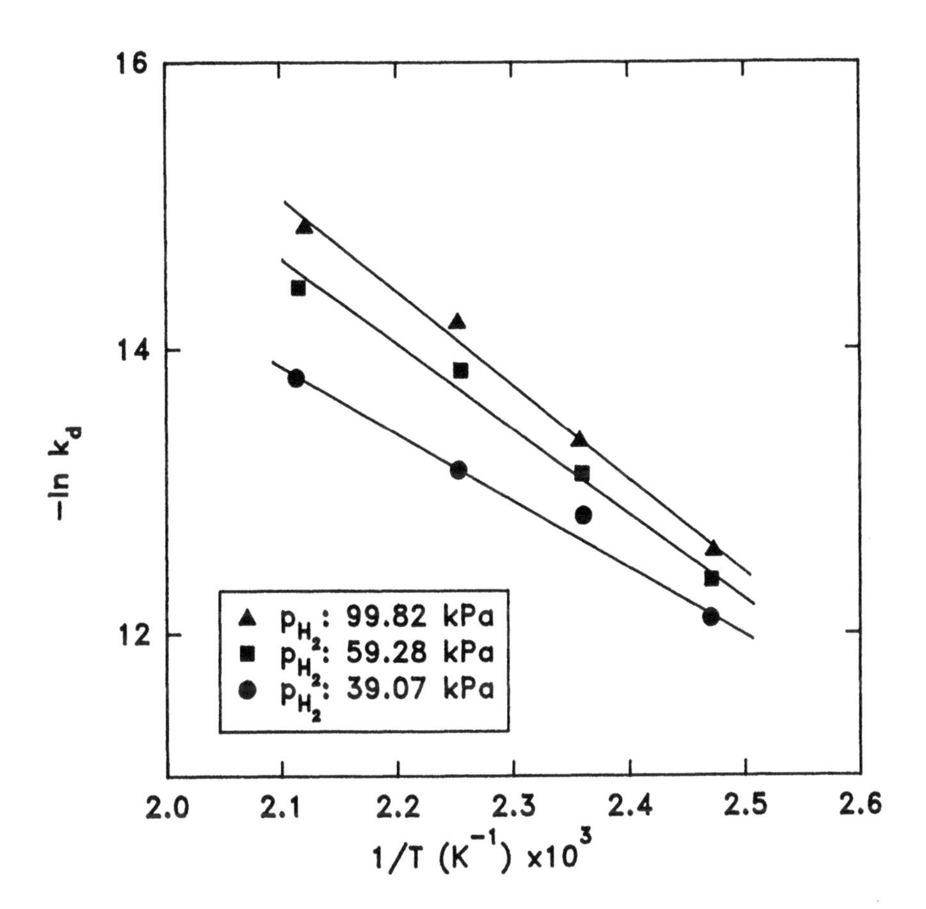

Figure 8—11
First—order rate constants for deactivation of Ni catalyst by thiophene
(Data from Zrncevic *et al.*, 1990)

pressures. The role of hydrogen would be to stabilize olefins and radicals on the surface of the catalyst before they can polymerize to form coke. For example, Furimsky [1983a] observed butadiene as a product of the hydrodeoxygenation of tetrahydrofuran. Higher hydrogen pressures would hydrogenate such diolefins and suppress their tendency to polymerize on the surface.

Diez *et al.* [1990] used model reactions to probe the activity of coked hydrotreating catalysts which contained 1—10% carbon by weight. Hydrogenation reactions of dibenzothiophene, quinoline, and phenanthrene followed similar trends as a function of carbon content, as illustrated by Figure 8—12. In this case the functional dependence of the activity ratio, Φ_C, would be

$$\Phi_c = k_{coked}/k_{fresh} = 1-\alpha C_c \tag{8.6}$$

where α is of the order of 0.08 $(wt\%)^{-1}$. The rate constants k_{coked} and k_{fresh} were measured for the reactions of the model compounds on coked and fresh catalysts respectively. Diez *et al.* [1990] suggest, on the basis of X—ray photoelectron spectroscopy, that the coke deposits most rapidly on the alumina support, leaving the metal crystallites relatively clear. As coke continues to accumulate, it would block the edges of the MoS_2 crystallites, and suppress activity. Hence, the thickness of the carbon deposit is critical, rather than the mass. The data of Figure 8—11 suggest that coke deposition is a first—order process, with some limiting level of accumulation (Figure 5—5), indicating that a suitable equation is

$$C_c = C_c^*(1-\exp(-\beta t)) \tag{8.7}$$

where C_C^* is the equilibrium carbon content (a function of feed, hydrogen pressure, and temperature) and β is the rate constant.

Unfortunately, the carbon content reaches a stable level but the catalyst activity continues to decline. Consequently measurements of carbon content alone are unable to correlate the observed activity. Aging of the catalyst may change the nature of the surface deposits, making them more aromatic. The aromaticity of the carbon surface deposits varies over a wide range depending on the feed and operating conditions, from about 30% to 100% [Egiebor *et al.*, 1989]. If the deposits became more aromatic with time, then hydrogen from the metal crystallites would be less able to keep the coke away from the Ni or Co edge sites.

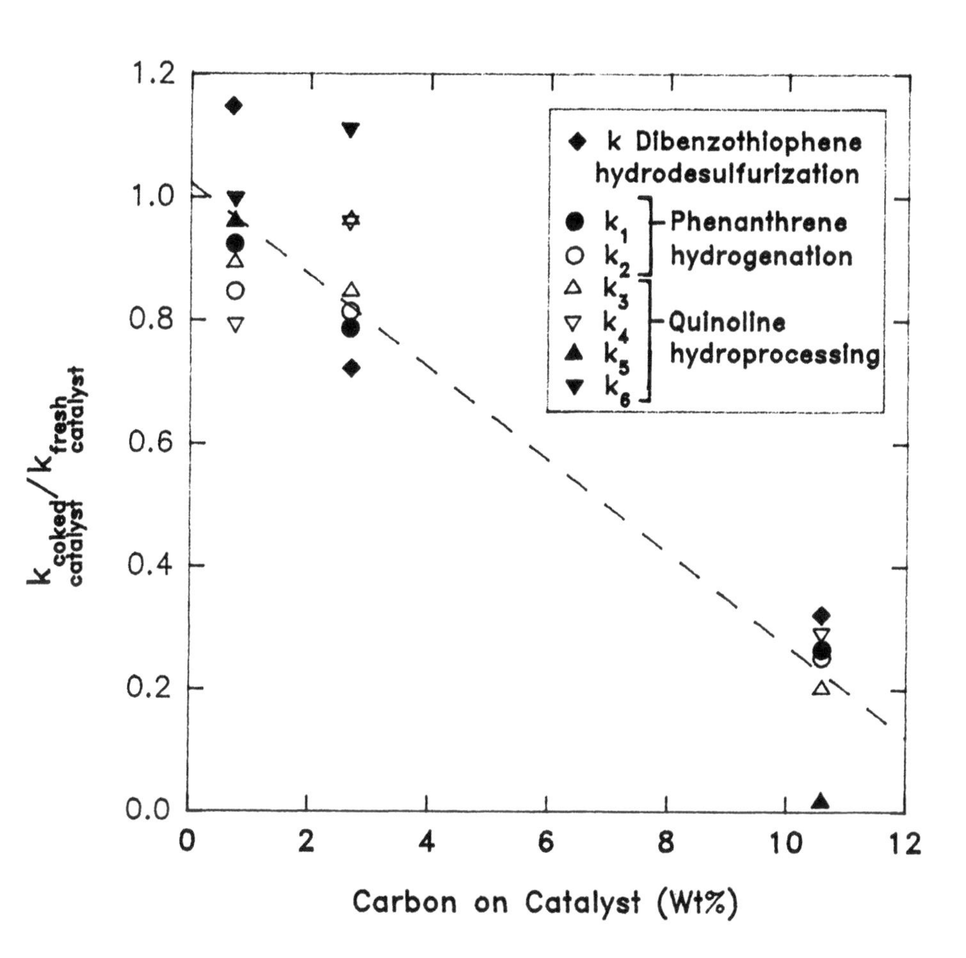

Figure 8—12

Activity of coked hydrotreating catalyst for hydrogenation reactions

Subscripted rate constants are for following reactions of model compounds: Phenanthrene k_1 = 9,10 hydrogenation, k_2 = 1,2,3,4 hydrogenation; Quinoline k_3 = 5,6,7,8 hydrogenation, k_4 = hydrogenation of 1,2,3,4 tetrahydro quinoline intermediate, k_5 = hydrogenation of 5,6,7,8 tetrahydro quinoline intermediate, k_6 = hydrogenation of o—propyl aniline intermediate (Data from Diez *et al.*, 1990)

8.6.3 Regeneration of spent hydrotreating catalysts

Loss of activity in spent catalysts can be partly reversed by burning off the carbonaceous deposits, and converting the metals back to the oxide form. Sintering, or redistribution of the metals, of catalysts does not seem to be significant in the sulfide form at the temperatures used to hydrotreat distillates, but sintering may occur during oxidative regeneration of spent catalyst [George *et al.*, 1988]. Careful control of the regeneration temperature and gas composition allow retrieval of initial activity for gas oil catalysts [Sanford, 1986; George *et al.*, 1988]. Unlike the Co/Mo catalysts in conventional refinery service, however, some catalysts used to treat some cracked distillates cannot be regenerated repeatedly due to a loss of crush strength and attrition of the particles. The activity of regenerated catalyst may be lower than the fresh catalyst, but in at least some cases subsequent regeneration of the catalyst gives no further loss of activity [Yui, 1991]. One mechanism for the irreversible loss of activity after the first cycle only may be migration of the nickel promoter during service. At higher temperatures, migration of nickel has been observed in coke on steam reforming catalysts [Trimm, 1977].

8.7 Secondary Hydrotreating of Distillates

The original development of hydrotreating processes was driven by the need to remove sulfur from sour distillates, but in processing cracked distillates from residues and bitumens other reactions are of interest. As discussed above, nitrogen removal can be a significant challenge, and oxygen removal may gain increased attention.

Product quality is often a significant issue with hydrotreated products from cracking processes. These fractions can be much more aromatic than a conventional crude oil distillate in the same boiling range, which may be of concern in reducing the total aromatics content of transportation fuels [Pauls and Weight, 1992]. Some hydrogenation of aromatics is achieved in hydrotreating gas oils and middle distillates on Ni/Mo catalysts at 350–400° C, changing the cracking characteristics of the oil by forming more naphthenes [Khorasheh *et al.*, 1989] and its product properties such as cetane number [Yui, 1989]. Secondary hydrotreating is of interest to help these distillates meet more stringent requirements for fuel formulation, such as very low sulfur and aromatics content.

A major problem with the middle distillates from hydroconversion of some residues, such as Alberta bitumens, is the low cetane number. Specifications for synthetic crude usually require a minimum value of 40 for cetane, which is insufficient for normal use without blending or further treatment. Straight run distillate is usable in diesel engines, as demonstrated on site by Syncrude, but the cetane requirement remains a concern for conventional refineries. The low cetane number is intrinsic in the composition of the Alberta bitumens. The paraffin side chains are in the range 1 to 26 carbons, with maximal abundance at 9—12 carbons [Mojelsky *et al.*, 1986b]. Thermal cracking of these chains, therefore, gives paraffins mainly in the naphtha range, rather than in the middle distillates where the paraffins would improve the cetane number. The middle distillates are then dominated by naphthenic and aromatic types with poor characteristics for diesel engines.

One example of secondary hydrotreating is work by Wilson and coworkers. Wilson and Kriz [1984] and Wilson *et al.* [1986] found that 97% of the aromatic carbon could be removed from middle distillates derived from Athabasca bitumen, using Ni/W catalyst at 380°C and 17.3 MPa hydrogen pressure. Tungsten—based catalysts were superior to Mo—based materials when the process objective was maximal conversion of aromatics. They suggested that conversion levels were limited by equilibrium between aromatic and naphthenics, depending on the reactor conditions (Figure 8—13). Moore and Akgerman [1985] pointed out that conversion should converge to a single value at higher temperatures, regardless of LHSV (see section 4.2.3 and Figure 4—11), and suggested that liquid—vapor phase equilibrium may contribute to the observed trends with temperature. One explanation for the discrepancy between the trends in the data of Figure 4—11, as compared to Figure 8—13, is the cracking of polycyclic naphthenes at 420—440°C which would prevent the observation of a single equilibrium—limited conversion [Wilson and Kriz, 1985]. Using gas oil feeds and a Ni/Mo catalyst, Yui and Sanford [1991] did observe convergence to a single conversion level at high temperatures, in support of Wilson and Kriz. Yui and Sanford also presented a kinetic analysis similar to section 4.2.3.

The feed oil used by Wilson and coworkers was produced by fluid and delayed coking of Athabasca bitumen, followed by primary hydrotreating. It had a cetane number of 31, while the products ranged as high as 42.5 [Wilson *et al.*, 1986]. Hydrogen consumption was approximately 120 m^3/m^3 feed to obtain specification product (cetane number = 40). Wilson *et al.* [1986] recommended development of catalysts to selectively open naphthenic rings as a means of improving the soot—forming potential of the product as jet or diesel fuel.

Deep hydrogenation of the middle distillates is an effective means of improving the cetane number, but it tends to be expensive due to the high consumption of hydrogen. These technologies were recently reviewed by Scubnik *et al.* [1990]. One possible way to modify the process lies in the naphthenic groups. In simple hydrogenation, these

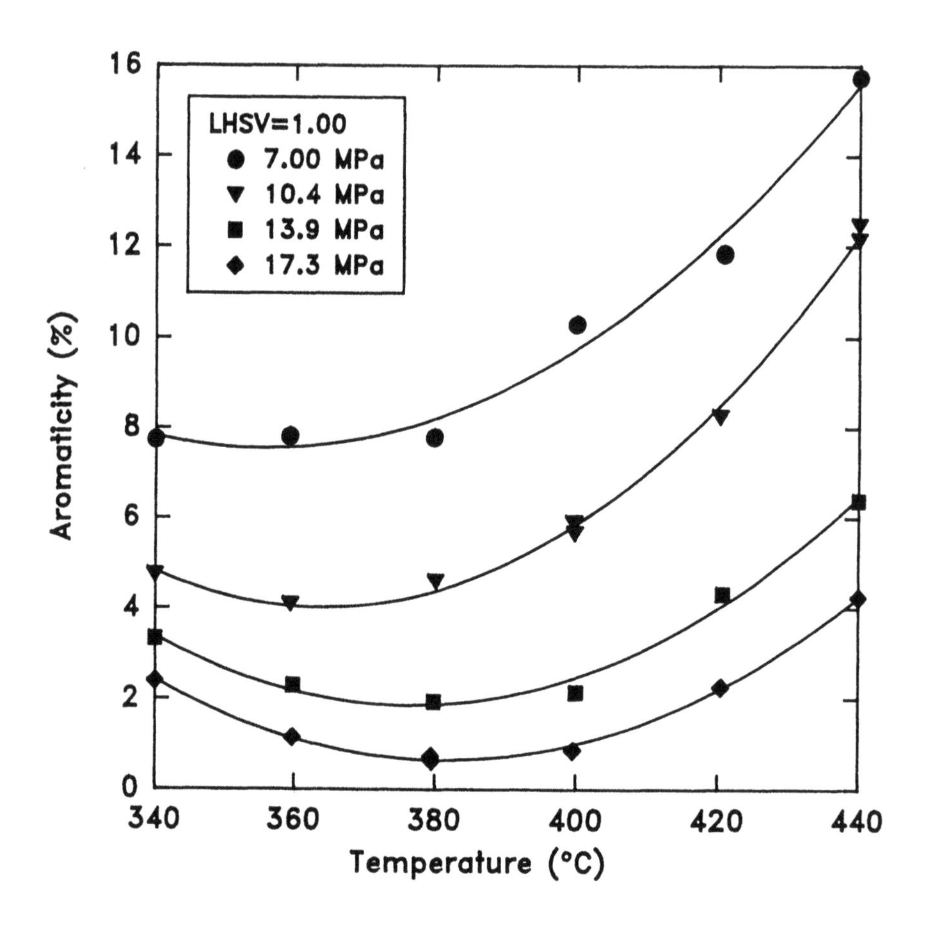

Figure 8—13
Effect of hydrogen pressure on aromatics saturation over sulfided Ni/W catalyst; LHSV 1.00, A = 7 MPa, B = 10.4 MPa, C = 13.9 MPa, D = 17.3 MPa (Data from Wilson and Kriz, 1984)

ring compounds approach the thermodynamic limits due to equilibrium with the aromatic compounds (Figures 3—4 and 4—11). Catalysts with selectivity for ring opening reactions would yield more paraffins, and overcome the thermodynamic limitations of simple hydrogenation. Payzant *et al.* [1988] identified a variety of substituted thiophenes and benzothiophenes in asphaltene from Athabasca bitumen. The pattern of substitution was interesting, in that both thiophenes and benzothiophenes were consistently substituted in the 2 and 5 positions (Table 8—3; numbering is the same as for carbazole). Sulfur removal and selective ring opening of these compounds would, therefore, yield paraffinic compounds. As an example, consider reaction (8.8):

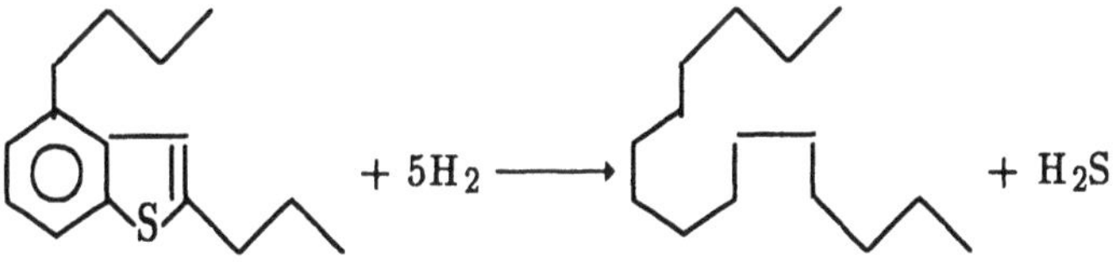

$$+ 5H_2 \longrightarrow \quad + H_2S \qquad (8.8)$$

Similar selective reactions could yield paraffins from a variety of naphthenics. Some evidence exists for enhanced formation of paraffins by selective reactions. Kirchen *et al.* [1989] found that hydrocracking in an ebullated bed, followed by coking, gave a product with more long—chain paraffins and side chains than straight coking of bitumen. The saturates fraction from hydrocracking/coking was a white solid at room temperature. The product from coking of hydrocracker bottoms was more aromatic and contained less saturates overall than did coking of uncracked bitumen. Further improvements in selectivity are unlikely during primary upgrading because of the dominance of thermal reactions.

Notation

C	concentration of reactant, $kmol/m^3$
C_C	concentration of coke, kg/kg
C_C^*	equilibrium coke concentration, kg/kg
d_p	catalyst pellet diameter, m
h	liquid hold—up, m^3/m^3
k	reaction rate constant, h^{-1}
L	lenght of reactor, m
LHSV	liquid hourly space velocity, h^{-1}
p	partial pressure, kPa
Q_0	feed flow rate, m^3/h

r	rate of reaction, kmol/(m^3—s) or kg/(m^3—s)
t	time, h
V	reactor volume, m^3
X_e	conversion at exit of reactor
α	parameter in equation 8.6
β	parameter in equation 8.1 and rate constant in equation 8.7
ν	kinematic viscosity, m^2/s
σ	stoichiometric ratio
Φ_c	activity ratio of coked catalyst

Subscripts

ar	aromatic
coked	coked catalyst
c	carbon
d	deactivation
e	exit or product condition
fresh	fresh catalyst
H_2	hydrogen
i,j	counters
o	initial or inlet condition
O,N,S	total heteroatom content

Superscripts

n	reaction order with respect to liquid—phase reactants
β	reaction order with respect to hydrogen

Further reading

Ho, T.C. 1988. "Hydrodenitrogenation catalysts", *Catal. Rev.–Sci. Engin. 30*, 117–160.

Satterfield, C.N. 1975. "Trickle—bed reactors". *AIChE J.* 21, 209—228.

Speight, J.G. 1981. *The Desulfurization of Heavy Oils and Residua*, Marcel Dekker, New York.

Problems

8.1 The following data are available for a bench—scale catalytic hydrotreating reactor:

Reactor pressure	13.9 MPa	
Reactor temperature	400° C	
Flow rates		
	Liquid m^3/min	Gas m^3/min (NTP)
Feed	3.33	2.114×10^3
Product	3.42	1.668×10^3

Liquid streams

Stream	Density kg/m^3	Composition, wt% S	N	324° C+
Feed	983	4.30	0.31	74.36
Product	925	1.02	0.19	54.36

Gas streams

Stream	H_2	H_2S	C_1	C_2	C_3	C_4+
Feed	100					
Product	95.0	3.50	0.39	0.26	0.33	0.63

a) Check the material balance of these reactor data.

b) What measures of performance would you apply to this reactor? Calculate these conversion, consumption and yield parameters.

c) What chemical reactions would give rise to the observed results?

d) How would performance change if the reactor temperature were increased?

References

Aalund, L. 1983. "Guide to export crudes of the 80's". *Oil Gas J.* Middle east 81(15), 93—95; 81(18), 204—215; North Sea 81(21), 69—76; 81(23), 75—79; West Africa 81(25), 93—95; 81(27), 96—99; North African 81(30), 146—150; Asian 81(34), 124—133; Minas, Soviet, and Chinese 81(36), 144—149; Latin America 81(43), 88—90; 81(45), 105—109; North America 81(50), 120—125; 81(51), 71—74.

Adam, P.; Mycke, B.; Schmid, J.C.; Connan, J.; Albrecht, P. 1992. "Steroid moieties attached to macromolecular petroleum fraction via di— or polysulfide bridges". *Energy Fuels* 6, 553—559.

Aiba, T.; Kaji, H.; Suzuki, T.; Wakamatsu, T. 1981. Residue thermal cracking by the Eureka Process. *Chem. Eng. Prog.* 77(2), 37—44.

Alberta Community and Environmental Health, 1988. *Report on H_2S Toxicity.* Alberta Ministry of Community and Environmental Health, Edmonton, AB, 64 pp.

Allan, D.E.; Martinez, C.H.; Eng, C.C.; Barton, W.J. 1983. "Visbreaking gains renewed interest". Chem. Eng. Progr. 79(1), 85—89.

Altgelt, K.H.; Boduszynski, M.M. 1992. "Composition of heavy petroleums. 3. An improved boiling point—molecular weight relation". *Energy Fuels* 6, 68—72.

Ammus, J.M.; Androutsopoulos, G.P. 1987. "HDS kinetic studies on Greek oil residue in a spinning basket reactor". *Ind. Eng. Chem.*

Res., 26, 494—501.
API 1983. *API Technical Data Book, Petroleum Refining*. American Petroleum Institute, Washington, D.C.
Aris, R.; Gavalas, G.R. 1966. "Theory of reactions in continuous mixtures" *Royal Soc. London, Phil. Trans.* 260, 351—393.
Asaoka, S.; Nakata, S.; Shiroto, Y.; Tekeuchi, C. 1987. "Characteristics of vanadium complexes in petroleum before and after hydrotreating". in *Metal Complexes in Fossil Fuels*, Filby, R.H.; Branthaver, J.F. eds., ACS Symposium Series 344, American Chemical Society, Washington DC, 275—289.
ASTM 1990. Annual Book of ASTM Standards, Volume 05, Petroleum Products and Lubricants, American Society for Testing of Materials, Philadelphia, PA.
Baltus, R.E.; Anderson, J.L. 1983. "Hindered diffusion of asphaltenes through microporous membranes". *Chem. Eng. Sci.* 38, 1959—1969.
Banerjee, D.K.; Laidler, .J.; Nandi, B.N.; Patmore, D.J. 1986. "Kinetic studies of coke formation in hydrocarbon fractions of heavy crudes". *Fuel* 65, 480—484.
Bartholdy, J.; Cooper, B.H. 1993. "Metal and coke deactivation of resid hydroprocessing catalysts". *ACS Div. Petrol. Chem. Prepr.* 38(2), 386—390.
Beaton W.I.; Bertolacini, R.J. 1991. "Resid hydroprocessing at Amoco". *Catal. Rev.-Sci. Eng.* 33, 281—317.
Benson, S.W. 1976. *Thermochemical Kinetics, 2nd Ed.*, John Wiley and Sons, New York.
Beret, S.; Reynolds, J.G. 1985. "Hydrogen incorporation in residuum conversion". *ACS Div. Petrol. Chem. Prepr.* 30, 664—671.
Beret, S.; Reynolds, J.G. 1990. "Effect of prehydrogenation on hydroconversion of Maya residuum, Part II: Hy incorporation". *Fuel Sci. Technol. Int.* 8, 191—219.
Birkholtz, D.A.; Hrudey, S.E.; Kimble, B.J.; Rawluk, M.; Gray, M. 1987. "Characterization of water soluble components of waste water oil sample from an oil sands bitumen upgrading plant". in *Oil in Freshwater: Chemistry, Biology, Countermeasure Technology*, F. Vermeulen and S.E. Hrudey (eds.), Pergamon, London, pp. 42—57.
Bishop, W. 1990. "LC—Finer operating experience at Syncrude". *Proc. Symp. Heavy Oil: Upgrading to Refining,* Can. Soc. Chem. Eng., Calgary, AB, pp. 14—27.
Boduszynski, M.M. 1984. "Limitations of average structural determination for heavy ends in fossil fuels". *Liquid Fuels Technol.* 2, 211—232.
Boduszynski, M.M. 1987. "Composition of heavy petroleums. 1. Molecular weight, hydrogen deficiency, and heteroatom

concentration as a function of atmospheric equivalent boiling point up to 1400° F (760° C)". *Energy Fuels* 1, 2—11.

Boduszynski, M.M. 1988. "Composition of heavy petroleums. 2. Molecular characterization". *Energy Fuels* 2, 597—613.

Bright, P.E.; Chipperfield, E.H.; Eyres, A.R.; Frati, E.; Hoehr, D.; Rettien, A.R.; Simpson, B.J. 1982. *Health Aspects of Bitumens.* CONCAWE Report, 7/82, CONCAWE, The Hague, Netherlands.

Broderick, D.H. 1980. "High—Pressure Reaction Chemistry and Kinetics Studies of Hydrodesulfurization of Dibenzothiophene Catalyzed by Sulfided Cobalt-Molybdenum Trioxide/γ—Alumina". Ph.D. Thesis, University of Delaware.

Busch, R.A.; Kociscin, J.J.; Schroeder, H.F.; Shah, G.N. 1979. "Coke heavy stocks, then treat". *Hydrocarbon Process.* 58(9), 136—142.

Carbongnani, L.; Garcia, C.; Izquierdo, A.; DiMarco, M.P.; Perez, C.; Rengel, A. ; Sanchez, V. 1987. "Correlation between physical properties and hydroprocessing of Venezeulan heavy ends". *ACS Pet. Div. Prepr.* 32, 406—412.

Carey, F.A.; Sundberg, R.J. 1980. *Advanced Organic Chemistry, Part A Structure and Mechanisms.* Plenum, New York.

Carlson, C.S.; Langer, A.W.; Stewart, J.; Hill, R.M. 1958. "Thermal hydrogenation. Transfer of hydrogen from tetralin to cracked residua". *Ind. Eng. Chem.* 53, 1067—1070.

Chase, S. 1990. "The Bi—Provincial Upgrader". *Proc. Symp. Heavy Oil: Upgrading to Refining,* Can. Soc. Chem. Eng., Calgary, AB, pp. 42—59.

Chmielowiec, J.; Fischer, P.; Pyburn, C.M. 1987. "Characterization of precursors which cause instability in hydroprocessed gas oils". *Fuel* 66, 1358—1363.

Choi, J.H.K.; Gray, M.R. 1988. "Structural Analysis of Extracts from Spent Hydroprocessing Catalysts". *Ind. Eng. Chem. Res.* 27, 1587—1595.

Chou, G.F.; Prausnitz, J.M. 1989. "A phenomenological correction to an equation of state for the critical region". *AIChE J.* 35, 1487—1496.

Christensen, H.; Cooper, B.H. 1990. "The influence of catalyst and feedstock properties in FCC pretreatment". AIChE Spring National Meeting, March, paper 44b.

Chung, S.Y.K. 1982. "Thermal Processing of Heavy Gas Oils". M.Sc. Thesis, Department of Chemical Engineering, University of Alberta.

Colyar, F.; Harris, E.; Popper, G. 1989. "Demonstration of the high—conversion H—Oil process on Cold Lake residuum", *4th UNITAR/UNDP Conf. Heavy Crude Tar Sands,* Edmonton, paper 5.

Coxson and Bischoff 1987. "Lumping strategy. 1. Introductory techniques and applications of cluster analysis". *Ind. Eng. Chem. Res.* 26, 1239—1248.

Cyr, N.; McIntyre, D.D.; Toth, G.; Strausz, O.P. 1987. "Hydrocarbon structural group analysis of Athabasca asphaltene and its g.p.c. fractions by ^{13}C n.m.r." *Fuel* 66, 1709—1714.

DeBruijn, T.J.W.; Chase, J.D.; Dawson, W.H. 1988. "Gas holdup in a two—phase vertical tubular reactor at high pressure". *Can. J. Chem. Eng.* 66, 330—333.

Decroocq, D. 1984. *Catalytic cracking of heavy petroleum fractions.* Editions Technip, Paris, 123 pp.

DeJong, E. 1980. "The effect of a crude oil spill on cereals". *Environ. Pollut. (Ser. A)* 22, 187—196.

Dejonghe, S.; Hubaut, R.; Des Courieres, T.; Grimblot, J. 1990. "Influence of the competitive adsorption of polyaromatic and vanadyl porphyrin molecules on the deactivation of a Ni—MoS_2—Al_2O_3 catalyst". *Appl. Catal.* 61, L9—L14.

Dickie, J.P.; Yen, T.F. 1967 "Macrostructures of the asphaltic fractions by various instrumental methods". *Anal. Chem.* 39, 1847—1852.

Diez, F.; Gates, B.C.; Miller, J.T.; Sajkowski, D.J.; Kukes, S.G. 1990. "Deactivation of a Ni—Mo/γ—Al_2O_3 catalyst: Influence of coke on the hydroprocessing activity". *Ind. Eng. Chem. Res.* 29, 1999—2004.

Dolbear, G.E.; Tang, A.; Moorehead, E.L. 1987. "Upgrading studies with Californian, Mexican, and Middle Eastern crude oils". in *Metal Complexes in Fossil Fuels*, Filby, R.H.; Branthaver, J.F. eds., ACS Symposium Series 344, American Chemical Society, Washington DC, 220—232.

Dorbon, M.; Ignatiadis, I.; Schmitter, J.—M.; Arpino, P.; Guiochon, G.; Toulhoat, H.; Huc A. 1984. "Identification of carbazoles and benzocarbazoles in a coker gas oil and influence of catalytic hydrotreatment on their distribution". *Fuel* 63, 565—570.

Edelman, A.M.; Lipuma, C.R.; Turpin, F.G. 1979. "Developments in thermal and catalytic cracking processes for heavy feeds". *Proc. Tenth World Pet. Congr., Bucharest, 1979,* Heyden, London, Vol. 4 pp. 167—174.

Edmister, W.C. 1958. "Applied hydrocarbon thermodynamics. 4. Compressibility factors and equations of state". *Petroleum Refiner* 37(4), 173—178.

Edmister, W.C. 1988. *Applied Hydrocarbon Thermodynamics, Vol 2, 2nd Ed.*, Gulf Publishing, Houston, TX.

Egiebor, N.O.; Gray, M.R.; Cyr, N. 1989. "^{13}C—NMR characterization of organic residues on spent hydroprocessing, hydrocracking, and demetallization catalysts". *Appl. Catal.* 55, 81—91.

Elvin, F.J.; Otterstedt, J.E.; Serte, J. 1988. "Processes for demetalization of fluid cracking catalysts". in *Fluid Catalytic Cracking: Role in modern refining*, M.L. Occelli, (ed), ACS Symp. Ser. 375, ACS, Washington DC, pp. 229—236.
Environment Canada 1984. *Technical Information for Problem Spills: Hydrogen Sulfide*, Environment Canada Environmental Protection Service, Ottawa, Ont., 92 pp.
Fabuss, B. M.; Smith, J. O.; Satterfield, C. N. 1964. "Thermal cracking of pure saturated hydrocarbons". *Adv. Pet. Chem. Ref.* 9, 157—201.
Fogler, H.S. 1986. *Elements of Chemical Reactor Design*, Prentice—Hall, Englewood Cliffs, NJ.
Ford, T.J. 1986. "Liquid Phase Thermal Decomposition of Hexadecane: Reaction Mechanisms". *Ind. Eng. Chem. Fundam.* 25, 240—243.
Fouda, S.A.; Kelly, J.F.; Rahimi, P.M. 1989. "Effects of coal concentration on coprocessing performance". *Energy Fuels* 3, 154—160.
Freund, H. ; Matturro, M.G.; Olmstead, W.N.; Reynolds, R.P.; Upton, T.H. 1991. "Anomalous side—chain cleavage in alkylaromatic thermolysis". *Energy Fuels* 5, 840—846.
Froment, G.F.; Bischoff, K.B. 1979. *Chemical Reactor Analysis and Design*, John Wiley, New York.
Frye, C.G.; Weitkamp, A.W. 1969. "Equilibrium hydrogenations of multi—ring aromatics" *J. Chem. Eng. Data* 14, 372—376.
Funk, G.A.; Harold, M.P.; Ng, K.M. 1990. "A novel model for trickle beds with flow maldistribution". *Ind. Eng. Chem. Res.* 29, 738—748.
Furimsky, E.; Ranganathan, R.; Parsons, B.I. 1978. "Catalytic hydrodenitrogenation of basic and non—basic nitrogen compounds in Athabasca bitumen distillates". *Fuel* 57, 427—430.
Furimsky, E. 1978a. "Catalytic deoxygenation of heavy gas oil". *Fuel* 57, 494—496.
Furimsky, E. 1978b. "Chemical origin of coke deposited on catalyst surface", *Ind. Eng. Chem. Prod. Res. Dev.* 17, 329—331.
Furimsky, E. 1979. "Deactivation and Regeneration of Refinery Catalysts". *Erdol & Kohle* 32, 383—390.
Furimsky, E. 1983a. "Chemistry of catalytic hydrodeoxygenation". *Catal. Rev. Sci.–Eng.* 25, 421—458.
Furimsky, E. 1983b. "Deactivation of molybdate catalyst durimg hydrodeoxyenation of tetrahydrofuran". *Ind. Eng. Chem. Prod. Res. Dev.* 22, 34—38.
Galbreath, R.B.; van Driesen, R.P. 1967 "Hydrocracking of residual petroleum stocks". *Proc. 8th World Petrol. Congr.*, vol 4, pp. 129—137.
Galiasso, R.; Garcia, W.; Ramirez de Agudelo, M.M.; Andreu, P. 1984.

"Hydrotreatment of cracked light gas oil". *Catal. Rev. Sci. Eng.* 26, 445—480.

Gary, J.H.; Handwerk, G.E. 1984. *Petroleum Refining: Technology and Economics*, Marcel Dekker, New York.

Gates, B.C. 1992. *Catalytic Chemistry*, John Wiley and Sons, New York.

George, Z.M.; Mohammed, P.; Tower, R. 1988. "Regeneration of spent hydroprocessing catalyst". *Proc. 9th Int. Congr. Catal, Calgary, 1988*. Vol. 1, pp 230—237.

Girgisz, M.J.; Gates, B.C. 1991. "Reactivities, reaction networks, and kinetics in high—pressure catalytic hydroprocessing" *Ind. Eng. Chem. Res.* 30, 2021—2058.

Golikeri, S.V.; Luss, D. 1972. "Analysis of activation energy of grouped parallel reactions". *AIChE J.* 18, 227—282.

Gosselink, J.W.; Stork, W.H.J.; de Vries, A.F.; Smit, C.H. 1987. "Mild hydrocracking: Coping with catalyst deactivation". in *Catalyst Deactivation 1987*, B. Delnom and G.F. Froment, eds., Elsevier, Amsterdam, pp 279—287.

Gough, M.A.; Rowland, S.J. 1990. "Characterization of unresolved complex mixtures of hydrocarbons in petroleum". *Nature* 344, 648—650.

GPSA 1987. *GPSA Engineering Data Book, vol II.*, Gas Processors Suppliers Association, Tulsa, OK.

Gray, M.R.; Choi, J.H.K.; Egiebor, N.O.; Kirchen, R.P.; Sanford, E.C. 1989. "Structural group analysis of residues from Athabasca bitumen". *Fuel Sci. Technol. Int.* 7, 599—610.

Gray, M.R. 1990. "Lumped Kinetics of Structural Groups: Hydrotreating of Heavy Distillate". *Ind. Eng. Chem. Res.* 29, 505—512.

Gray, M.R.; Jokuty, P.; Yeniova, H.; Nazarewycz, L.; Wanke, S.E.; Achia, U.; Krzywicki, A.; Sanford, E.C.; Sy, O.K.Y. 1991. "The relationship between chemical structure and reactivity of Alberta bitumens", *Can J. Chem. Eng.* 69, 833—843.

Gray, M.R.; Khorasheh, F.; Wanke, S.E.; Achia, U.; Krzywicki, A.; Sanford, E.C.; Sy, O.K.Y.; Ternan, M. 1992. "The role of catalyst in hydrocracking residues from Alberta bitumens", *Energy Fuels* 6, 478—485.

Gray, M.R.; Krzywicki, A.; Wanke, S.E. 1992b. "Chemical transformation during resid upgrading: Catalytic and Thermal". Report to CANMET, SSC # 06SQ.23440—7—9011.

Gray, R.D.; Heidman, J.L.; Springer, R.D.; Tsonopoulos, C. 1985. "VLE Predictions for multicomponent H2 systems with cubic equations of state". *Proc. Ann. Conv., Gas Process. Assoc. Tech.*

Papers 64, 289—298.

Green, J.B.; Yu, S.K.–T.; Pearson, C.D.; Reynolds, J.W. 1993. "Analysis of sulfur compound types in asphalt". *Energy Fuels* 7, 119—126.

Greenly, W.B.; Barta, .D.; Eddinger, R.T. 1982. "Literature review of toxicological properties of petroleum and coal–derived products". *Fuel Process. Technol.* 6, 9—25.

Groot, C.K.; de Beer, V.H.J.; Prins, R.; Stolarski, M.; Niedzweidz, W.S. 1986. "Comparative study of alumina and carbon supported catalysts for hydrogenolysis and hydrogenation of model compounds and coal–derived liquids". *Ind. Eng. Chem. Prod. Res. Dev.* 25, 522—530.

Gudin, C.; Syratt, W.J. 1975. "Biological aspects of land rehabilitation following hydrocarbon contamination". *Environ. Pollut.* 8, 107—112.

Guerin, M.R. et al. 1978. "Polycyclic hydrocarbons from fossil fuel conversion processes". in *Carcinogenesis, Vol. 3 PAH's*, P.W. Jones and R.I. Freundenthal (eds.), Raven Press, New York, pp. 21—33.

Guitian, J.; Souto, A.; Ramirez, R.; Marzin, R.; Solari, B. 1992. "Commercial design of a new upgrading process, HDH", *Proc. Int. Symp. on Heavy Oil and Residue Upgrading and Utilization*, C. Han and C. Hsi, eds, International Academic, Beijing, pp. 237—247.

Habermehl, R. 1988. "Safe handling and disposal of spent catalysts". *Chem. Eng. Progr.* 84(2) 16—19.

Han, C.; Liao, S.; Liu, Z. 1992. "Hydrocracking for high quality oil products and petrochemical feedstocks", *Proc. Int. Symp. on Heavy Oil and Residue Upgrading and Utilization*, C. Han and C. Hsi, eds, International Academic, Beijing, pp. 67—72.

Heck, R.H.; DiGuiseppi, F.T. 1993. "Kinetic and mechanistic effects in resid hydrocracking". *ACS Div. Petrol. Chem. Prepr.* 38(2), 417—421.

Henderson, J.H.; Weber, L. 1965. "Physical upgrading of heavy crude oils by the application of heat". *J. Can. Pet. Tech.* 4, 206—212.

Henry, H.C.; Gilbert, J.B. 1973. "Scale–up of pilot plant data for catalytic hydroprocessing". Ind. Eng. Chem. Process Des. Dev. 12, 328—334.

Hettinger, W.P. 1988. "Development of a reduced crude cracking catalyst". in *Fluid Catalytic Cracking: Role in modern refining*, M.L. Occelli, (ed), ACS Symp. Ser. 375, ACS, Washington DC, pp. 308—340.

Hirschberg, E.H.; Bertolacini, R.J. 1988. "Catalytic control of SO_X emissions from fluid catalytic cracking units". in *Fluid Catalytic Cracking: Role in modern refining*, M.L. Occelli, (ed), ACS Symp. Ser. 375, ACS, Washington DC, pp. 114—145.

Ho, T.C. 1988. "Hydrodenitrogenation catalysts", *Catal. Rev. –Sci. Engin.* 30, 117–160.

Ho, T.C.; Aris, R. 1987. "On apparent second–order kinetics". *AIChE J.* 33, 1050–1051.

Hutchinson, P.; Luss, D. 1971. "Lumping of mixtures with many n–th order reactions". *Chem. Eng. J.* 2, 172–178.

Hwang, S.–C.; Tsonopolous, C.; Cunningham, J.R.; Wilson, G.M. 1982. "Density, viscosity, and surface tension of coal liquids at high temperatures and pressures". *Ind. Eng. Chem. Process Des. Dev.* 21, 127–134.

Hyndman, A.W.; Liu, J.K. 1987. "How to select an upgrading process to suit your needs", *China–Canada Heavy Oil Technol. Symp. Proc., Zhuo'Zhou, October 26–30, 1987*, paper 50.

Inoue, K.; Zhang, P.; Tsyuyama, H. 1993. "Recovery of Mo, V, Ni, and Co from spent hydrodesulfurization catalysts". *ACS Div. Petrol. Chem. Prepr.* 38(1), 77–80.

Jaffe, S.B. 1974. "Kinetics of heat release in petroleum hydrogenation". *Ind. Eng. Chem. Process Des. Develop.* 13, 34–39.

Jeffries, R.B.; Gupta, R.K. 1986. "The Bi–Provincial Project primary upgrading technology: Why H–Oil was chosen". *Proc. CSChE Symp. Heavy/Synthetic Oil Upgrading, Edmonton, Alberta, February 1986*, 17 pp.

Jiang, P.; Arters, D.; Fan, L.–S. 1992. "Pressure effects on the hydrodynamic behavior of gas–liquid–solid fluidized beds". *Ind. Eng. Chem. Res.* 31, 2322–2327.

Jocker, J.S.M. 1993. "The Metrex process, full recycling of spent hydroprocessing catalysts". *ACS Div. Petrol. Chem. Prepr.* 38(1), 74–76.

Johnson, C.A.; Nongbri, G.; Lehman, L.M.; Chervenak, M.C. 1977. "Conversion of bitumen and heavy crude oils to more valuable products by the H–Oil process". *Oil Sands of Canada–Venezuela 1977*, Eds. D.A. Redford, A.G. Winestock, Canadian Institute of Mining and Metallurgy, Special volume 17, p. 224–230.

Katti, S.S.; Westerman, D.W.B.; Gates, B.C.; Youngless, T.; Petrakis, L. 1984. "Catalytic hydroprocessing of SRC–II heavy distillate fractions. 3. Hydrodesulfurization of the neutral oils". *Ind. Eng. Chem. Process Des. Dev.* 23, 773–778.

Kesler, M.G.; Lee, B.I. 1976. "Improve prediction of enthalpy of fractions". *Hydrocarbon Process.* 55(3), 153–158.

Khorasheh, F.; Gray, M.R.; Dalla Lana, I.G. 1987. "Structural analysis of Alberta heavy gas oils". *Fuel* 66, 505–511.

Khorasheh, F.; Rangwala, H.; Gray, M.R.; Dalla Lana, I.G. 1989. "Interactions between thermal and catalytic reactions in mild hydrocracking of gas oil". *Energy Fuels* 3, 716—722.

Khorasheh, F. 1992. "Thermal Cracking and Hydrocracking of n—Hexadecane in Aromatic Solvents", Ph.D. Thesis, University of Alberta.

Kim, H.; Curtis, C.W.; Cronauer, D.C.; Sajkowski, D.J. 1989. "Characterization of catalysts from molybdenum naphthenate". *Prepr. ACS Div. Fuel Chem.* 34(4), 1431—1434.

Kirchen, R.P.; Sanford, E.C.; Gray, M.R.; George, Z.M. 1989. "Coking of Athabasca—derived feedstocks". *AOSTRA J. Res.* 5, 225—235.

Kirchen, R.P.; Sanford, E.C. 1989. "Kinetics of Ni—Mo/Al_2O_3, Co—Mo/Al_2O_3 and layered catalyst charges for hydrotreating of Athabasca bitumen derived coker gas oil". *AOSTRA J. Res.* 5, 287—301.

Kissin, Y.V. 1987. "Free—radical reactions of high molecular weight isoalkanes". *Ind. Eng. Chem. Res.* 26, 1633—1638.

Koseoglu, R.O.; Phillips, C.R. 1988. "Kinetic models for the non—catalytic hydrocracking of Athabasca bitumen". *Fuel* 67, 906—915.

Kossiakoff, A.; Rice, F. O. 1943. "Thermal decomposition of hydrocarbons, resonance stabilization and isomerization of free radicals". *J. Am. Chem. Soc.* 65, 590—595.

Kroschwitz, J.I. 1985. *Encyclopedia of Polymer Science and Technology*, Wiley, New York, Chapter 1.

LaMarca, C.; Libanati, C.; Klein, M.T. 1990. "Design of kinetically coupled complex reactions systems". *Chem. Eng. Sci.* 45, 2059—2065.

Langer, A.W.; Stewart, J.; Thompson, C.E.; White, .T.; Hil, R.M. 1961. "Thermal hydrogenation of crude residua". *Ind. Eng. Chem.* 53, 27—30.

Lavopa, V.; Satterfield, C.N. 1988. "Response of dibenzothiophene hydrodesulfurization to presence of nitrogen compounds". *Chem. Eng. Commun.* 70, 171—176.

Le Page, J.F.; Davidson, M. 1986. "La conversion des residus et huiles lourdes: Au carrefour du thermique et du catalytique", *Rev. L'Inst. Francais Petrole* 41, 131—143.

Lee, S.Y.; Seader, J.D.; Tsai, C.H.; Massoth, F.E. 1991a. "Solvent and temperature effects on restrictive diffusion under reaction conditions". *Ind. Eng. Chem. Res.* 30, 607—613.

Lee, S.Y.; Seader, J.D.; Tsai, C.H.; Massoth, F.E. 1991b. "Restrictive liquid—phase diffusion and reaction in bidispersed catalysts". *Ind. Eng. Chem. Res.* 30, 1683—1693.

Lott, R.; Cyr, T.J. 1992. "Study of mechanism of coking in heavy oil processes". *Proc. Int. Symp. on Heavy Oil and Residue Upgrading and Utilization*, C. Han and C. Hsi, eds, International Academic, Beijing, pp. 309—316.

Lu, B.C.—Y.; Fu, C.—T. 1989. "Phase equilibria and PVT properties", in *AOSTRA Technical Handbook on Oil Sands, Bitumens, and Heavy Oils*, L.C. Hepler and C. Hsi, eds., AOSTRA, Edmonton, pp. 129.

MacKinnon, M.D. 1989. "Development of the tailings pond at Syncrude's oil sands plant: 1978—1987". *AOSTRA J. Res.* 5, 109—133.

McKnight, C.A.; Nowlan, V. 1993. "Metals accumulation and particle mixing in a commercial residue hydroprocessor with continuous catalyst addition". *ACS Div. Petrol. Chem. Prepr.* 38(2), 391—397.

Malhotra, R.; McMillen, D.F. 1990. "A mechanistic numerical model for coal liquefaction involving hydrogenolysis of strong bonds. Rationalization of interactive effects of solvent aromaticity and hydrogen pressure". *Energy Fuels* 4, 184—193.

Malhotra, R.; McMillen, D.F. 1993. "Relevance of cleavage of strong bonds in coal liquefaction". *Energy Fuels* 7, 227—233.

Maxwell, J.B.; Bonnett, L.S. 1955. *Vapor Pressure Charts for Petroleum Hydrocarbons*, Exxon Research and Engineering Co., Florham Park, NJ, 1955; reprinted 1974.

Mears, D.E. 1974. "Role of liquid holdup and effective wetting in the performance of trickle—bed reactors". *Adv. Chem. Ser.* 133, 218—227.

Mehrotra, A.K.; Eastick, R.R.; Svrcek, W.Y. 1989. "Viscosity of Cold Lake bitumen and its fractions", *Can. J. Chem. Eng.* 67, 1004—1009.

Menon, K.R.; Mink, B.H. 1992. "Residue conversion options for European refineries". *Hydrocarbon Process.* 71(5), 100I—100N.

Miki, Y. 1975. "The thermodynamic properties of C_9H_{18} naphthenes. I. The determination of the equilibirum constants for the hydrogenation of propyl— and isopropylbenzene and ethyltoluenes". *Bull. Chem. Soc. Jap.* 48, 201—208.

Miki, Y.; Yamadaya, S.; Oba, M.; Sugimoto, Y. 1983. "Role of catalyst in hydrocracking of heavy oil". *J. Catal.* 83, 371—383.

Mills, G.A.; Boedeker, E.A.; Oblad, A.G. 1950. "Chemical chracterization of catalysts. 1. Poisoning of cracking catalysts by nitrogen compounds and potassium ion". *J. Am. Chem. Soc.* 72, 1554—1560.

Mitchell, D.L.; Speight, J.G. 1973. "The solubility of asphaltenes in hydrocarbon solvents". *Fuel* 52, 149—152.

Mochida, I.; Inoue, S.—I.; Maeda, K.; Takeshita, K. 1977. "Carbonization of aromatic hydrocarbons". *Carbon* 15, 9—16.

Mochida, I.; Zhao, X.–Z.; Sakanishi, K.; Yamamoto, S.; Takashima, H.; Uemura, S. 1989. "Structure and proerties of sludges produced in the catalytic hydrocracking of vacuum residue". *Ind. Eng. Chem. Res.* 28, 418–421.

Mochida, I.; Zhao, X.–Z.; Sakanishi, K. 1990. "Suppression of sludge formation by two–stage hydrocracking of vacuum residue at high conversion". *Ind. Eng. Chem. Res.* 29, 2324–2327.

Mojelsky, T.W.; Montgomery, D.S.; Strausz, O.P. 1986a. "The basic nitrogen compounds in Athacasca bitumen", *AOSTRA J. Res.* 3, 25–33.

Mojelsky, T.W.; Montgomery, D.S.; Strausz, O.P. 1986b. "The side chains associated with the undistillable aromatic and resin components of Athabasca bitumen". *AOSTRA J. Res.* 3, 177–184.

Moore, P.K.; Akgerman, A. 1985. "Comments on upgrading of middle distillate fractions of syncrude from Athabasca oil sands". *Fuel* 64, 721–722.

Moreau, C.; Bekakra, L.; Olive, J.–L.; Geneste, P. 1988. "Influence of the aromaticity of the benzene ring on HDN of aniline–like compounds over a sulfided NiO-MoO_3/γ-Al_2O_3 catalyst". *Proc. 9th Int. Congr. Catal, Calgary, 1988,* Vol. 1, pp 58–65.

Morris, C.G.; Sim, W.D.; Vysniauskas, T.; Svrcek, W.Y. 1988. "Crude tower simulation on a personal computer". *Chem. Eng. Progr.* 84(11) 63–68.

Morrison, R.T.; Boyd, R.N. 1973. *Organic Chemistry, 3rd Edition.* Allyn and Bacon, Boston, MA.

Mosby, J.F.; Buttke, R.D.; Cox, J.A.; Nikolaides, C. 1986. "Process characterization of expanded–bed reactors in series". *Chem. Eng. Sci.* 41, 989–995.

Murcia, A.A. 1992. "Numerous changes mark FCC technology advance". *Oil Gas J.* 90(5), 68–71.

Mushrush, G.W.; Hazlett, R.N. 1984. "Pyrolysis of organic compounds containing long unbranched alkyl groups". *Ind. Eng. Chem. Fundam.* 23, 288–294.

Myers, T.E.; Lee, F.S.; Myers, B.L.; Fleisch, T.H.; Zajac, G.W. 1989. "Resid catalyst deactivation in expanded–bed service". in *Fundamentals of Resid Upgrading*, R.H. Heck, T.F. Degnan, eds., AIChE Symp. Ser. 273, AIChE, New York, pp. 21–31.

Nelson, W.L. 1958. *Petroleum Refinery Engineering, 4th Edition,* McGraw–Hill, New York, NY.

Neurock, M.; Libanati, C.; Nigam, A.; Klein, M.T. 1990. "Monte Carlo simulation of complex reaction systems: molecular structure and reactivity in modeling heavy oils". *Chem. Eng. Sci.* 45, 2083–2088.

Nishijima, A.; Shimada, H.; Yoshimura, Y.; Sato, T.; Matsubayashi, N.

1987. "Deactivation of molybdenum catalysts by metal and carbonaceous deposits during hydrotreating of coal derived liquids and heavy petroleums". hydrocracking: Coping with catalyst deactivation". in *Catalyst Deactivation 1987*, B. Delmon and G.F. Froment, eds., Elsevier, Amsterdam, pp 39—58.

Oballa, M.C.; Herrera, P.S.; Somogyvari, A.F. 1991. "Effect of catalyst in hydrotreating gas oil from residue hydrocracking", AIChE Annual Meeting, November 5—17, 1991, Los Angleles, CA, paper 130e.

Oballa, M.; Wong, C.; Krzywicki, A.; Chase, S.; Dennis, G.; Vandenhengel, W.; Jeffries, R. 1992. "Hydroprocessing catalyst research to support the operation of the Bi—Provincial Upgrader in Lloydminster", *Proc. Int. Symp. on Heavy Oil and Residue Upgrading and Utilization*, C. Han and C. Hsi, eds, International Academic, Beijing, pp. 67—72.

Occelli, M.L. 1988a. "Recent trends in fluid catalytic cracking technology". in *Fluid Catalytic Cracking: Role in modern refining*, M.L. Occelli, (ed), ACS Symp. Ser. 375, ACS, Washington DC, pp. 1—16.

Occelli, M.L. 1988b. "Cracking metal—contaminated oils with catalysts containing metal scavengers: Effects of sepiolite addition on vanadium passivation". in *Fluid Catalytic Cracking: Role in modern refining*, M.L. Occelli, (ed), ACS Symp. Ser. 375, ACS, Washington DC, pp. 162—181.

Otterstedt, J.E.; Gevert, S.B.; Jaras, S.G.; Menon, P.G. 1986. "Fluid catalytic cracking of heavy (residual) oil fractions: A review". *Appl. Catal.* 22, 159—179.

Overfield, R.E.; Sheu, E.Y.; Sinha, S.K.; Kiang, K.S. 1988. "SANS study of asphaltene aggregation". *ACS Div. Petrol. Chem. Prepr.* 33, 308—313.

Özüm, B.; Lewkowicz, L.; Cyr, T.; Oğuztöreli, M. 1989. "Hydrodynamics of a bitumen upgrader". *AIChE J.* 35, 1032—1038.

Papayannakos, N.; Georgiou, G. 1988. "Kinetics of hydrogen consumption during catalytic hydrodesulfurization of a residue in a trickle—bed reactor". *J. Chem. Eng. Japan* 21, 244—249.

Pasquini, R.; Taningher, M.; Monarca, S.; Pala, M.; Angeli, G. 1989. "Chemical composition nad genotoxic activity of petroleum derivatives collected in two working environments". *J. Toxicol. Environ. Health.* 27, 225—238.

Pauls, R.E.; Weight, G.J. 1992. "Methods for determining aromatics and benzene in reformulated gasolines". *Prepr. ACS Div. Fuel Chem.* 37(1), 9—16.

Payzant, J.D.; Montgomery, D.S.; Strausz, O.P. 1988. "The

identification of a homologous series of benzo[b]thiophenes, thiophenes, thiolanes, and thianes posessing a linear framework in the pyrolysis oil of Athabasca asphaltene". *AOSTRA J. Res.* 4, 117–129.

Pedersen, K.S.; Thomassen, P.; Fredenslund, A. 1984. "Thermodynamics of petroleum mixtures containing heavy hydrocarbons. 1. Phase envelope calculations by use of the Soave–Redlich Kwong equation of state". *Ind. Eng. Chem. Process Des. Dev.* 23, 163–170.

Peng, D.–Y.; Robinson, D.B. 1976. "A new two–constant equation of state". *Ind. Eng. Chem. Fundam.* 15, 59–64.

Pereira, C.J.; Beeckman, J.W.; Cheng, W.–C.; Suarez, W. 1990. "Metal deposition in hydrotreating catalysts. 2. Comparison with experiment". *Ind. Eng. Chem. Res.* 29, 520–521.

Pereira, C.J. 1990. "Metal deposition in hydrotreating catalysts. 1. A regular perturbation solution approach". *Ind. Eng. Chem. Res.* 29, 512–519.

Perry, R.H.; Green, D.W.; Maloney, J.O. (eds) 1984. *Perry's Chemical Engineer's Handbook, 6th Ed.*, McGraw–Hill, New York.

Peters, K.E.; Scheuerman, G.L.; Lee, C.Y.; Moldowan, J.M.; Reynolds, R.N.; Pefia, M.M. 1992. "Effects of refinery processes on biological markers". *Energy and Fuels* 6, 560–577.

Petrakis, L.; Allen, D.T. 1987. *NMR for Fossil Fuels*, Elsevier, Amsterdam.

Poutsma, M.L. 1990. "Free radical thermolysis and hydrogenolysis of model hydrocarbons relevant to processing of coal". *Energy Fuels* 4, 113–131.

Prausnitz, J.M. 1988. *Thermodynamic Properties and Characterization of Petroleum Fractions*, Amer. Petrol. Inst., Washington, DC.

Prins, R.; De Beer, V.H.J.; Somorjai, G.A. 1989. "Structure and function of the catalyst and the promoter in Co–Mo hydrodesulfurization catalysts", *Catal. Rev. Sci. Eng.* 31, 1–41.

Proust, P.; Vera, J. 1989. "PRSV: The Stryjek–Vera modification of the Peng–Robinson equation of state. Parameters for other pure compounds of industrial interest". *Can. J. Chem. Eng.* 67, 170–173.

Pruden, B.B.; Denis, J.M. 1976. "Heat of reaction and vaporization of feed and product in the thermal hydrocracking of Athabasca bitumen". CANMET Report 76–30, EMR Canada, Ottawa, 25 pp.

Pruden, B.B.; Denis, J.M.; Muir, G. 1989. "Upgrading of Cold Lake heavy oil in the CANMET hydrocracking demonstration plant", *4th UNITAR/UNDP Conf. Heavy Crude Tar Sands, Edmonton, 1989*, paper 80.

Puttagunta, V.R.; Miadonye, A. 1991. "Correlation and prediction of viscosity of heavy oils and bitumens containing dissolved gases Part 1: Alberta bitumens and heavy oils containing pure gases". *AOSTRA J. Res.* 7, 241—250.

Rangwala, H.A., Dalla Lana, I.G., Otto, F.O. and Wanke, S.E., 1984. "Hydroprocessing of Syncrude Coker Gas Oil", *Stud. Surf. Sci. Catal.* 19, 537—544.

Rangwala, H.A.; Wanke, S.E.; Otto, F.D.; Dalla Lana, I.G. 1986. "The hydrotreating of coker gas oil: Effects of operating conditions and catalyst properties". Prepr. 10th Can. Symp. Catal., Kinston, Ont., June 15—18, 1986, pp. 20—28.

Rangwala, H.A.; George, Z.M.; Hardin, A.H. 1991. "Upgrading of coprocessed liquid: Kinetics and internal mass—transfer effects". *Energy Fuels* 5, 835—839.

Reid, R.C.; Prausnitz, J.M.; Poling, B.E. 1987. *The Properties of Gases and Liquids, 4th Ed.*, McGraw—Hill, New York, NY.

Reis, T. 1975. "To coke, desulfurize, and calcine. Part 2: Coke quality and its control". *Hydrocarbon Process* 54(6), 97—104.

Reynolds, B.E.; Brown, E.C.; Silverman, M.A. 1992. "Clean gasoline via VRDS/RFCC". *Hydrocarbon Process.* 71(4), 43—51.

Rheaume, L.; Ritter, R.E. 1988. "Use of catalysts to reduce SO_X emissions from fluid catalytic cracking units". in *Fluid Catalytic Cracking: Role in modern refining*, M.L. Occelli, (ed), ACS Symp. Ser. 375, ACS, Washington DC, pp. 146—161.

Rhoe, A.; de Blignieres, C. 1979. "Visbreaking: a flexible process". *Hydrocarbon Process.* 58(1), 131—136.

Robschlager, K.W.; Deelen, W.J.; Naber, J.E. 1992. "The Shell residue hydroconversion process: Development and future applications". *Proc. Int. Symp. on Heavy Oil and Residue Upgrading and Utilization*, C. Han and C. Hsi, eds, International Academic, Beijing, pp. 249—254.

Sanchez, V.; Murgia, E.; Lubkowitz, J.A. 1984. "Size exclusion chromatographic apporach for the evaluation of processes for upgrading heavy petroleum". *Fuel* 63, 612—616.

Sanford, E. 1986. "Regeneration of Catalysts from Hydrotreating Bitumen Derived Coker Gas Oil: Correlation of Catalyst Activity with Regeneration Conditions and Measured Catalyst Properties", *Prepr. 10th Can. Symp. Catal., Kingston Ont.*, pp. 589—598.

Sanford, E.C.; Chung, K.H. 1991. "The mechanism of pitch conversion during coking, hydrocracking, and catalytic hydrocracking of Athabsca bitumen". *AOSTRA J. Res.* 7, 37—46.

Sanford, E.C. 1993. "Mechanism of coke prevention by hydrogen cracking during residuum hydrocracking", *Prepr. Div. Petrol. Chem. ACS* 38(2), 413—416.

Sankey, B.M.; Wu, F.S. 1989. "Mild thermal conversion of Cold Lake bitumen". *4th UNITAR/UNDP Conf. Heavy Crude Tar Sands, Edmonton, 1989*, paper 91.

Sapre, A.V.; Broderick, D.H.; Frankael, D.; Gates, B.C. Nag, N.K. 1980. "Hydrodesulfurization of benzo[b]naphtho[2,3–d]thiophene catalyzed by sulfided CoMo–MoO_3/γ–Al_2O_3; The reaction network". *AIChE J.* 26, 690–694.

Satterfield, C.N. 1975. "Trickle–bed reactors". *AIChE J.* 21, 209–228.

Satterfield, C.N.; Smith, C.M.; Ingalis, M. 1985. "Catalytic hydrodenitrogenation of quinoline. Effect of water and hydrogen sulfide". *Ind. Eng. Chem. Process Des. Dev.* 24, 1000–1004.

Savage, P.E.; Klein, M.T. 1987. "Asphaltene reaction pathways. 2. Pyrolysis of n-pentadecylbenzene". *Ind. Eng. Chem. Res.* 26, 488–493.

Savage, P.E.; Klein, M.T.; Kukes, S.G. 1988. "Asphaltene reaction pathways. 3. Effect of reaction environment". *Energy Fuels* 2, 619–628.

Savage, P.E.; Klein, M.T. 1988. "Asphaltene reaction pathways. 4. Pyrolysis of tridecylcyclohexane and 2–ethyltetralin". *Ind. Eng. Chem. Res.* 27, 1348–1355.

Savage, P.E.; Jacobs, G.E.; Javanmardian, M. 1989. "Autocatalysis and aryl–alkyl bond cleavage in 1–dodecylpyrene pyrolysis". *Ind. Eng. Chem. Res.* 28, 645–653.

Schmitter, J.–M.; Ignatiadis, I.; Dorbon, M.; Arpino, P.; Guischon, G.; Toulhoat, H.; Huc A. 1984. "Identification of nitrogen bases in a coker gas oil and influence of catalytic hydrotreatment on their composition". *Fuel* 63, 557–564.

Scubnik, M.; Wilson, M.F.; McCann, T.J. 1990. "Processing alternaitves for the 1990's to upgrade Canadian middle distillates from sands and conventional sources — A review". *AOSTRA J. Res.* 6, 1–16.

Seyer, F.A.; Gyte, C.W. 1989. "Viscosity". in *AOSTRA Technical Handbook on Oil Sands, Bitumens, and Heavy Oils*, L.G. Hepler and C. Hsi (eds), AOSTRA, Edmonton, AB, pp 153–184.

Shaw, J.M. 1987. "Correlation for hydrogen solubility in alicyclic and aromatic solvents". *Can. J. Chem. Eng.* 65, 293–298.

Shaw, J.M.; Gaikwad, R.P.; Stowe, D.A. 1988. "Phase splitting of pyrene–tetralin mixtures under coal liquefaction conditions". *Fuel* 67, 1554–1559.

Sim, W.J.; Daubert, T.E. 1980. "Prediction of vapor–liquid equilibria of undefined mixtures". *Ind. Eng. Chem. Process Des. Dev.* 16, 13–20.

Smith, J.M.; Van Ness, H.C. 1987. *Introduction to Chemical*

Engineering Thermodynamics, Fourth Edition, McGraw–Hill, New York, pp. 471–495.
Soave, G.S. 1972. "Equilibrium constants from a modified Redlich–Kwong equation of state". *Chem. Eng. Sci.* 27, 1197–1203.
Speight, J.G. 1981. *The Desulfurization of Heavy Oils and Residua*, Marcel Dekker, New York, NY.
Speight, J.G. 1991. *The Chemistry and Technology of Petroleum, 2nd Ed.*, Marcel Dekker, New York, NY.
Stangeland, B.E. 1974. "A kinetic model for the prediction of hydrocracker yields". *Ind. Eng. Chem. Process Des. Dev.* 13, 71–76.
Stanislaus, A.; Marafi, M.; Absi–Halabi, M. 1993. "Comparative evaluation of rejeuvenation of spent residue hydroprocessing catalysts". *ACS Div. Petrol. Chem. Prepr.* 38(1), 62–66.
Steer, J.G.; Muehlenbachs, K.; Gray, M.R. 1992. "Stable isotope analysis of hydrogen transfer during catalytic hydrocracking of residues". *Energy Fuels*, 6, 540–544.
Stohl, F.V.; Stephens, H.P. 1986. "The impact of the chemical constituents of hydrocracker feed on catalyst activity", *ACS Fuel Div. Prepr.* 31(4), 215–226.
Stokes, G.M.; Mott, R.W. 1989. "FCC resid processing: An overview". in *Fundamentals of Resid Upgrading*, R.H. Heck, T.F. Degnan, eds., AIChE Symp. Ser. 273, AIChE, New York, pp. 58–77.
Strausz, O.P. 1989. "Bitumen and heavy oil chemistry". in *AOSTRA Technical Handbook on Oil Sands, Bitumens, and Heavy Oils*, L.G. Hepler and C. Hsi (eds), AOSTRA, Edmonton, AB, pp 35–73.
Strausz, O.P. 1989b. "Structural studies on resids: Correlation between structure and activity". in *Fundamentals of Resid Upgrading*, R.H. Heck and T.F. Degnan (eds), AIChE, New York, NY, pp 1–6.
Stripinis, V.J. 1991. "Long residue processing in a riser pilot plant". in *Fluid Catalytic Cracking II*, M.L. Occelli, (ed), ACS Symp. Ser. 452, ACS, Washington DC, pp. 308–317.
Strong, D.; Filby, R.H. 1987. "Vandyl porphyrin distribution in the Alberta oil–sand bitumens". in *Metal Complexes in Fossil Fuels*, Filby, R.H.; Branthaver, J.F. eds., ACS Symposium Series 344, American Chemical Society, Washington DC, 154–172.
Stryjek, R.; Vera, J.H. 1986a. "PRSV: An improved Peng–Robinson equation of state for pure compounds and mixtures". *Can. J. Chem. Eng.* 64, 323–333.
Stryjek, R.; Vera, J.H. 1986b. "PRSV: An improved Peng–Robinson equation of state with new mixing rules for strongly non–ideal mixtures" *Can. J. Chem. Eng.* 64, 334–340.
Stubblefield, W.A.; McKee, R.H.: Kapp, R.W.; Hinz, J.P. 1989. "An evaluation of the acute toxic properties of liquids derived from oil

sands". *J. Appl. Toxicol.* 9, 59—65.

Sue, H.; Yoshimoto, M.; Armstrong, R.B.; Klein, B. 1988. "Mild resid hydrocracking for heavy oil upgrading". *Prepr. Fourth UNITAR/UNDP Conf. Heavy Crude Tar Sands*, Edmonton, Alberta, August 1988, paper no. 86.

Sullivan, R.F.; Boduszynski, M.M. 1989. "Molecular transformations in hydrotreating and hydrocracking". *Energy Fuels* 3, 603—612.

Sumida, Y.; Watari, R.; Bailey, R.T. 1984. Upgrading of oil sand bitumen by the Eureka process. *Energy Process. Canada* 77(1), Sept—Oct, 28—35.

Sundaram, K.M.; Katzer, J.R.; Bischoff, K.B. 1988. "Modeling of hydroprocessing reactions". *Chem. Eng. Comm.* 71, 53—71.

Sundaresan, S. 1993. "Liquid distribution in trickle—bed reactors". *Prepr. ACS Div. Petrol. Chem.* 38, 382—385.

Takatsuka, T.; Wada, Y.; Nakata, S.; Komatsu, S. 1988. "Reaction mechanism in VisABC process". *Prepr. Fourth UNITAR/UNDP Conf. Heavy Crude Tar Sands*, Edmonton, Alberta, August 1988, paper no. 123.

Takeuchi, C.; Fukui, Y.; Nakamura, M.; Shiroto, Y. 1983. "Asphaltene cracking in catalytic hydrotreating of heavy oils. 1. Processing of heavy oils by catalytic hydroprocessing and solvent deasphalting". *Ind. Eng. Chem. Process Des. Dev.* 22, 236—242.

Ternan, M.; Kriz, J.F. 1980. Some effects of catalyst composition on deactivation and coke formation when hydrocracking Athabasca bitumen". in *Catalyst Deactivation*, B. Delmon and G.F. Froment, eds., Elsevier, Amsterdam, pp. 283—293.

Ternan, M. 1986. "The effective diffusivity of residuum molecules in hydrocracking catalysts". Prepr. 10th Can. Symp. Catal., Kingston, Ont., June 15—18, 1986.

Ternan, M. 1987. "The diffusion of liquids in pores". *Can. J. Chem. Eng.* 65, 244—249.

Ternan, M.; Menashi, J. 1993. "Hydrocracking Boscan heavy oil with unimodal and bimodal catalysts". *Stud. Surf. Sci. Catal.* 75, 2387—2390.

Thakur, D.S.; Thomas, M.G. 1985. "Catalyst Deactivation in Heavy Hydrocarbon and Synthetic Crude Processing: A Review". *Appl. Catal.* 15, 197—225.

Trauth, D.M.; Stark, S.M.; Petti, T.F.; Neurock, M.; Yasar, M.; Klein, M.T. 1993. "Representation of the molecular structure of petroleum resid through characterization and Monte Carlo modeling. *ACS Petrol. Div. Prepr.* 38, 434—439 .

Trimm, D.L. 1977. "The formation and removal of coke from nickel catalyst". *Catal. Rev. Sci.-Eng.* 16, 155—189.

Trytten, L.C. 1989. "Hydroprocessing of gas oil fractions", M.Sc. Thesis, University of Alberta.

Trytten, L.C.; Gray, M.R.; Sanford, E.C. 1990. "Hydroprocessing of narrow—boiling gas oil fractions: Dependence of reaction kinetics on molecular weight". *Ind. Eng. Chem. Res.* 29, 725—730.

Trytten, L.C.; Gray, M.R. 1990. "Estimation of hydrocracking of C—C bonds during hydroprocessing of oils". *Fuel* 69, 397—399.

Tsonopoulos, C.; Heidman, J.L.; Hwang, S.—C. 1986. *Thermodynamic and Transport Properties of Coal Liquids*, Wiley, New York, NY.

Twu, C.H. 1984. "An internally consistent correlation for predicting the critical properties and molecular weights of petroleum and coal—tar liquids". *Fluid Phase Equil.* 16, 137—150.

Twu, C.H. 1985. "Internally consistent correlation for predicting liquid viscosities of petroleum fractions". *Ind. Eng. Chem. Process Des. Dev.* 24, 1287—1293.

Van Dreisen, R.P.; Caspers, J.; Campbell, A.R.; Lumin, G. 1979. "LC—Fining upgrades heavy crudes", *Hydrocarbon Process.* 58(5), 107—111.

Van Driesen, R.P.; Strangio, V.A.; Rhoe, A.; Kolstad, J.J. 1987. "The high—conversion LC—Fining process", *Energy Processing/Canada*, July/August, 13—19.

Voets, J.P.; Meerschman, M.; Verstreate, W. 1973. "Microbiological and biochemical effects of the application of bituminous emulsifiers as soil conditioners". *Plant Soil* 39, 433—436.

von Borstel, R.C.; Mehta, R.D. 1987. "Evaluation of the potential carcinogenicity and mutagenicity of process streams and other substances associated with oil sands development". AOSTRA Report, Agreement 204, AOSTRA, Edmonon, AB., 70 pp.

Walas, S.M. 1985. *Phase Equilibrium in Chemical Engineering*, Butterworth, Stoneham, MA.

Weekman, V.W. 1979. "Lumps, Models, and Kinetics in Practice", AIChE Monograph Series 11(75), 29 pp..

Wei, J.; Prater, C.D. 1962. "The structure and analysis of complex reaction systems", *Advances in Catalysis*, 13, Academic Press, New York.

Weisser, O.; Landa, S. 1973. *Sulphide Catalysts, Their Properties and Applications*, Pergamon, Oxford.

Wenzel., F.W. 1992. "VEBA—COMBI—Cracking, A commercial route for bottom of the barrel upgrading", *Proc. Int. Symp. on Heavy Oil and Residue Upgrading and Utilization*, C. Han and C. Hsi, eds, International Academic, Beijing, pp. 185—201.

Wiehe, I.A. 1993. "A phase separation kinetic model for coke formation". *ACS Div. Pet. Chem. Prepr.*, 38, 428—433.

Wilson, M.F.; Kriz, J.F. 1984. "Upgrading of middle distillate fractions of a syncrude from Athabasca oil sands". *Fuel* 63, 190—196.

Wilson, M.F.; Fisher, I.P.; Kriz J.F. 1985. "Hydrogenation of aromatic compounds in synthetic crude distillates catalyzed by sulfided Ni—W/γ—Al_2O_3". *J. Catal.* 95, 155—166 .

Wilson, M.F.; Kriz, J.F. 1985. "Response to comments on upgrading of middle distillate fractions of syncrude from Athabasca oil sands". *Fuel* 64, 1179—1180.

Wilson, M.F.; Fisher, I.P.; Kriz, J.F. 1986. "Cetane improvement of middle distillates from Athabasca syncrudes by catalytic hydroprocessing". *Ind. Eng. Chem. Prod. Res. Dev.* 25, 505—511.

World Health Organization 1982. "Selected petroleum products". *Environmental Health Criteria 20, IPCS Int. Prog. Chem. Safety,* WHO, Geneva, 139 pp.

Wukasch, J.E.; Rase, H.F. 1982. "Some characteristics of deposits on a commercially aged, gas oil hydrotreating catalyst". *Ind. Eng. Chem. Prod. Res. Dev.* 21, 558—565.

Young, D.C.; Galya, L.G. 1984. "Determination of paraffinic, naphthenic, and aromatic carbon in petroleum—derived materials by carbon—13 NMR spectroscopy". *Liq. Fuels Technol.* 2, 307—326.

Yui, S.M. 1989. "Hydrotreating of bitumen—derived coker gas oil: Kinetics of hydrodesulfurization, hydrodenitrogenation, and mild hydrocracking, and correlations to predict product yields and properties". *AOSTRA J. Res.* 5, 211—224.

Yui, S.M.; Sanford, E.C. 1989. "Mild hydrocracking of bitumen—derived coker and hydrocracker heavy gas oils: Kinetics, product yields, and product properties". *Ind. Eng. Chem. Res.* 28, 1278—1284.

Yui, S.M.; Sanford, E.C. 1991. "Kinetics of aromatics hydrogenation of bitumen—derived gas oils". *Can J. Chem. Eng.* 69, 1087—1095.

Yui, S.M. 1991. "Effect of using fresh or regenerated NiMo catalysts on coker gas oil hydrotreating". 1991 NPRA Annual Meeting, San Antonio TX, March 17—19, paper AM—91—60.

Zhou, P.; Crynes, B. L. 1986. "Thermolytic reactions of dodecane". *Ind. Eng. Chem. Process Des. Dev.* 25, 508—514.

Zhou, P.; Hollis, O. L.; Crynes, B. L. 1987. "Thermolysis of higher molecular weight straight—chain alkanes". *Ind. Eng. Chem. Res.* 26, 846—852.

Zrncevic, S.; Gomzi, Z.; Kotur, E. 1990. "Thiophene poisoning of Ni—SiO_2—Al_2O_3 in benzene hydrogenation: Deactivation kinetics". *Ind. Eng. Chem. Res.* 29, 774—777.

Index

Milton Keynes UK
Ingram Content Group UK Ltd.
UKHW020017071024
449327UK00032B/2820

9 780367 402075